I0034392

PROBLEMS AND SOLUTIONS ON THERMODYNAMICS AND STATISTICAL MECHANICS

Major American Universities Ph.D.
Qualifying Questions and Solutions

PROBLEMS AND SOLUTIONS ON THERMODYNAMICS AND STATISTICAL MECHANICS

Compiled by:
The Physics Coaching Class
University of Science and
Technology of China

Edited by:
Yung-Kuo Lim

World Scientific

NEW JERSEY • LONDON • SINGAPORE • BEIJING • SHANGHAI • HONG KONG • TAIPEI • CHENNAI

Published by

World Scientific Publishing Co. Pte. Ltd.

5 Toh Tuck Link, Singapore 596224

USA office: 27 Warren Street, Suite 401-402, Hackensack, NJ 07601

UK office: 57 Shelton Street, Covent Garden, London WC2H 9HE

British Library Cataloguing-in-Publication Data
A catalogue record for this book is available from the British Library.

First published 1990
Reprinted 1996, 2001, 2005, 2007

Major American Universities Ph.D. Qualifying Questions and Solutions
PROBLEMS AND SOLUTIONS ON THERMODYNAMICS AND STATISTICAL MECHANICS

Copyright © 1990 by World Scientific Publishing Co. Pte. Ltd.

All rights reserved. This book, or parts thereof, may not be reproduced in any form or by any means, electronic or mechanical, including photocopying, recording or any information storage and retrieval system now known or to be invented, without written permission from the Publisher.

For photocopying of material in this volume, please pay a copying fee through the Copyright Clearance Center, Inc., 222 Rosewood Drive, Danvers, MA 01923, USA. In this case permission to photocopy is not required from the publisher.

ISBN-13 978-981-02-0055-8
ISBN-10 981-02-0055-2
ISBN-13 978-981-02-0056-5 (pbk)
ISBN-10 981-02-0056-0 (pbk)

Printed in Singapore by World Scientific Printers (S) Pte Ltd

PREFACE

This series of physics problems and solutions which consists of seven parts – Mechanics, Electromagnetism, Optics, Atomic Nuclear and Particle Physics, Thermodynamics and Statistical Physics, Quantum Mechanics, Solid State Physics – contains a selection of 2550 problems from the graduate school entrance and qualifying examination papers of seven U.S. universities – California University Berkeley Campus, Columbia University, Chicago University, Massachusetts Institute of Technology, New York State University Buffalo Campus, Princeton University, Wisconsin University – as well as the CUSPEA and C.C. Ting's papers for selection of Chinese students for further studies in U.S.A. and their solutions which respresent the effort of more than 70 Chinese physicists.

The series is remarkable for its comprehensive coverage. In each area the problems span a wide spectrum of topics while many problems overlap several areas. The problems themselves are remarkable for their versatility in applying the physical laws and principles, their up-to-date realistic situations, and their scanty demand on mathematical skills. Many of the problems involve order of magnitude calculations which one often requires in an experimental situation for estimating a quantity from a simple model. In short, the exercises blend together the objectives of enhancement of one's understanding of the physical principles and practical applicability.

The solutions as presented generally just provide a guidance to solving the problems rather than step by step manipulation and leave much to the student to work out for him/herself, of whom much is demanded of the basic knowledge in physics. Thus the series would provide an invaluable complement to the textbooks.

In editing no attempt has been made to unify the physical terms and symbols. Rather, they are left to the setters' and solvers' own preferences so as to reflect the realistic situation of the usage today.

The present volume for Thermodynamics and Statistical Physics comprises 367 problems.

Lim Yung Kuo
Editor

INTRODUCTION

Solving problems in school work is the exercise of mental faculties, and examination problems are usually picked from the problems in school work. Working out problems is a necessary and important aspect in studies of Physics.

The *Major American University Ph.D. Qualifying Questions and Solutions* is a series of books which consists of seven volumes. The subjects of each volume and their respective referees (in parentheses) are as follows:

1. Mechanics (Qiang Yuan-qi, Gu En-pu, Cheng Jia-fu, Li Ze-hua, Yang De-tian)
2. Electromagnetism (Zhao Shu-ping, You Jun-han, Zhu Jun-jie)
3. Optics (Bai Gui-ru, Guo Guang-can)
4. Atomic, Nuclear and Particle Physics (Jin Huai-cheng, Yang Bao-zhong, Fan Yang-mei)
5. Thermodynamics and Statistical Physics (Zheng Jiu-ren)
6. Quantum Mechanics (Zhang Yong-de, Zhu Dong-pei, Fan Hong-yi)
7. Solid Physics and Comprehensive Topics (Zhang Jia-lu, Zhou You-yuan, Zhang Shi-ling)

The books cover almost all aspects of University Physics and contain 2550 problems, most of which are solved in detail.

These problems are carefully chosen from 3100 problems, some of which came from the China-U.S. Physics Examination and Application Program, others were selected from the Ph.D. Qualifying Examination on Experimental High Energy Physics, sponsored by Chao Chong Ting. The rest came from the graduate entrance examination questions of seven famous American universities during the last decade; they are: Columbia University, University of California at Berkeley, Massachusetts Institute of Technology, University of Wisconsin, University of Chicago, Princeton University and State University of New York, Buffalo.

In general, examination problems in physics in American universities do not involve too much Mathematics; however, they are to some extent characterized by the following three aspects: some problems involving various frontier subjects and overlapping domains of science are selected by professors directly from their own research work and show a "modern style", some problems involve a wider field and require a quick mind to analyse, and the methods used for solving the other problems are simple and practical which shows a full "touch of physics". From these, we think that these problems as a whole embody, to some extent, the characteristics of American science and culture and the features of the way of thinking of American education.

Just so, we believe it is worthwhile to collect and solve these problems and introduce them to students and teachers, even though the work is strenuous. About a hundred teachers and graduate students took part in this time-consuming job.

There are 367 problems in this volume which is divided into two parts: part I consists of 159 problems on Thermodynamics, part II consists of 208 problems on Statistical physics. Each part contains five sections.

The depth of knowledge involved in solving these problems is not beyond the contents of common textbooks on Thermodynamics and Statistical Physics used in colleges and universities in China, although the range of the knowledge and the techniques needed in solving some of these problems go beyond what we are familiar with. Furthermore, some new scientific research results are introduced into problems in school work, that will benefit not only the study of established theories and knowledge, but also the combination of teaching and research work by enlivening academic thoughts and making minds more active.

The people who contributed to solving these problems of this volume are Feng Ping, Wang Hai-da, Yao De-min and Jia Yun-fa. Associate professor Zheng Jiu-ren and Mr. Zheng Xin are referees of English of this volume.

15 October 1989

CONTENTS

PART I

THERMODYNAMICS

1. THERMODYNAMIC STATES AND THE FIRST LAW (1001-1030)

1001

Describe briefly the basic principle of the following instruments for making temperature measurements and state in one sentence the special usefulness of each instrument: constant-volume gas thermometer, thermo-couple, thermistor.

(Wisconsin)

Solution:

Constant-volume gas thermometer: It is made according to the principle that the pressure of a gas changes with its temperature while its volume is kept constant. It can approximately be used as an ideal gas thermometer.

Thermocouple thermometer: It is made according to the principle that thermoelectric motive force changes with temperature. The relation between the thermoelectric motive force and the temperature is

$$\varepsilon = a + bt + ct^2 + dt^3 ,$$

where ε is the electric motive force, t is the difference of temperatures of the two junctions, a, b, c and d are constants. The range of measurement of the thermocouple is very wide, from $-200°C$ to $1600°C$. It is used as a practical standard thermometer in the range from $630.74°C$ to $1064.43°C$.

Thermister thermometer: We measure temperature by measuring the resistance of a metal. The precision of a thermister made of pure platinum is very good, and its range of measurement is very wide, so it is usually used as a standard thermometer in the range from $13.81K$ to $903.89K$.

1002

Describe briefly three different instruments that can be used for the accurate measurement of temperature and state roughly the temperature range in which they are useful and one important advantage of each instrument. Include at least one instrument that is capable of measuring temperatures down to 1K.

(Wisconsin)

3

Solution:

1. *Magnetic thermometer:* Its principle is Curie's law $\chi = C/T$, where χ is the susceptibility of the paramagnetic substance used, T is its absolute temperature and C is a constant. Its advantage is that it can measure temperatures below 1K.

2. *Optical pyrometer:* It is based on the principle that we can find the temperature of a hot body by measuring the energy radiated from it, using the formula of radiation. While taking measurements, it does not come into direct contact with the measured body. Therefore, it is usually used to measure the temperatures of celestial bodies.

3. *Vapor pressure thermometer:* It is a kind of thermometer used to measure low temperatures. Its principle is as follows. There exists a definite relation between the saturation vapor pressure of a chemically pure material and its boiling point. If this relation is known, we can determine temperature by measuring vapor pressure. It can measure temperatures greater than 14K, and is the thermometer usually used to measure low temperatures.

1003

A bimetallic strip of total thickness x is straight at temperature T. What is the radius of curvature of the strip, R, when it is heated to temperature $T + \Delta T$? The coefficients of linear expansion of the two metals are α_1 and α_2, respectively, with $\alpha_2 > \alpha_1$. You may assume that each metal has thickness $x/2$, and you may assume that $x \ll R$.

(*Wisconsin*)

Solution:

We assume that the initial length is l_0. After heating, the lengths of the mid-lines of the two metallic strips are respectively

$$l_1 = l_0(1 + \alpha_1 \Delta T) , \tag{1}$$
$$l_2 = l_0(1 + \alpha_2 \Delta T) . \tag{2}$$

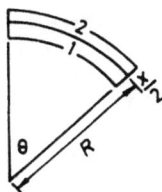

Fig. 1.1.

Assuming that the radius of curvature is R, the subtending angle of the strip is θ, and the change of thickness is negligible, we have

$$l_2 = \left(R + \frac{x}{4}\right)\theta , \quad l_1 = \left(R - \frac{x}{4}\right)\theta ,$$

$$l_2 - l_1 = \frac{x}{2}\theta = \frac{x}{2}\frac{l_1 + l_2}{2R} = \frac{xl_0}{4R}[2 + (\alpha_1 + \alpha_2)\Delta T] . \tag{3}$$

From (1) and (2) we obtain

$$l_2 - l_1 = l_0 \Delta T(\alpha_2 - \alpha_1) , \tag{4}$$

(3) and (4) then give

$$R = \frac{x}{4}\frac{[2 + (\alpha_1 + \alpha_2)\Delta T]}{(\alpha_2 - \alpha_1)\Delta T} .$$

1004

An ideal gas is originally confined to a volume V_1 in an insulated container of volume $V_1 + V_2$. The remainder of the container is evacuated. The partition is then removed and the gas expands to fill the entire container. If the initial temperature of the gas was T, what is the final temperature? Justify your answer.

(*Wisconsin*)

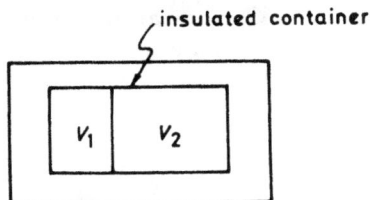

Fig. 1.2.

Solution:

This is a process of adiabatic free expansion of an ideal gas. The internal energy does not change; thus the temperature does not change, that is, the final temperature is still T.

1005

An insulated chamber is divided into two halves of volumes. The left half contains an ideal gas at temperature T_0 and the right half is evacuated. A small hole is opened between the two halves, allowing the gas to flow through, and the system comes to equilibrium. No heat is exchanged with the walls. Find the final temperature of the system.

(Columbia)

Solution:

After a hole has been opened, the gas flows continuously to the right side and reaches equilibrium finally. During the process, internal energy of the system E is unchanged. Since E depends on the temperature T only for an ideal gas, the equilibrium temperature is still T_0.

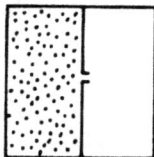

Fig. 1.3.

1006

Define heat capacity C_v and calculate from the first principle the numerical value (in calories/$^\circ$C) for a copper penny in your pocket, using your best physical knowledge or estimate of the needed parameters.

(UC, Berkeley)

Solution:

$C_v = (dQ/dT)_v$. The atomic number of copper is 64 and a copper penny is about 32 g, i.e., 0.5 mol. Thus $C_v = 0.5 \times 3R = 13$ J/K.

1007

Specific heat of granite may be: $0.02, 0.2, 20, 2000$ cal/g·K.

Solution:

The main component of granite is $CaCO_3$; its molecular weight is 100. The specific heat is $C = 3R/100 = 0.25$ cal/g· K. Thus the best answer is 0.2 cal/g·K.

1008

The figure below shows an apparatus for the determination of C_p/C_v for a gas, according to the method of Clement and Desormes. A bottle G, of reasonable capacity (say a few litres), is fitted with a tap H, and a manometer M. The difference in pressure between the inside and the outside can thus be determined by observation of the difference h in heights of the two columns in the manometer. The bottle is filled with the gas to be investigated, at a very slight excess pressure over the outside atmospheric pressure. The bottle is left in peace (with the tap closed) until the temperature of the gas in the bottle is the same as the outside temperature in the room. Let the reading of the manometer be h_i. The tap H is then opened for a very short time, just sufficient for the internal pressure to become equal to the atmospheric pressure (in which case the manometer reads $h = 0$). With the tap closed the bottle is left in peace for a while, until the inside temperature has become equal to the outside temperature. Let the final reading of the manometer be h. From the values of h_i and h_f it is possible to find C_p/C_v. (a) Derive an expression for C_p/C_v in terms of h_i and h_f in the above experiment. (b) Suppose that the gas in question is oxygen. What is your theoretical prediction for C_p/C_v at 20°C, within the framework of statistical mechanics?

Fig. 1.4.

Solution:

(a) The equation of state of ideal gas is $pV = nkT$. Since the initial and final T, V of the gas in the bottle are the same, we have $p_f/p_i = n_f/n_i$.

Meanwhile, $n_f/n_i = V/V'$, where V' is the volume when the initial gas in the bottle expands adiabatically to pressure p_0. Therefore

$$\frac{V}{V'} = \left(\frac{p_0}{p_i}\right)^{\frac{1}{7}}, \quad \frac{p_f}{p_i} = \left(\frac{p_0}{p_i}\right)^{\frac{1}{7}},$$

$$\gamma = \frac{\ln\dfrac{p_i}{p_0}}{\ln\dfrac{p_i}{p_f}} = \frac{\ln\left(1 + \dfrac{h_i}{h_0}\right)}{\ln\left(1 + \dfrac{h_i}{h_0}\right) - \ln\left(1 + \dfrac{h_f}{h_0}\right)}.$$

Since $h_i/h_0 \ll 1$ and $h_f/h_0 \ll 1$, we have $\gamma = h_i/(h_i - h_f)$.

(b) Oxygen consists of diatomic molecules. When $t = 20°C$, only the translational and rotational motions of the molecules contribute to the specific heat. Therefore

$$C_v = \frac{5R}{2}, \quad C_p = \frac{7R}{2}, \quad \gamma = \frac{7}{5}.$$

1009

(a) Starting with the first law of thermodynamics and the definitions of c_p and c_v, show that

$$c_p - c_v = \left[p + \left(\frac{\partial U}{\partial V}\right)_T\right]\left(\frac{\partial V}{\partial T}\right)_p$$

where c_p and c_v are the specific heat capacities per mole at constant pressure and volume, respectively, and U and V are energy and volume of one mole.

(b) Use the above results plus the expression

$$p + \left(\frac{\partial U}{\partial V}\right)_T = T\left(\frac{\partial p}{\partial T}\right)_V$$

to find $c_p - c_v$ for a Van der Waals gas

$$\left(p + \frac{a}{V^2}\right)(V - b) = RT .$$

Use that result to show that as $V \to \infty$ at constant p, you obtain the ideal gas result for $c_p - c_v$.

(SUNY, Buffalo)

Solution:

(a) From $H = U + pV$, we obtain

$$\left(\frac{\partial H}{\partial T}\right)_p = \left(\frac{\partial U}{\partial T}\right)_p + p\left(\frac{\partial V}{\partial T}\right)_p .$$

Let $U = U[T, V(T, p)]$. The above expression becomes

$$\left(\frac{\partial H}{\partial T}\right)_p = \left(\frac{\partial U}{\partial T}\right)_V + \left[p + \left(\frac{\partial U}{\partial V}\right)_T\right]\left(\frac{\partial V}{\partial T}\right)_p .$$

Hence

$$c_p - c_v = \left[p + \left(\frac{\partial U}{\partial V}\right)_T\right]\left(\frac{\partial V}{\partial T}\right)_p .$$

(b) For the Van der Waals gas, we have

$$\left(\frac{\partial p}{\partial T}\right)_V = \frac{R}{(V - b)} ,$$

$$\left(\frac{\partial V}{\partial T}\right)_p = R \bigg/ \left[\frac{RT}{V - b} - \frac{2a(V - b)}{V^3}\right] .$$

Hence,

$$c_p - c_v = \frac{R}{1 - 2a(1 - b/V)^2/VRT} ,$$

When $V \to \infty$, $c_p - c_v \to R$, which is just the result for an ideal gas.

1010

One mole of gas obeys Van der Waals equation of state. If its molar internal energy is given by $u = cT - a/V$ (in which V is the molar volume, a is one of the constants in the equation of state, and c is a constant), calculate the molar heat capacities C_v and C_p.

(*Wisconsin*)

Solution:

$$C_v = \left(\frac{\partial u}{\partial T}\right)_V = c ,$$

$$C_p = \left(\frac{\partial u}{\partial T}\right)_p + p\left(\frac{\partial V}{\partial T}\right)_p = \left(\frac{\partial u}{\partial T}\right)_V + \left[\left(\frac{\partial u}{\partial V}\right)_T + p\right]$$

$$\times \left(\frac{\partial V}{\partial T}\right)_p = c + \left(\frac{a}{V^2} + p\right)\left(\frac{\partial V}{\partial T}\right)_p .$$

From the Van der Waals equation

$$(p + a/V^2)(V - b) = RT \ ,$$

we obtain

$$\left(\frac{\partial V}{\partial T}\right)_p = R \bigg/ \left(p - \frac{a}{V^2} + \frac{2ab}{V^3}\right) \ .$$

Therefore

$$C_p = c + \frac{R\left(p + \frac{a}{V^2}\right)}{p - \frac{a}{V^2} + \frac{2ab}{V^3}} = c + \frac{R}{1 - \frac{2a(V - b)^2}{RTV^3}} \ .$$

1011

A solid object has a density ρ, mass M, and coefficient of linear expansion α. Show that at pressure p the heat capacities C_p and C_v are related by

$$C_p - C_v = 3\alpha M p/\rho \ .$$

(*Wisconsin*)

Solution:

From the first law of thermodynamics $dQ = dU + pdV$ and $\left(\dfrac{dU}{dT}\right)_p \approx \left(\dfrac{dU}{dT}\right)_v$ (for solid), we obtain

$$C_p - C_v = \left(\frac{dQ}{dT}\right)_p - \left(\frac{dU}{dT}\right)_v = p\frac{dV}{dT} \ . \qquad (*)$$

From the definition of coefficient of linear expansion $\alpha = \alpha_{\text{solid}}/3 = \dfrac{1}{3V}\dfrac{dV}{dT}$, we obtain

$$\frac{dV}{dT} = 3\alpha V = 3\alpha\frac{M}{\rho} \ .$$

Substituting this in (*), we find

$$C_p - C_v = 3\alpha\frac{M}{\rho}p \ .$$

1012

One mole of a monatomic perfect gas initially at temperature T_0 expands from volume V_0 to $2V_0$, (a) at constant temperature, (b) at constant pressure.

Calculate the work of expansion and the heat absorbed by the gas in each case.

(*Wisconsin*)

Solution:

(a) At constant temperature T_0, the work is

$$W = \int_A^B p dV = RT_0 \int_{V_0}^{2V_0} dV/V = RT_0 \ln 2 .$$

As the change of the internal energy is zero, the heat absorbed by the gas is

$$Q = W = RT_0 \ln 2 .$$

(b) At constant pressure p, the work is

$$W = \int_{V_0}^{2V_0} p dV = pV_0 = RT_0 .$$

The increase of the internal energy is

$$\Delta U = C_v \Delta T = \frac{3}{2} R \Delta T = \frac{3}{2} p \Delta V = \frac{3}{2} pV_0 = \frac{3}{2} RT_0 .$$

Thus the heat absorbed by the gas is

$$Q = \Delta U + W = \frac{5}{2} RT_0 .$$

1013

For a diatomic ideal gas near room temperature, what fraction of the heat supplied is available for external work if the gas is expanded at constant pressure? At constant temperature?

(*Wisconsin*)

Solution:

In the process of expansion at constant pressure p, assuming that the volume increases from V_1 to V_2 and the temperature changes from T_1 to T_2, we have

$$\begin{cases} pV_1 = nRT_1 \\ pV_2 = nRT_2 \ . \end{cases}$$

In this process, the work done by the system on the outside world is $W = p(V_2 - V_1) = nR\Delta T$ and the increase of the internal energy of the system is

$$\Delta U = C_v \Delta T \ .$$

Therefore

$$\frac{W}{Q} = \frac{W}{\Delta U + W} = \frac{nR}{C_v + nR} = \frac{2}{7} \ .$$

In the process of expansion at constant temperature, the internal energy does not change. Hence

$$W/Q = 1 \ .$$

1014

A compressor designed to compress air is used instead to compress helium. It is found that the compressor overheats. Explain this effect, assuming the compression is approximately adiabatic and the starting pressure is the same for both gases.

(Wisconsin)

Solution:

The state equation of ideal gas is

$$pV = nRT \ .$$

The equation of adiabatic process is

$$p\left(\frac{V}{V_0}\right)^\gamma = p_0 \ ,$$

where $\gamma = c_p/c_v$, p_0 and p are starting and final pressures, respectively, and V_0 and V are volumes. Because $V_0 > V$ and $\gamma_{He} > \gamma_{Air}$ ($\gamma_{He} = 7/5, \gamma_{Air} = 5/3$), we get

$$p_{He} > p_{Air} \quad \text{and} \quad T_{He} > T_{Air} \ .$$

1015

Calculate the temperature after adiabatic compression of a gas to 10.0 atmospheres pressure from initial conditions of 1 atmosphere and 300K (a) for air, (b) for helium (assume the gases are ideal).

(Wisconsin)

Solution:

The adiabatic process of an ideal gas follows the law

$$T_B = (p_B/p_A)^{(\gamma-1)/\gamma} T_A = 10^{(\gamma-1)/\gamma} \times 300 \text{ K} .$$

(a) For air, $\gamma = C_p/C_v = 1.4$, thus $T_B = 5.8 \times 10^2\text{K}$.
(b) For helium, $\gamma = C_p/C_v = 5/3$, thus $T_B = 7.5 \times 10^2\text{K}$.

1016

(a) For a mole of ideal gas at $t = 0°\text{C}$, calculate the work W done (in Joules) in an isothermal expansion from V_0 to $10V_0$ in volume.

(b) For an ideal gas initially at $t_i = 0°\text{C}$, find the final temperature t_f (in $°\text{C}$) when the volume is expanded to $10V_0$ reversibly and adiabatically.

(UC, Berkeley)

Solution:

(a) $W = \displaystyle\int_{V_0}^{10V_0} pdV = \int_{V_0}^{10V_0} \frac{RT}{V}dV = RT\ln 10 = 5.2 \times 10^3\text{J}$.

(b) Combining the equation of adiabatic process $pV^\gamma = \text{const}$ and the equation of state $pV = RT$, we get $TV^{\gamma-1} = \text{const}$. Thus

$$T_f = T_i \left(\frac{V_i}{V_f}\right)^{\gamma-1} .$$

If the ideal gas molecule is monatomic, $\gamma = 5/3$, we get $t_f = 59\text{K}$ or $-214°\text{C}$.

1017

(a) How much heat is required to raise the temperature of 1000 grams of nitrogen from $-20°\text{C}$ to $100°\text{C}$ at constant pressure?
(b) How much has the internal energy of the nitrogen increased?
(c) How much external work was done?

(d) How much heat is required if the volume is kept constant?

Take the specific heat at constant volume $c_v = 5$ cal/mole$\cdot°$C and $R = 2$ cal/mole$\cdot°$C.

(Wisconsin)

Solution:

(a) We consider nitrogen to be an ideal gas. The heat required is

$$Q = n(c_v + R)\Delta T = \frac{1000}{28}(5 + 2) \times 120 = 30 \times 10^3 \text{cal} .$$

(b) The increase of the internal energy is

$$\Delta U = nc_v \Delta T = \frac{100}{28} \times 5 \times 120$$
$$= 21 \times 10^3 \text{cal} .$$

(c) The external work done is

$$W = Q - \Delta U = 8.6 \times 10^3 \text{ cal} .$$

(d) If it is a process of constant volume, the required heat is

$$Q = nc_v \Delta T = 21 \times 10^3 \text{cal} .$$

1018

10 litres of gas at atmospheric pressure is compressed isothermally to a volume of 1 litre and then allowed to expand adiabatically to 10 litres.

(a) Sketch the processes on a pV diagram for a monatomic gas.

(b) Make a similar sketch for a diatomic gas.

(c) Is a net work done on or by the system?

(d) Is it greater or less for the diatomic gas?

(Wisconsin)

Solution:

We are given that $V_A = 10l, V_B = 1l, V_C = 10l$ and $p_A = 1$ atm.

$A \rightarrow B$ is an isothermal process, thus

$$pV = \text{const. or } p_A V_A = p_B V_B ,$$

hence

$$p_B = \frac{V_A}{V_B} p_A = 10 \text{ atm} .$$

(The curve AB of the two kinds of gas are the same).

$B \to C$ is an adiabatic process, thus

$$pV^\gamma = \text{const, or } p_B V_B^\gamma = p_C V_C^\gamma , \quad \text{hence}$$

$$p_C = \left(\frac{V_B}{V_C}\right)^\gamma p_B = 10^{1-\gamma} \text{atm.}$$

(a) For the monatomic gas, we have

$$\gamma = 5/3, p_C = 10^{-2/3} = 0.215 \text{ atm} .$$

(b) For the diatomic gas, we have

$$\gamma = 7/5, p_C = 10^{-2/5} = 0.398 \text{ atm.}$$

The two processes are shown in the figures 1.5. (The curve BC of the monatomic gas (a) is lower than that of the diatomic gas (b)).

(c) In each case, as the curve AB for compression is higher than the curve BC for expansion, net work is done on the system. As p_C (monatomic gas) $< p_C$ (diatomic gas) the work on the monatomic gas is greater than that on the diatomic gas.

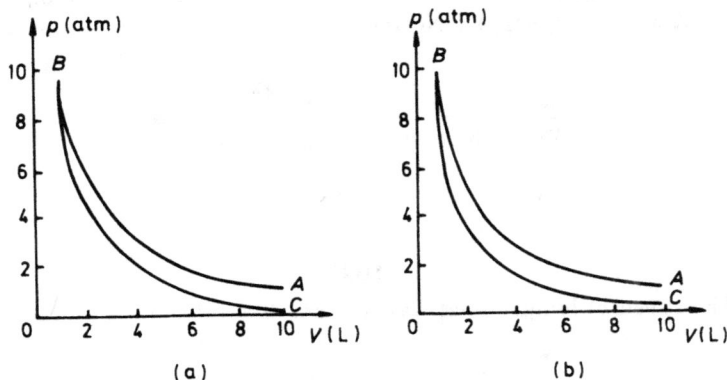

Fig. 1.5.

1019

An ideal gas is contained in a large jar of volume V_0. Fitted to the jar is a glass tube of cross-sectional area A in which a metal ball of mass M fits snugly. The equilibrium pressure in the jar is slightly higher than atmospheric pressure p_0 because of the weight of the ball. If the ball is displaced slightly from equilibrium it will execute simple harmonic motion (neglecting friction). If the states of the gas represent a quasistatic adiabatic process and γ is the ratio of specific heats, find a relation between the oscillation frequency f and the variables of the problem.

(UC, Berkeley)

Fig. 1.6.

Solution:

Assume the pressure in the jar is p. As the process is adiabatic, we have

$$pV^\gamma = \text{const} ,$$

giving

$$\frac{dp}{p} + \gamma \frac{dV}{V} = 0 .$$

This can be written as $F = Adp = -kx$, where F is the force on the ball, $x = dV/A$ and $k = \gamma A^2 p/V$. Noting that $p = p_0 + mg/A$, we obtain

$$f = \frac{\omega}{2\pi} = \frac{1}{2\pi}\sqrt{\frac{k}{m}} = \frac{1}{2\pi}\sqrt{\frac{\gamma A^2 \left(p_0 + \dfrac{mg}{A}\right)}{Vm}} .$$

1020

The speed of longitudinal waves of small amplitude in an ideal gas is

$$C = \sqrt{\frac{dp}{d\rho}}$$

where p is the ambient gas pressure and ρ is the corresponding gas density. Obtain expressions for

(a) The speed of sound in a gas for which the compressions and rarefactions are isothermal.

(b) The speed of sound in a gas for which the compressions and rarefactions are adiabatic.

(*Wisconsin*)

Solution:

The isothermal process of an ideal gas follows $pV = \text{const}$; the adiabatic process of an ideal gas follows $pV^\gamma = \text{const}$. We shall use $pV^t = \text{const}$ for a general process, its differential equation being

$$\frac{dp}{p} + t\frac{dV}{V} = 0 .$$

Thus

$$\left(\frac{dp}{dV}\right) = -t\frac{p}{V} .$$

With $\rho = M/V$, we have

$$\frac{dp}{d\rho} = \frac{dp}{dV}\left(\frac{dV}{d\rho}\right) = \left(-t\frac{p}{V}\right)\left(-\frac{M}{\rho^2}\right) = t\frac{RT}{M} ,$$

Therefore

$$c = \sqrt{\frac{dp}{d\rho}} = \sqrt{\frac{tRT}{M}} .$$

(a) The isothermal process: $t = 1$, thus $c = \sqrt{RT/M}$.

(b) The adiabatic process: $t = \gamma$, thus $c = \sqrt{\gamma RT/M}$.

1021

Two systems with heat capacities C_1 and C_2, respectively, interact thermally and come to a common temperature T_f. If the initial temperature of system 1 was T_1, what was the initial temperature of system 2? You may assume that the total energy of the combined systems remains constant.

(*Wisconsin*)

Solution:

We assume that the initial temperature of system 2 is T_2. According to the conservation of energy, we know the heat released from system 1 is equal to that absorbed by the other system, i.e.,

$$C_1(T_f - T_1) = C_2(T_2 - T_f) .$$

The solution is

$$T_2 = \frac{C_1}{C_2}(T_f - T_1) + T_f .$$

1022

A large solenoid coil for a physics experiment is made of a single layer of conductor of cross section 4cm × 2cm with a cooling water hole 2 cm × 1cm in the conductor. The coil, which consists of 100 turns, has a diameter of 3 meters, and a length of 4 meters (the insulation thickness is negligible). At the two ends of the coil are circular steel plates to make the field uniform and to return the magnetic flux through a steel cylindrical structure external to the coil, as shown in the diagram. A magnetic field of 0.25 Tesla is desired. The conductor is made of aluminium.

(a) What power (in kilowatts) must be supplied to provide the desired field, and what must be the voltage of the power supply?

(b) What rate of water flow (litres/second) must be supplied to keep the temperature rise of the water at 40°C? Neglect all heat losses from the coil except through the water.

(c) What is the outward pressure exerted on the coil by the magnetic forces?

(d) If the coil is energized by connecting it to the design voltage calculated in (a), how much time is required to go from zero current to 99% of the design current? Neglect power supply inductance and resistance. The resistivity of aluminium is 3×10^{-8} ohm-meters. Assume that the steel is far below saturation.

(*CUSPEA*)

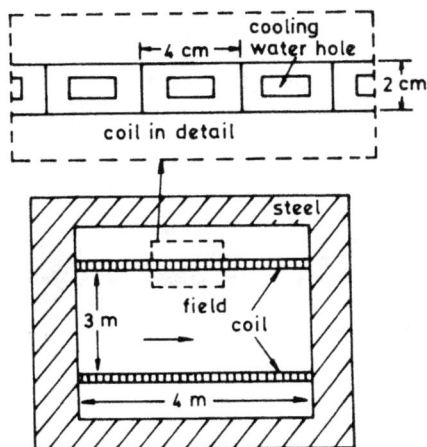

Fig. 1.7.

Solution:

(a) The magnetic field is $B = \mu_0 N I / L$,
where N is the number of turns, L is the length of the solenoid coil. The current is therefore

$$I = \frac{BL}{\mu_0 N} = \frac{0.25 \times 4}{4\pi \times 10^{-7} \times 100} = 7960 \text{A} .$$

The total resistance of the coil is $R = \rho L / A$. Therefore, the resistance, the voltage and the power are respectively

$$R = \frac{(3 \times 10^{-8})(100 \times 2\pi \times 1.5)}{(4 \times 2 - 2 \times 1) \times 10^{-4}} = 0.0471\Omega$$

$$V = RI = 375 \text{V}$$

$$P = VI = 2.99 \times 10^3 \text{ kW} .$$

(b) The rate of flow of the cooling water is W. Then $\rho W C \Delta T = P$, where ρ is the density, C is the specific heat and ΔT is the temperature rise of the water. Hence

$$W = \frac{p}{\rho C \Delta T} = \frac{2.99 \times 10^3 \times 10^3}{1 \times 4190 \times 40} = 17.8 \text{ } l/s .$$

(c) The magnetic pressure is

$$p = \frac{B^2}{2\mu_0} = \frac{(0.25)^2}{2(4\pi \times 10^{-7})} = 2.49 \times 10^4 \text{ N/m}^2 .$$

(d) The time constant of the circuit is

$$\tau = L/R, \quad \text{with} \quad L = N\Phi/I ,$$

where L is the inductance, R is the resistance, N is the number of turns, I is the current and Φ is the magnetic flux. Thus we have

$$L = 100 \times 0.25\pi \times (1.5)^2/7960 = 0.0222 \text{ H}$$

and

$$\tau = 0.0222/0.0471 = 0.471 \text{ s} .$$

The variation of the current before steady state is reached is given by

$$I(t) = I_{\max}[1 - \exp(-t/\tau)] .$$

When $I(t)/I_{\max} = 99\%$,

$$t = \tau \ln 100 = 4.6\tau \approx 2.17 \text{ s} .$$

1023

Consider a black sphere of radius R at temperature T which radiates to distant black surroundings at $T = 0$K.

(a) Surround the sphere with a nearby heat shield in the form of a black shell whose temperature is determined by radiative equilibrium. What is the temperature of the shell and what is the effect of the shell on the total power radiated to the surroundings?

(b) How is the total power radiated affected by additional heat shields? (Note that this is a crude model of a star surrounded by a dust cloud.)

(*UC, Berkeley*)

Solution:

(a) At radiative equilibrium, $J - J_1 = J_1$ or $J_1 = J/2$. Therefore $T_1^4 = T^4/2$, or $T_1 = \sqrt[4]{\dfrac{T^4}{2}} = \dfrac{T}{\sqrt[4]{2}} .$

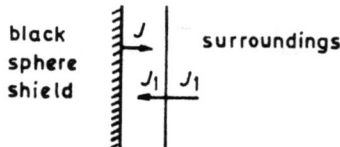

Fig. 1.8.

(b) The heat shield reduces the total power radiated to half of the initial value. This is because the shield radiates a part of the energy it absorbs back to the black sphere.

1024

In vacuum insulated cryogenic vessels (Dewars), the major source of heat transferred to the inner container is by radiation through the vacuum jacket. A technique for reducing this is to place "heat shields" in the vacuum space between the inner and outer containers. Idealize this situation by considering two infinite sheets with emissivity $= 1$ separated by a vacuum space. The temperatures of the sheets are T_1 and T_2 $(T_2 > T_1)$. Calculate the energy flux (at equilibrium) between them. Consider a third sheet (the heat shield) placed between the two which has a reflectivity of R. Find the equilibrium temperature of this sheet. Calculate the energy flux from sheet 2 to sheet 1 when this heat shield is in place.

For $T_2 =$ room temperature, $T_1 =$ liquid He temperature (4.2 K) find the temperature of a heat shield that has a reflectivity of 95%. Compare the energy flux with and without this heat shield.
$(\sigma = 0.55 \times 10^{-7} \text{ watts/m}^2\text{K})$

(*UC, Berkeley*)

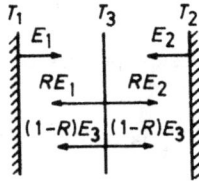

Fig. 1.9.

Solution:

When there is no "heat shield", the energy flux is

$$J = E_2 - E_1 = \sigma(T_2^4 - T_1^4) .$$

When "heat shield" is added, we have

$$J^* = E_2 - RE_2 - (1 - R)E_3 ,$$
$$J^* = (1 - R)E_3 + RE_1 - E_1 .$$

These equations imply $E_3 = (E_1 + E_2)/2$, or $T_3 = [(T_2^4 + T_1^4)/2]^{1/4}$. Hence

$$J^* = (1 - R)(E_2 - E_1)/2 = (1 - R)J/2 .$$

With $T_1 = 4.2$ K, $T_2 = 300$K and $R = 0.95$, we have

$$T_3 = 252 \text{ K and } J^*/J = 0.025 .$$

1025

Two parallel plates in vacuum, separated by a distance which is small compared with their linear dimensions, are at temperatures T_1 and T_2 respectively $(T_1 > T_2)$.

(a) If the plates are non-transparent to radiation and have emission powers e_1 and e_2 respectively, show that the net energy W transferred per unit area per second is

$$W = \frac{E_1 - E_2}{\dfrac{E_1}{e_1} + \dfrac{E_2}{e_2} - 1} .$$

where E_1 and E_2 are the emission powers of black bodies at temperatures T_1 and T_2 respectively.

(b) Hence, what is W if T_1 is 300 K and T_2 is 4.2 K, and the plates are black bodies?

(c) What will W be if n identical black body plates are interspersed between the two plates in (b)?
($\sigma = 5.67 \times 10^{-8} \text{W/m}^2\text{K}^4$).

(SUNY, Buffalo)

Solution:

(a) Let f_1 and f_2 be the total emission powers (thermal radiation plus reflection) of the two plates respectively. We have

$$f_1 = e_1 + \left(1 - \frac{e_1}{E_1}\right) f_2 , \quad f_2 = e_2 + \left(1 - \frac{e_2}{E_2}\right) f_1 .$$

The solution is

$$f_1 = \frac{\dfrac{E_1 E_2}{e_1 e_2}(e_1 + e_2) - E_2}{\dfrac{E_1}{e_1} + \dfrac{E_2}{e_2} - 1} ,$$

$$f_2 = \frac{\dfrac{E_1 E_2}{e_1 e_2}(e_1 + e_2) - E_1}{\dfrac{E_1}{e_1} + \dfrac{E_2}{e_2} - 1} .$$

Hence

$$W = f_1 - f_2 = \frac{E_1 - E_2}{\dfrac{E_1}{e_1} + \dfrac{E_2}{e_2} - 1} .$$

(b) For black bodies, $W = E_1 - E_2 = \sigma(T_1^4 - T_2^4) = 460$ W/m^2.

(c) Assume that the n interspersed plates are black bodies at temperatures $t_1, t_2, \ldots, t_n$. When equilibrium is reached, we have

$$T_1^4 - t_1^4 = t_1^4 - T_2^4 , \quad \text{for } n = 1 ,$$

with solution

$$t_1^4 = \frac{T_1^4 + T_2^4}{2} , \quad W = \sigma(T_1^4 - t_1^4) = \frac{\sigma}{2}(T_1^4 - T_2^4) .$$
$$T_1^4 - t_1^4 = t_1^4 - t_2^4 = t_2^4 - T_2^4 , \quad \text{for } n = 2 ,$$

with solution

$$t_1^4 = \frac{4}{3}\left(\frac{T_1^4}{2} + \frac{T_2^4}{4}\right) , \quad W = \frac{\sigma}{3}(T_1^4 - T_2^4) .$$

Then in the general we have

$$T_1^4 - t_1^4 = t_1^4 - t_2^4 = \ldots = t_n^4 - T_2^4 ,$$

with solution

$$t_1^4 = \frac{n}{n+1}T_1^4 - \frac{1}{n+1}T_2^4 ,$$
$$W = \sigma(T_1^4 - T_2^4) = \frac{\sigma}{n+1}(T_1^4 - T_2^4) .$$

1026

A spherical black body of radius r at absolute temperature T is surrounded by a thin spherical and concentric shell of radius R, black on both sides. Show that the factor by which this radiation shield reduces the rate of cooling of the body (consider space between spheres evacuated, with no thermal conduction losses) is given by the following expression: $aR^2/(R^2 + br^2)$, and find the numerical coefficients a and b.

(*SUNY, Buffalo*)

Solution:

Let the surrounding temperature be T_0. The rate of energy loss of the black body before being surrounded by the spherical shell is

$$Q = 4\pi r^2 \sigma (T^4 - T_0^4) \ .$$

The energy loss per unit time by the black body after being surrounded by the shell is

$$Q' = 4\pi r^2 \sigma (T^4 - T_1^4), \quad \text{where } T_1 \text{ is temperature of the shell} \ .$$

The energy loss per unit time by the shell is

$$Q'' = 4\pi R^2 \sigma (T_1^4 - T_0^4) \ .$$

Since $Q'' = Q'$, we obtain

$$T_1^4 = (r^2 T^4 + R^2 T_0^4)/(R^2 + r^2) \ .$$

Hence $Q'/Q = R^2/(R^2 + r^2)$, i.e., $a = 1$ and $b = 1$.

1027

The solar constant (radiant flux at the surface of the earth) is about $0.1 \ \text{W/cm}^2$. Find the temperature of the sun assuming that it is a black body.

(*MIT*)

Solution:

The radiant flux density of the sun is

$$J = \sigma T^4 \ , \quad \text{where } \sigma = 5.7 \times 10^{-8} \ \text{W/m}^2\text{K}^4. \text{ Hence } \sigma T^4 (r_S/r_{SE})^2 = 0.1 \ ,$$

where the radius of the sun $r_S = 7.0 \times 10^5$km, the distance between the earth and the sun $r_{SE} = 1.5 \times 10^8$km. Thus

$$T = \left[\frac{0.1}{\sigma} \left(\frac{r_{SE}}{r_S} \right)^2 \right]^{\frac{1}{4}} \approx 5 \times 10^3 \text{ K} .$$

1028

(a) Estimate the temperature of the sun's surface given that the sun subtends an angle θ as seen from the earth and the earth's surface temperature is T_0. (Assume the earth's surface temperature is uniform, and that the earth reflects a fraction, ε, of the solar radiation incident upon it). Use your result to obtain a rough estimate of the sun's surface temperature by putting in "reasonable" values for all parameters.

(b) Within an unheated glass house on the earth's surface the temperature is generally greater than T_0. Why? What can you say about the maximum possible interior temperature in principle?

(*Columbia*)

Solution:

(a) The earth radiates heat while it is absorbing heat from the solar radiation. Assume that the sun can be taken as a black body. Because of reflection, the earth is a grey body of emissivity $1 - \varepsilon$. The equilibrium condition is

$$(1 - \varepsilon) J_S 4\pi R_S^2 \cdot \pi R_E^2 / 4\pi r_{S-E}^2 = J_E \cdot 4\pi R_E^2 ,$$

where J_S and J_E are the radiated energy flux densities on the surfaces of the sun and the earth respectively, R_S, R_E and r_{S-E} are the radius of the sun, the radius of the earth and the distance between the earth and the sun respectively. Obviously $R_S / r_{S-E} = \tan(\theta/2)$. From the Stefan-Boltzman law, we have

for the sun, $J_S = \sigma T_S^4$;
for the earth $J_E = (1 - \varepsilon) \sigma T_E^4$.

Therefore

$$T_S = T_E \sqrt{\frac{2 r_{S-E}}{R_S}} \approx 300 \text{ K} \times \left(2 \times \frac{1.5 \times 10^8 \text{ km}}{7 \times 10^6 \text{ km}} \right)^{1/2}$$

$$\approx 6000 \text{ K} .$$

(b) Let T be temperature of the glass house and t be the transmission coefficient of glass. Then

$$(1 - t)T^4 + tT_0^4 = tT^4 \; ,$$

giving

$$T = \left[\frac{t}{(2t - 1)} \right]^{1/4} T_0 \; .$$

Since $t < 1$, we have $t > 2t - 1$, so that

$$T > T_0 \; .$$

1029

Consider an idealized sun and earth, both black bodies, in otherwise empty flat space. The sun is at a temperature of $T_S = 6000$ K and heat transfer by oceans and atmosphere on the earth is so effective as to keep the earth's surface temperature uniform. The radius of the earth is $R_E = 6 \times 10^8$ cm, the radius of the sun is $R_S = 7 \times 10^{10}$ cm, and the earth-sun distance is $d = 1.5 \times 10^{13}$ cm.

(a) Find the temperature of the earth.

(b) Find the radiation force on the earth.

(c) Compare these results with those for an interplanetary "chondrule" in the form of a spherical, perfectly conducting black-body with a radius of $R = 0.1$cm, moving in a circular orbit around the sun with a radius equal to the earth-sun distance d.

(*Princeton*)

Solution:

(a) The radiation received per second by the earth from the sun is approximately

$$q_{SE} = 4\pi R_S^2 (\sigma T_S^4) \frac{\pi R_E^2}{4\pi d^2} \; .$$

The radiation per second from the earth itself is

$$q_E = 4\pi R_E^2 \cdot (\sigma T_E^4) \; .$$

Neglecting the earth's own heat sources, energy conservation leads to the relation $q_E = q_{SE}$, so that

$$T_E^4 = \frac{R_S^2}{4d^2} T_S^4 \; ,$$

i.e.,

$$T_E = \sqrt{R_S/2d} \cdot T_S = 290 \text{ K} = 17°C \; .$$

(b) The angles subtended by the earth in respect of the sun and by the sun in respect of the earth are very small, so the radiation force is

$$F_E = \frac{q_E}{c} = \frac{1}{c} \frac{R_S^2}{d^2} \cdot \pi R_E^2 \cdot (\sigma T_S^4) = 6 \times 10^8 \text{ N} \; .$$

(c) As $R_E \rightarrow R, T = T_E = 17°C$

$$F = (R/R_E)^2 F_E = 1.7 \times 10^{-11} \text{ N} \; .$$

1030

Making reasonable assumptions, estimate the surface temperature of Neptune. Neglect any possible internal sources of heat. What assumptions have you made about the planet's surface and/or atmosphere?

Astronomical data which may be helpful: radius of sun=7×10^5 km; radius of Neptune = 2.2×10^4 km; mean sun-earth distance = 1.5×10^8 km; mean sun-Neptune distance = 4.5×10^9 km; $T_S = 6000$ K; rate at which sun's radiation reaches earth = 1.4 kW/m^2; Stefan-Boltzman constant = 5.7×10^{-8} W/m^{2}K^4.

(*Wisconsin*)

Solution:

We assume that the surface of Neptune and the thermodynamics of its atmosphere are similar to those of the earth. The radiation flux on the earth's surface is

$$J_E = 4\pi R_S^2 \sigma T_S^4 / 4\pi R_{SE}^2$$

The equilibrium condition on Neptune's surface gives

$$4\pi R_S^2 \sigma T_S^4 \cdot \pi R_N^2 / 4\pi R_{SN}^2 = \sigma T_N^4 \cdot 4\pi R_N^2 \; .$$

Hence

$$R_{SE}^2 J_E / R_{SN}^2 = 4\sigma T_N^4 \ ,$$

and we have

$$
\begin{aligned}
T_N &= (R_{SE}^2 J_E / 4\sigma R_{SN}^2)^{1/4} \\
&= \left[\frac{(1.5 \times 10^8)^2}{(5.7 \times 10^9)^2} \cdot \frac{1.4 \times 10^3}{4 \times 5.7 \times 10^{-8}} \right]^{\frac{1}{4}} \\
&= 52 \text{ K} \ .
\end{aligned}
$$

2. THE SECOND LAW AND ENTROPY (1031-1072)

1031

A steam turbine is operated with an intake temperature of 400°C, and an exhaust temperature of 150°C. What is the maximum amount of work the turbine can do for a given heat input Q? Under what conditions is the maximum achieved?

(*Wisconsin*)

Solution:

From the Clausius formula

$$\frac{Q_1}{T_1} - \frac{Q_2}{T_2} \leq 0 \ ,$$

we find the external work to be

$$W = Q_1 - Q_2 \leq \left(1 - \frac{T_2}{T_1} \right) Q_1 \ .$$

Substituting $Q_1 = Q, T_1 = 673$ K and $T_2 = 423$ K in the above we have

$$W_{max} = \left(1 - \frac{T_2}{T_1} \right) Q = 0.37 Q \ .$$

As the equal sign in the Clausius formula is valid if and only if the cycle is reversible, when and only when the steam turbine is a reversible engine can it achieve maximum work.

1032

What is a Carnot cycle? Illustrate on a pV diagram and an ST diagram. Derive the efficiency of an engine using the Carnot cycle.

(*Wisconsin*)

Solution:

A Carnot cycle is a cycle composed of two isothermal lines and two adiabatic lines (as shown in Fig. 1.10 (a)).

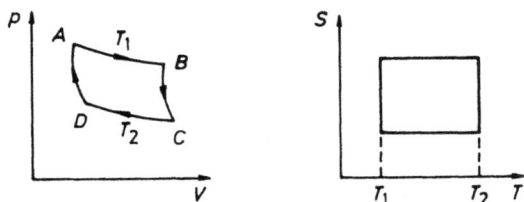

Fig. 1.10.

Now we calculate the efficiency of the Carnot engine. First, we assume the cycle is reversible and the gas is 1 mole of an ideal gas. As $A \to B$ is a process of isothermal expansion, the heat absorbed by the gas from the heat source is

$$Q_1 = RT_1 \ln(V_B/V_A) .$$

As $C \to D$ is a process of isothermal compression, the heat released by the gas is

$$Q_2 = RT_2 \ln(V_C/V_D) .$$

The system comes back to the initial state through the cycle $ABCDA$. In these processes, the relations between the quantities of state are

$$p_A V_A = p_B V_B , \quad p_B V_B^\gamma = p_C V_C^\gamma ,$$
$$p_C V_C = p_D V_D , \quad p_D V_D^\gamma = p_A V_A^\gamma .$$

Thus we find

$$\frac{V_B}{V_A} = \frac{V_C}{V_D} .$$

Therefore the efficiency of the engine is

$$\eta = \frac{Q_1 - Q_2}{Q_1} = \frac{T_1 - T_2}{T_1} = 1 - \frac{T_2}{T_1} .$$

If the engine (or the cycle) is not reversible, its efficiency is

$$\eta' < \eta = 1 - T_2/T_1 \ .$$

1033

A Carnot engine has a cycle pictured below.

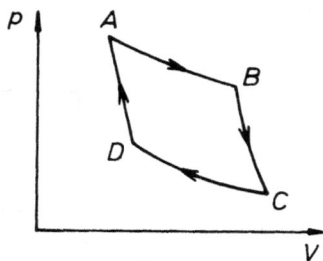

Fig. 1.11.

(a) What thermodynamic processes are involved at boundaries AD and BC; AB and CD?

(b) Where is work put in and where is it extracted?

(c) If the above is a steam engine with $T_{in} = 450$ K, operating at room temperature, calculate the efficiency.

(*Wisconsin*)

Solution:

(a) DA and BC are adiabatic processes, AB and CD are isothermal processes.

(b) Work is put in during the processes CD and DA; it is extracted in the processes AB and BC.

(c) The efficiency is

$$\eta = \frac{1 - T_{out}}{T_{in}} = 1 - \frac{300}{450} = \frac{1}{3} \ .$$

1034

A Carnot engine has a cycle as shown in Fig. 1.12. If W and W' represent work done by 1 mole of monatomic and diatomic gas, respectively, calculate W'/W.

(*Columbia*)

Solution:

For the Carnot engine using monatomic gas, we have

$$W = R(T_1 - T_2)\ln(V_2/V_1) \, ,$$

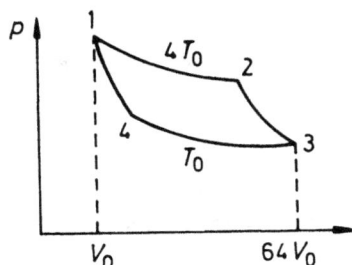

Fig. 1.12.

where $T_1 = 4T_0$, and $T_2 = T_0$ are the temperatures of the respective heat sources, $V_1 = V_0$, and V_2 is the volume at state 2. We also have $V_3 = 64\,V_0$. With $W' = R(T_1 - T_2)\ln\left(\dfrac{V_2'}{V_1}\right)$ for the diatomic gas engine, we obtain

$$\frac{W'}{W} = \frac{\ln(V_2'/V_1)}{\ln(V_2/V_1)} \, .$$

Then, using the adiabatic equations $4T_0 V_2^{\gamma-1} = T_0 V_3^{\gamma-1}$,

$$4T_0 V_2'^{\gamma'-1} = T_0 V_3^{\gamma'-1} \, ,$$

we obtain

$$\frac{W'}{W} = \frac{3 + (1 - \gamma')^{-1}}{3 + (1 - \gamma)^{-1}} \, .$$

For a monatomic gas $\gamma = 5/3$; for a diatomic gas, $\gamma' = 7/5$. Thus

$$\frac{W'}{W} = \frac{1}{3} \, .$$

1035

Two identical bodies have internal energy $U = NCT$, with a constant C. The values of N and C are the same for each body. The initial temperatures of the bodies are T_1 and T_2, and they are used as a source of work by connecting them to a Carnot heat engine and bringing them to a common final temperature T_f.

(a) What is the final temperature T_f?

(b) What is the work delivered?

(*CUSPEA*)

Solution:

(a) The internal energy is $U = NCT$. Thus $dQ_1 = NCdT_1$ and $dQ_2 = NCdT_2$. For a Carnot engine, we have $\dfrac{dQ_1}{T_1} = -\dfrac{dQ_2}{T_2}$. Hence

$$\frac{dT_1}{T_1} = -\frac{dT_2}{T_2} .$$

Thus $\displaystyle\int_{T_1}^{T_f} \frac{dT_1}{T_1} = -\int_{T_2}^{T_f} \frac{dT_2}{T_2}, \quad \ln\frac{T_f}{T_1} = -\ln\frac{T_f}{T_2},$

Therefore $T_f = \sqrt{T_1 T_2}$.

(b) Conservation of energy gives

$$W = (U_1 - U) - (U - U_2) = U_1 + U_2 - 2U$$
$$= NC(T_1 + T_2 - 2T_f) .$$

1036

Water powered machine. A self-contained machine only inputs two equal steady streams of hot and cold water at temperatures T_1 and T_2. Its only output is a single high-speed jet of water. The heat capacity per unit mass of water, C, may be assumed to be independent of temperature. The machine is in a steady state and the kinetic energy in the incoming streams is negligible.

(a) What is the speed of the jet in terms of T_1, T_2 and T, where T is the temperature of water in the jet?

(b) What is the maximum possible speed of the jet?

(*MIT*)

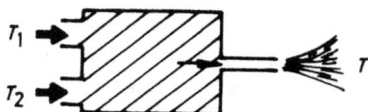

Fig. 1.13.

Solution:

(a) The heat intake per unit mass of water is

$$\Delta Q = [C(T_1 - T) - C(T - T_2)]/2 \; .$$

As the machine is in a steady state, $v^2/2 = \Delta Q$, giving

$$v = \sqrt{C(T_1 + T_2 - 2T)} \; .$$

(b) Since the entropy increase is always positive, i.e.,

$$\Delta S = \frac{1}{2}C\left[\ln\frac{T}{T_1} + \ln\frac{T}{T_2}\right] \geq 0 \; ,$$

we have

$$T \geq \sqrt{T_1 T_2} \; .$$

Thus $v \leq v_{\max} = \sqrt{C(T_1 + T_2 - 2\sqrt{T_1 T_2})}$.

1037

In the water behind a high power dam (110 m high) the temperature difference between surface and bottom may be 10°C. Compare the possible energy extraction from the thermal energy of a gram of water with that generated by allowing the water to flow over the dam through turbines in the conventional way.

(Columbia)

Solution:

The efficiency of a perfect engine is

$$\eta = 1 - T_{\text{low}}/T_{\text{high}} \; .$$

The energy extracted from one gram of water is then

$$W = \eta Q = \left(1 - \frac{T_{\text{low}}}{T_{\text{high}}}\right) \cdot C_v(T_{\text{high}} - T_{\text{low}}) \; ,$$

where Q is the heat extracted from one gram of water, C_v is the specific heat of one gram of water. Thus

$$W = C_v(T_{\text{high}} - T_{\text{low}})^2/T_{\text{high}} \; .$$

If T_{high} can be taken as the room temperature, then

$$W = 1 \times 10^2/300 = 0.3 \text{ cal} .$$

The energy generated by allowing the water to flow over the dam is

$$W' = mgh = 1 \times 980 \times 100 \times 10^2$$
$$= 10^7 \text{ erg} = 0.24 \text{ cal} .$$

We can see that under ideal conditions $W' < W$. However, the efficiency of an actual engine is much less than that of a perfect engine. Therefore, the method by which we generate energy from the water height difference is still more efficient.

1038

Consider an engine working in a reversible cycle and using an ideal gas with constant heat capacity c_p as the working substance. The cycle consists of two processes at constant pressure, joined by two adiabatics.

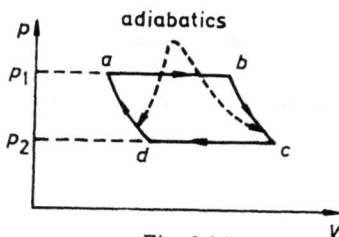

Fig. 1.14.

(a) Find the efficiency of this engine in terms of p_1, p_2.

(b) Which temperature of T_a, T_b, T_c, T_d is highest, and which is lowest?

(c) Show that a Carnot engine with the same gas working between the highest and lowest temperatures has greater effficiency than this engine.

(Columbia)

Solution:

(a) In the cycle, the energy the working substance absorbs from the source of higher temperature is

$$Q_{\text{ab}} = c_p(T_b - T_a) .$$

The energy it gives to the source of lower temperature is $Q_{gi} = c_p(T_c - T_d)$. Thus

$$\eta = 1 - Q_{gi}/Q_{ab} = 1 - \frac{T_c - T_d}{T_b - T_a} .$$

From the equation of state $pV = nRT$ and the adiabatic equations

$$p_2 V_d^\gamma = p_1 V_a^\gamma , \quad p_2 V_c^\gamma = p_1 V_b^\gamma ,$$

we have

$$\eta = 1 - \left(\frac{p_2}{p_1}\right)^{\frac{\gamma-1}{\gamma}} .$$

(b) From the state equation, we know $T_b > T_a, T_c > T_d$; from the adiabatic equation, we know $T_b > T_c, T_a > T_d$; thus

$$T_b = \max(T_a, T_b, T_c, T_d) ,$$
$$T_d = \min(T_a, T_b, T_c, T_d) .$$

(c) $\eta_c = 1 - \dfrac{T_d}{T_b} > 1 - \dfrac{T_d}{T_a} = 1 - \left(\dfrac{p_2}{p_1}\right)^{\frac{\gamma-1}{\gamma}} = \eta .$

1039

A building at absolute temperature T is heated by means of a heat pump which uses a river at absolute temperature T_0 as a source of heat. The heat pump has an ideal performance and consumes power W. The building loses heat at a rate $\alpha(T - T_0)$, where α is a constant.

(a) Show that the equilibrium temperature T_e of the building is given by

$$T_e = T_0 + \frac{W}{2\alpha}\left[1 + \left(1 + \frac{4\alpha T_0}{W}\right)^{\frac{1}{2}}\right] .$$

(b) Suppose that the heat pump is replaced by a simple heater which also consumes a constant power W and which converts this into heat with 100% efficiency. Show explicitly why this is less desirable than a heat pump.

(Columbia)

Solution:

(a) The rate of heat from the pump is

$$Q = \frac{W}{\eta} = \frac{W}{1 - (T_0/T)} \ .$$

At equilibrium, $T = T_e$ and $Q = Q_e = \alpha(T_e - T_0)$. Thus

$$T_e = T_0 + \frac{W}{2\alpha}\left[1 + \left(1 + \frac{4\alpha T_0}{W}\right)^{1/2}\right] \ .$$

(b) In this case, the equilibrium condition is

$$W = \alpha(T'_e - T_0) \ .$$

Thus

$$T'_e = T_0 + \frac{W}{\alpha} < T_e \ .$$

Therefore it is less desirable than a heat pump.

1040

A room at temperature T_2 loses heat to the outside at temperature T_1 at a rate $A(T_2 - T_1)$. It is warmed by a heat pump operated as a Carnot cycle between T_1 and T_2. The power supplied by the heat pump is dW/dT.

(a) What is the maximum rate dQ_m/dt at which the heat pump can deliver heat to the room? What is the gain dQ_m/dW? Evaluate the gain for $t_1 = 2°C$, $t_2 = 27°C$.

(b) Derive an expression for the equilibrium temperature of the room, T_2, in terms of T_1, A and dW/dt.

(UC, Berkeley)

Solution:

(a) From $dQ_m \cdot (T_2 - T_1)/T_2 = dW$, we get

$$\frac{dQ_m}{dt} = \frac{T_2}{T_2 - T_1}\frac{dW}{dt} \ ,$$

With $T_1 = 275K$, $T_2 = 300K$, we have $dQ_m/dW = 12$.

(b) When equilibrium is reached, one has

$$A(T_2 - T_1) = \frac{T_2}{T_2 - T_1}\frac{dW}{dt} \ ,$$

giving

$$T_2 = T_1 + \frac{1}{2A}\left(\frac{dW}{dt}\right) + \frac{1}{2A}\sqrt{\left(\frac{dW}{dt}\right)^2 + 4AT_1\left(\frac{dW}{dt}\right)} \ .$$

1041

A building at a temperature T (in K) is heated by an ideal heat pump which uses the atmosphere at $T_0(\mathrm{K})$ as heat source. The pump consumes power W and the building loses heat at a rate $\alpha(T - T_0)$. What is the equilibrium temperature of the building?

(MIT)

Solution:

Let T_e be the equilibrium temperature. Heat is given out by the pump at the rate $Q_1 = W/\eta$, where $\eta = 1 - T_0/T_e$. At equilibrium $Q_1 = \alpha(T_e - T_0)$, so that

$$W = \frac{\alpha}{T_e}(T_e - T_0)^2 \ ,$$

from which we get

$$T_e = T_0 + \frac{W}{2\alpha} + \sqrt{T_0\frac{W}{\alpha} + \left(\frac{W}{2\alpha}\right)^2} \ .$$

1042

Let M represent a certain mass of coal which we assume will deliver 100 joules of heat when burned – whether in a house, delivered to the radiators or in a power plant, delivered at 1000°C. Assume the plant is ideal (no waste in turbines or generators) discharging its heat at 30°C to a river. How much heat will M, burned at the plant to generate electricity, provide for the house when the electricity is:

(a) delivered to residential resistance-heating radiators?

(b) delivered to a residential heat pump (again assumed ideal) boosting heat from a reservoir at 0°C into a hot-air system at 30°C?

(Wisconsin)

Solution:

When M is burned in the power plant, the work it provides is

$$W = Q_1 \eta = Q_1 \left(1 - \frac{T_2}{T_1}\right) = 100 \left(1 - \frac{273 + 30}{273 + 1000}\right)$$

$$= 76.2 J .$$

This is delivered in the form of electric energy.

(a) When it is delivered to residential resistance-heating radiators, it will transform completely into heat: $Q' = W = 76.2 J$.

(b) When the electricity is delivered to a residential heat pump, heat flows from a source of lower temperature to a system at higher temperature, the working efficiency being

$$\varepsilon = \frac{T_1}{T_1 - T_2} = 273/30 = 9.1 .$$

Hence the heat provided for the house is

$$Q' = (1 + \varepsilon)W = 770 \text{ J} .$$

1043

An air conditioner is a device used to cool the inside of a home. It is, in essence, a refrigerator in which mechanical work is done and heat removed from the (cooler) inside and rejected to the (warmer) outside.

A home air conditioner operating on a reversible Carnot cycle between the inside, absolute temperature T_2, and the outside, absolute temperature $T_1 > T_2$, consumes P joules/sec from the power lines when operating continuously.

(a) In one second, the air conditioner absorbs Q_2 joules from the house and rejects Q_1 joules outdoors. Develop a formula for the efficiency ratio Q_2/P in terms of T_1 and T_2.

(b) Heat leakage into the house follows Newton's law $Q = A(T_1 - T_2)$. Develop a formula for T_2 in terms of T_1, P, and A for continuous operation of the air conditioner under constant outside temperature T_1 and uniform (in space) inside temperature T_2.

(c) The air conditioner is controlled by the usual on-off thermostat and it is observed that when the thermostat set at 20°C and an outside temperature at 30°, it operates 30% of the time. Find the highest outside temperature, in °C, for which it can maintain 20°C inside (use $-273°$C for absolute zero).

(d) In the winter, the cycle is reversed and the device becomes a heat pump which absorbs heat from outside and rejects heat into the house. Find the lowest outside temperature in °C for which it can maintain 20°C inside.

(*CUSPEA*)

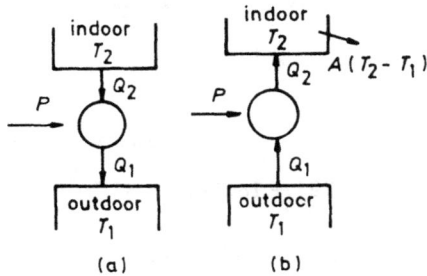

Fig. 1.15.

Solution:

(a) From the first and second thermodynamic laws, we have

$$Q_1 = P + Q_2, \quad Q_2/T_2 = Q_1/T_1 .$$

Hence

$$\frac{Q_2}{P} = \frac{T_2}{T_1 - T_2} .$$

(b) At equilibrium, heat leakage into the house is equal to the heat transfered out from the house, i.e., $Q_2 = A(T_1 - T_2)$. We obtain, using the result in (a),

$$\frac{T_2 P}{T_1 - T_2} = A(T_1 - T_2) .$$

Hence

$$T_2 = T_1 + \frac{1}{2}\left(\frac{P}{A} \pm \sqrt{\left(\frac{P}{A}\right)^2 + 4T_1\frac{P}{A}}\right) .$$

In view of the fact $T_2 < T_1$, the solution is

$$T_2 = T_1 + \frac{1}{2}\left(\frac{P}{A} - \sqrt{\left(\frac{P}{A}\right)^2 + 4T_1\frac{P}{A}}\right) .$$

(c) When the air conditioner works 30% of the time, we know from (b)

$$P_{30\%} = A \cdot \frac{(T_1 - T_2)^2}{T_2} = A \cdot \frac{10^2}{293} = A \cdot \frac{100}{293} \ .$$

When it operates continuously, we have

$$P = P_{30\%} \cdot \frac{100}{30} = A \cdot \frac{100}{30} \cdot \frac{100}{239} \approx 1.1377A \ .$$

With $T_2 = 20°C = 293K$, we get

$$T_1 = T_2 + \sqrt{T_2 \frac{P}{A}} = 293 + \sqrt{\frac{100^2}{30}}$$
$$= 293 + 18.26 \ K = 38.26°C \ .$$

(d) When the cycle is reversed in winter, we have $Q_2 = P + Q_1$ and $\frac{Q_2}{T_2} = \frac{Q_1}{T_1}$. At equilibrium, $Q_2 = A(T_2 - T_1)$, so that

$$T_2 - T_1 = \sqrt{T_2 \frac{P}{A}} \ .$$

Thus $T_1 = T_2 - \sqrt{\frac{P}{A} T_2} = 293 - (1.14 \times 293)^{1/2} = 275K = 2°C.$

1044

Calculate the change of entropy involved in heating a gram-atomic weight of silver at constant volume from 0° to 30°C. The value of C_v over this temperature may be taken as a constant equal to 5.85 cal/deg·mole.

(Wisconsin)

Solution:

The change of entropy is

$$\Delta S = n \int_{T_1}^{T_2} \frac{C_v dT}{T} = nC_v \ln \frac{T_2}{T_1} = 5.85 \ln \frac{30 + 273}{273}$$
$$= 0.61 \ cal/K \ .$$

1045

A body of constant heat capacity C_p and a temperature T_i is put into contact with a reservoir at temperature T_f. Equilibrium between the body and the reservoir is established at constant pressure. Determine the total entropy change and prove that it is positive for either sign of $(T_f - T_i)/T_f$. You may regard $|T_f - T_i|/T_f < 1$.

(*Wisconsin*)

Solution:

We assume $T_i \neq T_f$ (because the change of entropy must be zero when $T_i = T_f$). The change of entropy of the body is

$$\Delta S_1 = \int_{T_i}^{T_f} \frac{C_p dT}{T} = C_p \ln \frac{T_f}{T_i} .$$

The change of entropy of the heat source is

$$\Delta S_2 = \frac{\Delta Q}{T_f} = \frac{C_p(T_i - T_f)}{T_f} .$$

Therefore the total entropy change is

$$\Delta S = \Delta S_1 + \Delta S_2 = C_p \left(\frac{T_i}{T_f} - 1 + \ln \frac{T_f}{T_i} \right) .$$

When $x > 0$ and $x \neq 1$, the function $f(x) = x - 1 - \ln x > 0$. Therefore

$$\Delta S = C_p f \left(\frac{T_i}{T_f} \right) > 0 .$$

1046

One kg of H_2O at $0°C$ is brought in contact with a heat reservoir at $100°C$. When the water has reached $100°C$,

(a) what is the change in entropy of the water?

(b) what is the change in entropy of the universe?

(c) how could you heat the water to $100°C$ so the change in entropy of the universe is zero?

(*Wisconsin*)

Solution:

The process is irreversible. In order to calculate the change of entropy of the water and of the whole system, we must construct a reversible process which has the same initial and final states as the process in this problem.

(a) We assume the process is a reversible process of constant pressure. The change in entropy of the water is

$$\Delta S_{H_2O} = \int_{273}^{373} mC_{H_2O} dT/T = mC_{H_2O} \ln(373/273) \ .$$

We substitute $m = 1$kg, and $C_{H_2O} = 4.18$ J/g into it, and find

$$\Delta S_{H_2O} = 1305 \text{ J/K} \ .$$

(b) The change in entropy of the heat source is

$$\Delta S_{hs} = -|Q|/T = -1000 \times 4.18 \times 100/373$$
$$= -11121 \text{ J/K} \ .$$

Therefore the change of entropy of the whole system is

$$\Delta S = \Delta S_{H_2O} + \Delta S_{hs} = 184 \text{ J/K} \ .$$

(c) We can imagine infinitely many heat sources which have infinitesimal temperature difference between two adjacent sources from 0°C to 100°C. The water comes in contact with the infinitely many heat sources in turn in the order of increasing temperature. This process which allows the temperature of the water to increase from 0°C to 100°C is reversible; therefore $\Delta S = 0$.

1047

Compute the difference in entropy between 1 gram of nitrogen gas at a temperature of 20°C and under a pressure of 1 atm, and 1 gram of liquid nitrogen at a temperature -196°C, which is the boiling point of nitrogen, under the same pressure of 1 atm. The latent heat of vaporization of nitrogen is 47.6 cal/gm. Regard nitrogen as an ideal gas with molecular weight 28, and with a temperature-independent molar specific heat at constant pressure equal to 7.0 cal/mol·K.

(UC, Berkeley)

Solution:

The number of moles of 1g nitrogen is

$$n = 1/28 = 3.57 \times 10^{-2} \text{mol}.$$

The entropy difference of an ideal gas at 20°C and at −196°C is

$$\Delta S' = nC_p \ln(T_1/T_2) = 0.33 \text{ cal/K} ,$$

and the entropy change at phase transition is

$$\Delta S'' = nL/T_2 = 0.64 \text{ cal/K} .$$

Therefore $\Delta S = \Delta S' + \Delta S'' = 0.97 \text{ cal/K}.$

1048

A Carnot engine is made to operate as a refrigerator. Explain in detail, with the aid of (a) a pressure-volume diagram, (b) an enthalpy-entropy diagram, all the processes which occur during a complete cycle or operation.

This refrigerator freezes water at 0°C and heat from the working substance is discharged into a tank containing water maintained at 20°C. Determine the minimum amount of work required to freeze 3 kg of water.

(SUNY, Buffalo)

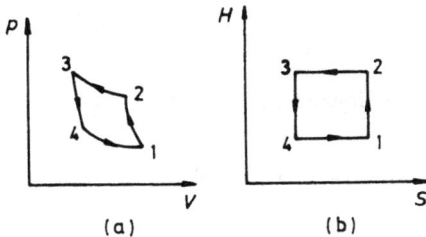

Fig. 1.16.

Solution:

(a) As shown in Fig. 1.16(a),

1-2: adiabatic compression,

2-3: isothermal compression,

3-4: adiabatic expansion,

4-1: isothermal expansion.

(b) As shown in Fig. 1.16(b):

1-2: Adiabatic compression. The entropy is conserved.

2-3: Isothermal compression. If the working matter is an ideal gas, the enthalpy is conserved.

3-4: Adiabatic expansion. The entropy is conserved.

4-1: Isothermal expansion. The enthalpy is conserved.

The refrigeration efficiency is

$$\eta = \frac{Q_2}{W} = \frac{T_2}{(T_1 - T_2)} \ .$$

Hence

$$W = Q_2 \frac{T_1 - T_2}{T_2} \ .$$

$Q_2 = ML$ is the latent heat for $M = 3$ kg of water at $T = 0°C$ to become ice. As

$$L = 3.35 \times 10^5 \text{ J/kg} \ ,$$

we find $W = 73.4 \times 10^3$ J.

1049

$n = 0.081$ kmol of He gas initially at $27°C$ and pressure $= 2 \times 10^5 \text{N/m}^2$ is taken over the path $A \to B \to C$. For He

$$C_v = 3R/2 \ , \quad C_p = 5R/2 \ .$$

Assume the ideal gas law.

(a) How much work does the gas do in expanding at constant pressure from $A \to B$?

(b) What is the change in thermal or internal energy of the helium from $A \to B$?

(c) How much heat is absorbed in going from $A \to B$?

(d) If $B \to C$ is adiabatic, what is the entropy change and what is the final pressure?

(*Wisconsin*)

Solution:

(a) For $A \to B$, the external work is

$$W = p_A(V_B - V_A) = 1.0 \times 10^5 \text{ J} \ .$$

(b) For $A \to B$, the increase of the internal energy is

$$\Delta U = nC_v \Delta T = C_v p_A (V_B - V_A)/R = 3W/2 = 1.5 \times 10^5 \text{ J} .$$

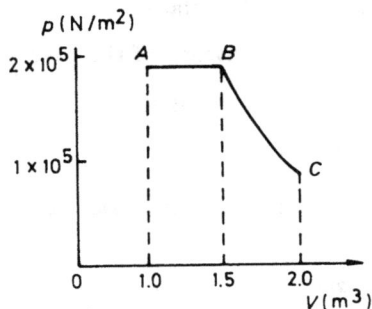

Fig. 1.17.

(c) By the first law of thermodynamics, the heat absorbed during $A \to B$ is $W + \Delta U = 2.5 \times 10^5$ J.

(d) For $B \to C$, the adiabatic process of an monatomic ideal gas satisfies the equation

$$pV^\gamma = \text{const.} , \quad \text{where } \gamma = C_p/C_v = 5/3 .$$

Thus $p_B V_B^\gamma = p_C V_C^\gamma$ and $p_C = (V_B/V_C)^\gamma p_B = 1.24 \times 10^5 \text{ N/m}^2$.
In the process of reversible adiabatic expansion, the change in entropy is $\Delta S = 0$. This is shown by the calculation in detail as follows:

$$\Delta S = nC_v \ln \frac{T_C}{T_B} + nR \ln \frac{V_C}{V_B}$$

$$= nC_v \ln \frac{T_C V_C^{\gamma-1}}{T_B V_B^{\gamma-1}} = 0 .$$

1050

A mole of an ideal gas undergoes a reversible isothermal expansion from volume V_1 to $2V_1$.

    (a) What is the change in entropy of the gas?

    (b) What is the change in entropy of the universe?

Suppose the same expansion takes place as a free expansion:

    (a) What is the change in entropy of the gas?

    (b) What is the change in the entropy of the universe?

(Wisconsin)

Solution:

    (a) In the process of isothermal expansion, the external work done by the system is

$$W = \int_{V_1}^{2V_1} p\,dV = RT \int_{V_1}^{2V_1} \frac{dV}{V} = RT \ln 2 \ .$$

Because the internal energy does not change in this process, the work is supplied by the heat absorbed from the external world. Thus the increase of entropy of the gas is

$$\Delta S_1 = \frac{\Delta Q}{T} = \frac{W}{T} = R \ln 2 \ .$$

    (b) The change in entropy of the heat source $\Delta S_2 = -\Delta S_1$, thus the total change in entropy of the universe is

$$\Delta S = \Delta S_1 + \Delta S_2 = 0 \ .$$

If it is a free expansion, the internal energy of the system is constant. As its final state is the same as for the isothermal process, the change in entropy of the system is also the same. In this case, the state of the heat source does not change, neither does its entropy. Therefore the change in entropy of the universe is $\Delta S = R \ln 2$.

1051

    N atoms of a perfect gas are contained in a cylinder with insulating walls, closed at one end by a piston. The initial volume is V_1 and the initial temperature T_1.

(a) Find the change in temperature, pressure and entropy that would occur if the volume were suddenly increased to V_2 by withdrawing the piston.

(b) How rapidly must the piston be withdrawn for the above expressions to be valid?

(*MIT*)

Fig. 1.18.

Solution:

(a) The gas does no work when the piston is withdrawn rapidly. Also, the walls are thermally insulating, so that the internal energy of the gas does not change, i.e., $dU = 0$. Since the internal energy of an ideal gas is only dependent upon temperature T, the change in temperature is 0, i.e., $T_2 = T_1$. As for the pressure, $p_2/p_1 = V_1/V_2$. The increase in entropy is

$$S_2 - S_1 = \int_{V_1}^{V_2} \frac{p}{T} dV = Nk \ln \frac{V_2}{V_1} .$$

(b) The speed at which the piston is withdrawn must be far greater than the mean speed of the gas molecules, i.e., $v \gg \bar{v} = (8kT_1/\pi m)^{1/2}$.

1052

A cylinder contains a perfect gas in thermodynamic equilibrium at p, V, T, U (internal energy) and S (entropy). The cylinder is surrounded by a very large heat reservoir at the same temperature T. The cylinder walls and piston can be either perfect thermal conductors or perfect thermal insulators. The piston is moved to produce a small volume change $\pm \Delta V$. "Slow" or "fast" means that during the volume change the speed of the piston is very much less than, or very much greater than, molecular speeds at temperature T. For each of the five processes below show (on your answer sheet) whether the changes (after the reestablishment of equilibrium) in the other quantities have been positive, negative, or zero.

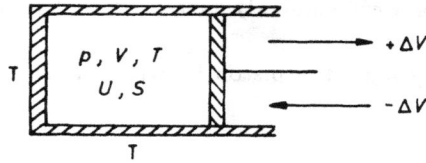

Fig. 1.19.

	ΔT	ΔU	ΔS	Δp

1. $(+\Delta V)$ (slow) (conduct)
2. $(+\Delta V)$ (slow) (insulate)
3. $(+\Delta V)$ (fast) (insulate)
4. $(+\Delta V)$ (fast) (conduct)
5. $(-\Delta V)$ (fast) (conduct)

(*Wisconsin*)

Solution:

(1) For isothermal expansion, $\Delta T = 0, \Delta U = 0$, and

$$\Delta S = R\frac{\Delta V}{V} > 0, \quad \Delta p = \frac{-p}{V}\Delta V < 0 \ .$$

(2) For adiabatic expansion, $\Delta Q = 0$. Because the process proceeds very slowly it can be taken as a reversible process of quasistatic states, then $\Delta S = 0$. The adiabatic process satisfies $pV^\gamma = $ const. While V increases, p decreases, i.e., $\Delta p < 0$; and the internal energy of the system decreases because it does work externally, thus $\Delta U < 0$, or $\Delta T < 0$.

(3) The process is equivalent to adiabatic free expansion of an ideal gas, thus $\Delta S > 0$, $\Delta U = 0$, $\Delta T = 0$, $\Delta p < 0$.

(4) The result is as the same as that of isothermal free expansion, thus $\Delta T = 0$, $\Delta U = 0$, $\Delta S > 0$, $\Delta p < 0$.

(5) The result is the same as that of isothermal free compression, thus $\Delta T = 0$, $\Delta U = 0$, $\Delta p > 0$, $\Delta S < 0$.

The above are summarized in the table below.

	ΔT	ΔU	ΔS	Δp
1	0	0	+	−
2	−	−	0	−
3	0	0	+	−
4	0	0	+	−
5	0	0	−	+

1053

A thermally insulated box is separated into two compartments (volumes V_1 and V_2) by a membrane. One of the compartments contains an ideal gas at temperature T; the other is empty (vacuum). The membrane is suddenly removed, and the gas fills up the two comparments and reaches equilibrium.

(a) What is the final temperature of the gas?

(b) Show that the gas expansion process is irreversible.

(*MIT*)

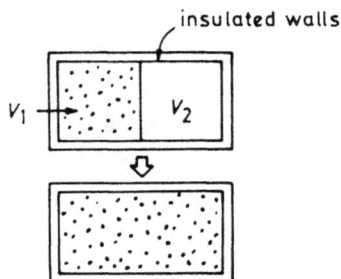

Fig. 1.20.

Solution:

(a) Freely expanding gas does no external work and does not absorb heat. So the internal energy does not change, i.e., $dU = 0$. The internal energy of an ideal gas is only a function of temperature; as the temperature does not change in the process, $T_f = T$.

(b) Assuming a quasi-static process of isothermal expansion, we can calculate the change in entropy resulting from the free expansion. In the process, we have $dS = pdV/T, pV = NkT$. Hence,

$$S_f - S = \int dS = \int_{V_1}^{V_1 + V_2} \frac{p}{T} dV = Nk \ln \frac{V_1 + V_2}{V_1} > 0 \; .$$

Thus the freely expanding process of the gas is irreversible.

1054

A thermally conducting, uniform and homogeneous bar of length L, cross section A, density ρ and specific heat at constant pressure c_p is brought to a nonuniform temperature distribution by contact at one end with a hot reservoir at a temperature T_H and at the other end with a cold reservoir at a temperature T_c. The bar is removed from the reservoirs, thermally insulated and kept at constant pressure. Show that the change in entropy of the bar is

$$\Delta S = C_p \left(1 + \ln T_f + \frac{T_c}{T_H - T_c} \ln T_c - \frac{T_H}{T_H - T_c} \ln T_H \right) \; ,$$

where $C_p = c_p \rho A L$, $T_f = (T_H + T_c)/2$.

(SUNY, Buffalo)

Solution:

As the temperature gradient in the bar is $(T_H - T_c)/L$, the temperature at the cross section at a distance x from the end at T_c can be expressed by $T_x = T_c + (T_H - T_c)x/L$. As the bar is adiabatically removed, we have

$$\int_0^L \rho c_p [T(x) - T_f] dx = 0 \; ,$$

from which we obtain $T_f = (T_H + T_c)/2$.
But $c_p = T(\partial S/\partial T)_p$

$$\Delta S = c_p \rho A \int_0^L dx \int_{T(x)}^{(T_H + T_c)/2} \frac{dT}{T} = C_p \left[1 + \ln T_f + \frac{T_c}{T_H - T_c} \ln T_c \right.$$
$$\left. - \frac{T_H}{T_H - T_c} \ln T_H \right] \; ,$$

where $C_p = c_p \rho A L$.

1055

A mixture of 0.1 mole of helium ($\gamma_1 = C_p/C_v = 5/3$) with 0.2 mole of nitrogen ($\gamma_2 = 7/5$), considered an ideal mixture of two ideal gases, is initially at 300K and occupies 4 litres. Show that the changes of temperature and pressure of the system which occur when the gas is compressed slowly and adiabatically can be described in terms of some intermediate value of γ. Calculate the magnitude of these changes when the volume is reduced by 1%.

(*UC, Berkeley*)

Solution:

The entropy change for an ideal gas is

$$\Delta S = nC_v \ln(T_f/T_i) + nR \ln(V_f/V_i) \ ,$$

where n is the mole number, i and f indicate initial and final states respectively. As the process is adiabatic the total entropy change in the nitrogen gas and helium gas must be zero, that is, $\Delta S_1 + \Delta S_2 = 0$. The expression for ΔS then gives

$$T_f = T_i \left(\frac{V_i}{V_f}\right)^g \ ,$$

where

$$g = \frac{(n_1 + n_2)R}{n_1 C_{v1} + n_2 C_{v2}} \ .$$

Together with the equation of state for ideal gas, it gives

$$p_f = p_i \left(\frac{V_i}{V_f}\right)^\gamma \ ,$$

where

$$\gamma = \frac{n_1 C_{p1} + n_2 C_{p2}}{n_1 C_{v1} + n_2 C_{v2}} \ .$$

Helium is monatomic, so that $C_{v1} = 3R/2$, $C_{p_1} = 5R/2$; nitrogen is diatomic, so that $C_{v2} = 5R/2$, $C_{p2} = 7/2$. Consequently, $\gamma = 1.46$.

When $V_f = 0.99V_i$, we have

$$T_f = 1.006T_i = 302 \text{ K} \ ,$$
$$p_f = 1.016p_i = 1.016nRT/V = 2.0 \times 10^5 \text{ N/m}^2$$

1056

Consider two ways to mix two perfect gases. In the first, an adiabatically isolated container is divided into two chambers with a pure gas A in the left hand side and a pure gas B in the right. The mixing is accomplished by opening a hole in the dividing wall.

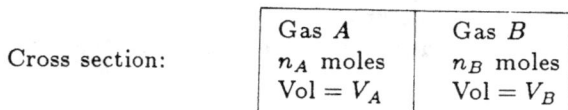

Cross section:

Gas A	Gas B
n_A moles	n_B moles
Vol $= V_A$	Vol $= V_B$

Fig. 1.21(a).

In the second case the chamber is divided by two rigid, perfectly selective membranes, the membrane on the left is perfectly permeable to gas A but impermeable to gas B. The membrane on the right is just the reverse. The two membranes are connected by rods to the outside and the whole chamber is connected to a heat reservoir at temperature T. The gases can be mixed in this case by pulling left hand membrane to the left and the right hand one to the right.

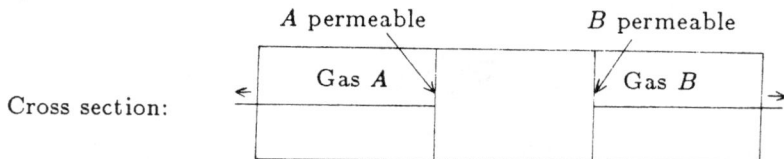

Cross section:

A permeable B permeable

Gas A Gas B

Fig. 1.21(b).

(a) Find the change in entropy of the container and its contents for second process.

(b) Find the change in entropy of the container and contents for the first process.

(c) What is the change in entropy of the heat reservoir in part (a)?

(*CUSPEA*)

Solution:

(a) Because the process is reversible, we have

$$\Delta S = \int \frac{dQ}{T} = \frac{1}{T} \left(\int_{V_A}^{V_A+V_B} p_A dV + \int_{V_A}^{V_A+V_B} p_B dV \right)$$

$$= R \left(n_A \ln \frac{V_A + V_B}{V_A} + n_B \ln \frac{V_A + V_B}{V_B} \right) ,$$

where we have made use of the equation of state $pV = nRT$.

(b) Because energy is conserved and the internal energy of an ideal gas is related only to its temperature, the temperatures of the initial and final states are the same. The initial and final states of the gas in this case are identical with those in case (a). As entropy is a function of state, ΔS is equal to that obtained in (a).

(c) $\Delta S_{\text{heat source}} = -\Delta S$, where ΔS is that given in (a).

1057

Consider a cylinder with a frictionless piston composed of a semi-permeable membrane permeable to water only. Let the piston separate a volume V of N moles of pure water from a volume V' of a dilute salt (NaCl) solution. There are N' moles of water and n moles of the salt in the solution. The system is in contact with a heat reservoir at temperature T.

(a) Evaluate an expression for entropy of mixing in the salt solution.

(b) If the piston moves so that the amount of water in the salt solution doubles, how much work is done?

(c) Derive an expression for the pressure π across the semipermeable membrane as a function of the volume of the salt solution.

(*Princeton*)

Solution:

(a) The entropy of mixing, i.e., the increase of entropy during mixing isothermally and isobarically is

$$\Delta S = -N'R \ln \frac{N'}{N'+n} - nR \ln \frac{n}{N'+n} \ .$$

(b) The osmotic pressure of a dilute solution is

$$\pi V' = nRT \quad \text{(Van't Hoff's law)} \ .$$

When the amount of water in the salt solution doubles, the work done is

$$W = \int_{V'}^{2V'} \pi dV = \int_{V'}^{2V'} \frac{nRT}{V} dV = nRT \ln 2 \ .$$

(c) $\pi = nRT/V'$. The osmotic pressure, i.e., the pressure difference across the membrane, is the net and effective pressure on the membrane.

1058

(a) In the big-bang theory of the universe, the radiation energy initially confined in a small region adiabatically expands in a spherically symmetric manner. The radiation cools down as it expands. Derive a relation between the temperature T and the radius R of the spherical volume of radiation, based purely on thermodynamic considerations.

(b) Find the total entropy of a photon gas as a function of its temperature T, volume V, and the constants $k, \hbar, c$.

(SUNY, Buffalo)

Solution:

(a) The expansion can be treated as a quasi-static process. We then have $dU = TdS - pdV$. Making use of the adiabatic condition $dS = 0$ and the expression for radiation pressure $p = U/3V$, we obtain $dU/U = -dV/3V$; hence $U \propto V^{-1/3}$. The black body radiation energy density is $u = U/V = aT^4$, a being a constant. The above give $T^4 \propto V^{-4/3} \propto R^{-4}$, so that $T \propto R^{-1}$, i.e., $RT =$ constant.

(b) $dS = \dfrac{dU}{T} + \dfrac{p}{T}dV = \dfrac{V}{T}du + \dfrac{4}{3}\dfrac{u}{T}dV = d\left(\dfrac{4}{3}aT^3V\right)$, from which we obtain $S = \dfrac{4}{3}aT^3V$. By dimensional analysis we find $a \sim k^4/(\hbar c)^3$. In fact, $a = \dfrac{\pi^2}{15}\dfrac{k^4}{(\hbar c)^3}$, so that $S = \dfrac{4\pi^2}{45}\dfrac{k^4}{(\hbar c)^3}T^3V$.

1059

(a) A system, maintained at constant volume, is brought in contact with a thermal reservoir at temperature T_f. If the initial temperature of the system is T_i, calculate ΔS, change in the total entropy of the system + reservoir. You may assume that c_v, the specific heat of the system, is independent of temperature.

(b) Assume now that the change in system temperature is brought about through successive contacts with N reservoirs at temperature $T_i + \Delta T, T_i + 2\Delta T, \ldots, T_f - \Delta T, T_f$, where $N\Delta T = T_f - T_i$. Show that in the limit $N \to \infty, \Delta T \to 0$ with $N\Delta T = T_f - T_i$ fixed, the change in entropy of the system + reservoir is zero.

(c) Comment on the difference between (a) and (b) in the light of the second law of thermodynamics.

(SUNY, Buffalo)

Solution:

(a) The change in entropy of the system is

$$\Delta S_1 = \int_{T_i}^{T_f} \frac{M c_v \, dT}{T} = M c_v \ln \frac{T_f}{T_i} \; .$$

The change in entropy of the heat source is

$$\Delta S_2 = -\frac{|Q|}{T_f} = M c_v \frac{T_i - T_f}{T_f} \; .$$

The total change in entropy is

$$\Delta S = \Delta S_1 + \Delta S_2 = M c_v \left(\ln \frac{T_f}{T_i} + \frac{T_i - T_f}{T_f} \right) \; .$$

(b)

$$\Delta S = \lim_{\substack{\Delta T \to 0 \\ \Delta N \to \infty}} \sum_{n=0}^{N-1} \Delta S_n \; ,$$

where

$$\Delta S_n = M c_v \left(\ln \frac{T_i + (n+1)\Delta T}{T_i + n\Delta T} - \frac{\Delta T}{T_i + (n+1)\Delta T} \right)$$

is the change in entropy of the $(n+1)$th contact. Thus

$$\Delta S = M c_v \left(\ln \frac{T_f}{T_i} - \int_{T_i}^{T_f} \frac{dT}{T} \right) = 0 \; .$$

(c) The function $f(x) = x - \ln x - 1 > 0$ if $x > 0$ and $x \neq 1$. Thus in (a) $\Delta S = M c_v f(T_i/T_f) > 0$, that is, the entropy is increased. We know the process is irreversible from the second law of thermodynamics. In (b) $\Delta S = 0$, the process is reversible.

1060

A material is brought from temperature T_i to temperature T_f by placing it in contact with a series of N reservoirs at temperatures $T_i + \Delta T, T_i +$

$2\Delta T, \ldots, T_i + N\Delta T = T_f$. Assuming that the heat capacity of the material, C, is temperature independent, calculate the entropy change of the total system, material plus reservoirs. What is the entropy change in the limit $N \to \infty$ for fixed $T_f - T_i$?

(Wisconsin)

Solution:

Consider the material at temperature $T_i + t\Delta T$ in contact with the reservoir at temperature $T_i + (t + 1)\Delta T$. When they come to thermal equilibrium, the change of entropy of the material is

$$\Delta S_1 = \int_{T_i + t\Delta T}^{T_i + (t+1)\Delta T} \frac{CdT}{T} = C \ln \frac{T_i + (t + 1)\Delta T}{T_i + t\Delta T} .$$

The change in entropy of the heat reservoir is

$$\Delta S_2 = -\frac{C\Delta T}{T_i + (t + 1)\Delta T} .$$

The total change in entropy is

$$\Delta S_t = \Delta S_1 + \Delta S_2 = C \left(\ln \frac{T_i + (t + 1)\Delta T}{T_i + t\Delta T} - \frac{\Delta T}{T_i + (t + 1)\Delta T} \right) .$$

Therefore, after the material of initial temperature T_i has had contacts with the series of reservoirs, the total change of entropy of the whole system is

$$\Delta S = \sum_{t=0}^{N-1} \Delta S_t = C \left(\ln \frac{T_f}{T_i} - \sum_{t=0}^{N-1} \frac{\Delta T}{T_i + (t + 1)\Delta T} \right) .$$

When $N \to \infty$, or $\Delta T \to 0$, the above sum can be written as an integration, so that

$$\Delta S = C \left(\ln \frac{T_f}{T_i} - \int_{T_i}^{T_f} \frac{dT}{T} \right) = C \left(\ln \frac{T_f}{T_i} - \ln \frac{T_f}{T_i} \right) = 0 .$$

1061

The specific heat of water is taken as 1 cal/g·K, independent of temperature, where 1 calorie = 4.18 joules.

(a) Define the specific heat of a substance at constant pressure in terms of such quantities as Q (heat), S (entropy), and T (temperature).

(b) One kg of water at 0°C is brought into sudden contact with a large heat reservoir at 100°C. When the water has reached 100°C, what has been the change in entropy of the water? Of the reservoir? Of the entire system consisting of both water and the heat reservoir?

(c) If the water had been heated from 0°C to 100°C by first bringing it into contact with a reservoir at 50°C and then another reservoir at 100°C, what would be the change in entropy of the entire system?

(d) Show how the water might be heated from 0°C to 100°C with negligible change in entropy of the entire system.

(UC, Berkeley)

Solution:

(a) $c_p = \left(\dfrac{\partial Q}{\partial T}\right)_p = T\left(\dfrac{\partial S}{\partial T}\right)_p.$

(b) The change in the entropy of the water is

$$\Delta S_1 = \int_{T_1}^{T_2} \frac{c_p}{T}\, dT = c_p \ln \frac{T_2}{T_1} = 0.312 \text{ cal/g·K} ,$$

and the change in entropy of the reservoir is

$$\Delta S_2 = -c_p \frac{T_2 - T_1}{T_2} = -0.268 \text{ cal/g·K} .$$

Thus $\Delta S = 0.044$ cal/g·K.

(c) In this process, the change in entropy of the water is still $\Delta S_1' = 0.312$ cal/g·K, while that of the reservoir is

$$\Delta S_2' = -\frac{1 \times (50 - 1)}{273 + 50} - \frac{1 \times (100 - 50)}{273 + 100}$$
$$= -0.289 \text{ cal/g·K} .$$

So that $\Delta S' = \Delta S_1' + \Delta S_2' = 0.023$ cal/g·K.

(d) Divide the range of temperature 0°C - 100°C into N equal parts, with $N \gg 1$.

At every temperature point, there exists a large heat reservoir. Let the water come into contact with them successively from low temperature to high temperature, to make the process of thermal contact quasi-static. Then $\Delta S = 0$ at every step and consequently for the entire process.

1062

Two finite, identical, solid bodies of constant total heat capacity per body, C, are used as heat sources to drive heat engine. Their initial temperatures are T_1 and T_2 respectively. Find the maximum work obtainable from the system.

(*MIT*)

Solution:

As energy is conserved, the work obtainable is $W = C(T_1 + T_2 - 2T_f)$, where T_f is the final temperature of the system. From the second law of thermodynamics, we have

$$\Delta S = C \ln \frac{T_f}{T_1} + C \ln \frac{T_f}{T_2} \geq 0 , \quad \text{so that} \quad T_f \geq \sqrt{T_1 T_2} .$$

Hence $W_{\max} = C(T_1 + T_2 - 2\sqrt{T_1 T_2})$.

1063

A rigid box containing one mole of air at temperature T_0 (in K) is initially in thermal contact with an "infinite' heat-capacity reservoir" at the same temperature T_0. The box is removed from the reservoir and a cyclic engine is used to take some heat from the reservoir and put some into the air in the box. What is the minimum amount of work from T_0 to T_1? Express W in terms of T_0, T_1 and the gas constant R, and state units. Ignore vibrational degrees-of-freedom in the air molecules and the heat capacity of the container. Would inclusion of vibrational degrees-of-freedom increase or reduce the value of W?

(*Columbia*)

Solution:

As $\Delta Q + W = C_v(T_1 - T_0)$, where ΔQ is the heat absorbed from the "infinite heat-capacity reservoir", we get

$$0 \leq \Delta S = \Delta S_{\text{source}} + \Delta S_{\text{air}} = -\Delta Q/T_0 + C_v \ln(T_1/T_0) .$$

Hence
$$W \geq C_v(T_1 - T_0) - C_v T_0 \ln(T_1/T_0) = W_{\min} .$$
With the inclusion of vibrational degrees-of-freedom, $W_{\min}$ increases as C_v increases.

1064

A reversible heat engine operates between two reservoirs, T_1 and T_2 $(T_2 > T_1)$. T_1 can be considered to have infinite mass, i.e., T_1 remains constant. However the warmer reservoir at T_2 consists of a finite amount of gas at constant volume (μ moles with a specific heat capacity C_v).

After the heat engine has operated for some long period of time, the temperature T_2 is lowered to T_1.

(a) What is the heat extracted from the warmer reservoir during this period?

(b) What is the change of entropy of the warmer reservoir during this period?

(c) How much work did the engine do during this period?

(Columbia)

Solution:

(a) $Q_{ab} = \mu C_v(T_2 - T_1)$.

(b) Because $dS = \dfrac{dQ}{T} = \dfrac{\mu C_v dT}{T}$, $\quad \Delta S = \mu C_v \ln \dfrac{T_1}{T_2}$.

(c) $\dfrac{dW}{dQ} = 1 - \dfrac{T_1}{T}$, $\quad dQ = -\mu C_v dT$, therefore the work done by the engine is

$$W = \int dW = -\int_{T_2}^{T_1} \left(1 - \frac{T_1}{T}\right) \mu C_v dT = \mu C_v(T_2 - T_1) - \mu C_v T_1 \ln\left(\frac{T_2}{T_1}\right) .$$

1065

Large heat reservoirs are available at 900 K (H) and 300 K (C).

(a) 100 cal of heat are removed from the reservoir H and added to C. What is the entropy change of the universe?

(b) A reversible heat engine operates between H and C. For each

100 cal of heat removed from H, what work is done and what heat is added to C?

(c) What is the entropy change of the universe in the process of part (b) above?

(d) A real heat engine is operated as a heat pump removing heat from C and adding heat to H. What can be said about the entropy change in the universe produced by the heat pump?

(Wisconsin)

Solution:

(a) The change of entropy of the universe is

$$\Delta S = Q \left(\frac{1}{T_C} - \frac{1}{T_H} \right) = 100 \left(\frac{1}{300} - \frac{1}{900} \right) = \frac{2}{9} \text{ cal/K} .$$

(b) The external work done by the engine for each 100 cal of heat is

$$W = \eta Q_1 = \left(1 - \frac{T_C}{T_H} \right) Q_1 = \left(1 - \frac{300}{900} \right) \times 100 = \frac{200}{3} \text{ cal} .$$

The heat absorbed by C is

$$Q_2 = Q_1 - W = \frac{100}{3} \text{ cal} .$$

(c) The change in entropy of the universe is

$$\Delta S = -\frac{Q_1}{T_H} + \frac{Q_2}{T_C} = -\frac{100}{900} + \frac{100/3}{300} = 0 .$$

(d) The change of entropy is

$$\Delta S = -\frac{Q_2}{T_C} + \frac{Q_1}{T_H} ,$$

where Q_2 is the heat released by the reservoir of lower temperature, Q_1 is the heat absorbed by the reservoir of higher temperature. As $\frac{Q_2}{T_C} - \frac{Q_1}{T_H} \leq 0, \Delta S \geq 0$.

1066

Consider an arbitrary heat engine which operates between two reservoirs, each of which has the same finite temperature-independent heat capacity c. The reservoirs have initial temperatures T_1 and T_2, where $T_2 > T_1$, and the engine operates until both reservoirs have the same final temperature T_3.

(a) Give the argument which shows that $T_3 > \sqrt{T_1 T_2}$.

(b) What is the maximum amount of work obtainable from the engine?

(*UC, Berkeley*)

Solution:

(a) The increase in entropy of the total system is

$$\Delta S = \int_{T_1}^{T_3} \frac{cdT}{T} + \int_{T_2}^{T_3} \frac{cdT}{T} = c \ln \frac{T_3^2}{T_1 T_2} \geq 0 .$$

Thus $T_3^2 \geq T_1 T_2$, or $T_3 \geq \sqrt{T_1 T_2}$.

(b) The maximum amount of work can be obtained using a reversible heat engine, for which $\Delta S = 0$.

$$W_{max} = c(T_1 + T_2 - 2T_{3\,min}) = c(T_1 + T_2 - 2\sqrt{T_1 T_2}) = c(\sqrt{T_1} - \sqrt{T_2})^2 .$$

1067

(a) What is the efficiency for a reversible engine operating around the indicated cycle, where T is temperature in K and S is the entropy in joules/K?

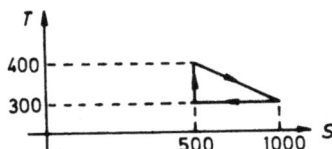

Fig. 1.22.

(b) A mass M of a liquid at a temperature T_1 is mixed with an equal mass of the same liquid at a temperature T_2. The system is thermally insulated. If c_p is the specific heat of the liquid, find the total entropy change. Show that the result is always positive.

(*UC, Berkeley*)

Solution:

(a) In the cycle, the heat absorbed by the engine is

$$Q = (1000 - 500) \frac{400 + 300}{2} = 1.75 \times 10^5 \text{ J} ,$$

and the work it does is

$$W = (1000 - 500) \frac{400 - 300}{2} = 2.5 \times 10^4 \text{ J} .$$

Thus the efficiency is $\eta = W/Q = 14.3\%$.

(b) Obviously the equilibrium temperature is $T_3 = (T_1 + T_2)/2$. Therefore

$$\Delta S_1 = \int_{T_1}^{T_3} \frac{c_p dT}{T} = c_p \ln \frac{T_3}{T_1} ,$$

and

$$\Delta S_2 = \int_{T_2}^{T_3} \frac{c_p dT}{T} = c_p \ln \frac{T_3}{T_2} ,$$

thus

$$\Delta S = \Delta S_1 + \Delta S_2 = c_p \ln \frac{(T_1 + T_2)^2}{4T_1 T_2} .$$

Since $(T_1 + T_2)^2 \geq 4T_1 T_2$, we have $\Delta S \geq 0$.

1068

(a) One mole of an ideal gas is carried from temperature T_1 and molar volume V_1 to T_2, V_2. Show that the change in entropy is

$$\Delta S = C_v \ln \frac{T_2}{T_1} + R \ln \frac{V_2}{V_1} .$$

(b) An ideal gas is expanded adiabatically from (p_1, V_1) to (p_2, V_2). Then it is compressed isobarically to (p_2, V_1). Finally the pressure is increased to p_1 at constant volume V_1. Show that the efficiency of the cycle is

$$\eta = 1 - \gamma(V_2/V_1 - 1)/(p_1/p_2 - 1) ,$$

where $\gamma = C_p/C_v$.

(*Columbia*)

Solution:

(a) From $dS = \dfrac{1}{T}(dU + pdV) = \dfrac{1}{T}(C_v dT + pdV)$ and

$$pV = RT ,$$

we obtain

$$\Delta S = C_v \ln \frac{T_2}{T_1} + R \ln \frac{V_2}{V_1} .$$

(b) The cycle is shown in the Fig. 1.23.

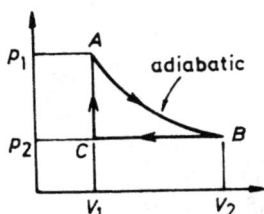

Fig. 1.23.

The work the system does in the cycle is

$$W = \oint pdV = \int_{AB} pdV + p_2(V_1 - V_2) .$$

Because AB is adiabatic and an ideal gas has the equations $pV = nkT$ and $C_p = C_v + R$, we get

$$\int_{AB} pdV = -\int_{AB} C_v dT = -C_v(T_2 - T_1)$$

$$= \frac{1}{1-\gamma}(p_2V_2 - p_1V_1) .$$

During the CA part of the cycle the gas absorbs heat

$$Q = \int_{CA} TdS = \int_{CA} C_v dT = C_v (T_1 - T_2)$$
$$= \frac{1}{1 - \gamma} V_1 (p_2 - p_1) .$$

Hence, the efficiency of the engine is

$$\eta = \frac{W}{Q} = 1 - \gamma \frac{\dfrac{V_2}{V_1} - 1}{\dfrac{p_1}{p_2} - 1} .$$

1069

(1) Suppose you are given the following relation among the entropy S, volume V, internal energy U, and number of particles N of a thermodynamic system: $S = A[NVU]^{1/3}$, where A is a constant. Derive a relation among:

(a) U, N, V and T;

(b) the pressure p, N, V, and T.

(c) What is the specific heat at constant volume c_v?

(2) Now assume two identical bodies each consists solely of a material obeying the equation of state found in part (1). N and V are the same for both, and they are initially at temperatures T_1 and T_2, respectively. They are to be used as a source of work by bringing them to a common final temperature T_f. This process is accomplished by the withdrawal of heat from the hotter body and the insertion of some fraction of this heat in the colder body, the remainder appearing as work.

(a) What is the range of possible final temperatures?

(b) What T_f corresponds to the maximum delivered work, and what is this maximum amount of work?

You may consider both reversible and irreversible processes in answering these questions.

(Princeton)

Solution:

(1)
$$U = \frac{S^3}{A^3 NV} \, ,$$

$$T = \left(\frac{\partial U}{\partial S} \right)_{V,N} = \frac{3S^2}{A^3 NV} = \frac{3U^{2/3}}{A(NV)^{1/3}} \, ,$$

$$p = - \left(\frac{\partial U}{\partial V} \right)_{S,N} = \frac{S^2}{A^3 NV^2} = \frac{U}{V} = \frac{1}{V} \sqrt{NV} \cdot \left(\frac{AT}{3} \right)^{3/2}$$

$$c_v = T \left(\frac{\partial S}{\partial T} \right)_{V,N} = \frac{1}{2} \sqrt{\frac{A^3 NV}{3}} \cdot \sqrt{T} \equiv \lambda \sqrt{T} \, .$$

(2) When no work is delivered, T_f will be maximum. Then

$$Q_1 = \int_{T_1}^{T_f} c_v \, dT = \int_{T_1}^{T_f} \lambda \sqrt{T} \, dT = \frac{2}{3} \lambda (T_f^{3/2} - T_1^{3/2}) \, ,$$

$$Q_2 = \int_{T_2}^{T_f} c_v \, dT = \frac{2}{3} \lambda (T_f^{3/2} - T_2^{3/2}) \, .$$

Since $Q_1 + Q_2 = 0$, we have

$$T_{f \, \max} = \left[\frac{T_1^{3/2} + T_2^{3/2}}{2} \right]^{2/3} \, .$$

The minimum of T corresponds to a reversible process; for which the change in entropy of the system is zero. As

$$\Delta S_1 = \int_{T_1}^{T_f} c_v \, dT/T = 2\lambda (T_f^{1/2} - T_1^{1/2}) \, ,$$

$$\Delta S_2 = \int_{T_2}^{T_f} c_v \, dT/T = 2\lambda (T_f^{1/2} - T_2^{1/2}) \, .$$

and $\Delta S_1 + \Delta S_2 = 0$, we have

$$T_{f \, \min} = \left(\frac{T_1^{1/2} + T_2^{1/2}}{2} \right)^2 \, .$$

Hence

$$T_{f \, \min} \le T_f \le T_{f \, \max}$$

$\overline{W}_{max}$ corresponds to $T_{f\,min}$, i.e., the reversible heat engine has the maximum delivered work

$$\overline{W}_{max} = -(Q_1 + Q_2) = \frac{2\lambda}{3}\left[T_1^{3/2} + T_2^{3/2} - 2\left(\frac{\sqrt{T_1} + \sqrt{T_2}}{2}\right)^3\right],$$

where $\lambda = \frac{1}{2}\sqrt{\frac{A^3 NV}{3}}$.

1070

One kilogram of water is heated by an electrical resistor from 20°C to 99°C at constant (atmospheric) pressure. Estimate:

(a) The change in internal energy of the water.

(b) The entropy change of the water.

(c) The factor by which the number of accessible quantum states of the water is increased.

(d) The maximum mechanical work achievable by using this water as heat reservoir to run an engine whose heat sink is at 20°C.

(UC, Berkely)

Solution:

(a) The change in internal energy of the water is
$\Delta U = Mc\Delta T = 1000 \times 1 \times 79 = 7.9 \times 10^4$ cal.

(b) The change in entropy is
$\Delta S = \int \frac{Mc}{T} dT = Mc\ln\frac{T_2}{T_1} = 239$ cal/K.

(c) From Boltzmann's relation $S = k\ln\Omega$, we get
$\frac{\Omega_2}{\Omega_1} = \exp\left(\frac{\Delta S}{k}\right) = \exp(7 \times 10^{25})$.

(d) The maximum mechanical work available is

$$W_{max} = \int_{T_1}^{T_2}\left(1 - \frac{T_1}{T}\right)McdT = Mc(T_2 - T_1) - T_1 Mc\ln\frac{T_2}{T_1}$$
$$= 9 \times 10^3 \text{ cal}.$$

1071

One mole of the paramagnetic substance whose TS diagram is shown below is to be used as the working substance in a Carnot refrigerator operating between a sample at 0.2 K and a reservoir at 1K:

(a) Show a possible Carnot cycle on the TS diagram and describe in detail how the cycle is performed.

(b) For your cycle, how much heat will be removed from the sample per cycle?

(c) How much work will be performed on the paramagnetic substance per cycle?

(*Columbia*)

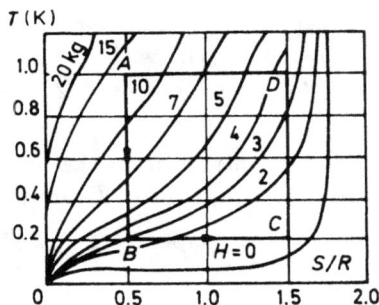

Fig. 1.24.

Solution:

(a) The Carnot cycle is shown in the Fig. 1.24;

$A \rightarrow B$, adiabatically decrease the magnetic field;

$B \rightarrow C$, isothermally decrease the magnetic field;

$C \rightarrow D$, adiabatically increase the magnetic field;

$D \rightarrow A$, isothermally increase the magnetic field;

(b) $Q_{abs} = T_{low} \Delta S_{B \rightarrow C} = 0.2 \times (1.5 - 0.5) R$
$\qquad = 1.7 \times 10^7$ ergs/mol.

(c) $Q_{rel} = T_{high} \Delta S_{D \rightarrow A} = 1 \times (1.5 - 0.5) R$
$\qquad = 8.3 \times 10^7$ ergs/mol.

The work done is

$$W = Q_{rel} - Q_{abs} = 6.6 \times 10^7 \text{ ergs/mol.}$$

1072

A capacitor with a capacity that is temperature sensitive is carried through the following cycle:

(1) The capacitor is kept in a constant temperature bath with a temperature T_1 while it is slowly charged (without any ohmic dissipation) to charge q and potential V_1. An amount of heat Q_1 flows into the capacitor during this charging.

(2) The capacitor is now removed from the bath while charging continues until a potential V_2 and temperature T_2 are reached.

(3) The capacitor is kept at a temperature T_2 and is slowly discharged.

(4) It is removed from the bath which kept it at temperature T_2 and discharged completely until it is returned to its initial uncharged state at temperature T_1.

(a) Find the net amount of work done in charging and discharging the capacitor.

(b) How much heat flows out of the capacitor in step (3)?

(c) For fixed capacitor charge q find dV/dT.
$\qquad$ Hint: Consider $V_2 = V_1 + dV$

$\qquad\qquad\qquad\qquad\qquad\qquad\qquad\qquad\qquad$ (*Columbia*)

Solution:

(a) The whole cycle can be taken as a reversible Carnot cycle.

(1) and (3) are isothermal processes; (2) and (4) are adiabatic processes.

In the whole cycle, the work done by the outside world is

$$W = Q_{rel} - Q_{abs} = T_2 |\Delta S_2| - T_1 |\Delta S_1| \ .$$

The total change of entropy in the whole cycle is 0. Thus

$$\Delta S_1 + \Delta S_2 = 0 , \quad \text{i.e.,} \quad |\Delta S_1| = |\Delta S_2| .$$

As $T_1|\Delta S_1| = Q_{abs}$,

$$W = \left(\frac{T_2}{T_1} - 1\right) Q_{abs}$$

(b) $Q_{rel} = T_2|\Delta S_2| = T_2 Q_{abs}/T_1$.

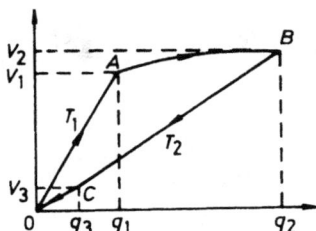

Fig. 1.25.

(c) We construct the V(Voltage)-q(charge) diagram for the cycle as shown in the Fig. 1.25. We have

$$W = \oint V\,dq .$$

Assume $V_2 = V_1 + dV$, where dV is an infinitesimal voltage change, and let the capacitance of the capacitor be $C(T)$. We then have

$$O \rightarrow A : \ V = q/C(T_1) , \quad B \rightarrow C : \ V = q/C(T_2) .$$

Consider the work done by the outside world in each process:

$$W_{O \rightarrow A} = \frac{1}{2}q_1 V_1 = \frac{1}{2}C(T_1)V_1^2 ,$$
$$W_{A \rightarrow B} \approx V_1[C(T_2)V_2 - C(T_1)V_1] ,$$
$$W_{B \rightarrow C} = -\frac{1}{2}(V_2 + V_3)(q_2 - q_3) = -\frac{1}{2}(V_2^2 - V_3^2)C(T_2)$$
$$W_{C \rightarrow O} \approx -V_3 q_3 = -C(T_2)V_3^2 .$$

Obviously the adiabatic line $B \rightarrow C$ crosses point O. Thus if dV is a small quantity, V_3 is also a small quantity. Then in the first-order approximation,

$$W_{C \rightarrow O} \approx 0, \quad W_{B \rightarrow C} \simeq -V_2^2 C(T_2)/2 .$$

Therefore

$$W = \sum W_i \approx \frac{1}{2}C(T_1)V_1^2 - \frac{1}{2}C(T_2)V_2^2 + V_1[C(T_2)V_2 - C(T_1)V_1]$$

$$\approx -d\left(\frac{1}{2}C(T)V^2\right) + V\,d(C(T)V)$$

$$= \frac{1}{2}V^2\frac{dC(T)}{dT} \cdot dT \ .$$

On the other hand, we know from (a)

$$W = \frac{Q_1}{T_1}(T_2 - T_1) = \frac{Q}{T}dT \ .$$

Thus

$$\frac{1}{2}V^2\frac{dC(T)}{dT} = \frac{Q}{T}$$

or

$$\frac{dC(T)}{dT} = \frac{2Q}{TV^2} \ .$$

Finally we have

$$\left(\frac{dV}{dT}\right)_q = \left[\frac{d}{dT} \cdot \frac{q}{C(T)}\right]_q = q\frac{d}{dT}\left(\frac{1}{C(T)}\right)$$

$$= -2Q/Tq \ ,$$

or

$$\left[\frac{\partial V(T, q)}{\partial T}\right]_q = -\frac{2Q(T, q)}{Tq} \ ,$$

where $Q(T, q)$ is the heat that the capacitor absorbs when it is charged from 0 to q while in contact with a heat source of constant temperature T.

3. THERMODYNAMIC FUNCTIONS AND EQUILIBRIUM CONDITIONS (1073-1105)

1073

For each of the following thermodynamic conditions, describe a system, or class of systems (the components or range of components, temperatures, etc.), which satisfies the condition. Confine yourself to classical, single component, chemical systems of constant mass. U is the internal energy and S is the entropy of the system.

(a) $\left(\dfrac{\partial U}{\partial V}\right)_T = 0$, (b) $\left(\dfrac{\partial S}{\partial V}\right)_p < 0$,

(c) $\left(\dfrac{\partial T}{\partial S}\right)_p = 0$, (d) $\left(\dfrac{\partial S}{\partial V}\right)_T = 0$,

(e) $\left(\dfrac{\partial T}{\partial V}\right)_S = -\left(\dfrac{\partial p}{\partial S}\right)_V$.

(*Wisconsin*)

Solution:

(a) The classical ideal gas.

(b) $\left(\dfrac{\partial S}{\partial V}\right)_p = \dfrac{\partial(S,p)}{\partial(V,p)} = \dfrac{\partial(S,p)}{\partial(T,p)} \Big/ \dfrac{\partial(V,p)}{\partial(T,p)} = \dfrac{C_p}{\alpha TV} < 0$

This requires $\alpha < 0$, i.e., the system has a negative coefficient of expansion at constant pressure.

(c) $\left(\dfrac{\partial T}{\partial S}\right)_p = \dfrac{T}{C_p} = 0$. This requires $C_p = \infty$. The system has two coexistent phases.

(d) $\left(\dfrac{\partial S}{\partial V}\right)_T = \left(\dfrac{\partial p}{\partial T}\right)_V = 0$. This requires $\beta = \dfrac{1}{p}\left(\dfrac{\partial p}{\partial T}\right)_V = 0$.

It is a system whose coefficient of pressure at constant volume is zero.

(e) All systems of a single component and constant mass satisfy this Maxwell relation.

1074

Consider an ideal gas whose entropy is given by

$$S = \frac{n}{2}\left[\sigma + 5R\ln\frac{U}{n} + 2R\ln\frac{V}{n}\right],$$

where n = number of moles, R = universal gas constant, U = internal energy, V = volume, and σ = constant.

(a) Calculate c_p and c_v, the specific heats at constant pressure and volume.

(b) An old and drafty house is initially in equilibrium with its surroundings at 32°F. Three hours after turning on the furnace, the house is at a cozy 70°F. Assuming that the air in the house is described by the above equation, show how the energy density (energy/volume) of the air inside the house compares at the two temperatures.

(Columbia)

Solution:

(a) The temperature T is determined by the following equation:

$$\frac{1}{T} = \left(\frac{\partial S}{\partial U}\right)_V = \frac{n}{2} 5R \frac{1}{U} \;, \quad \text{or} \quad U = \frac{5}{2} nRT \;.$$

Therefore, the specific heat at constant volume is

$$c_v = \left(\frac{\partial U}{\partial T}\right)_V = \frac{5}{2} nR \;.$$

The specific heat at constant pressure is

$$c_p = c_v + nR = \frac{7}{2} nR \;.$$

(b) $\dfrac{U}{V} = \dfrac{5}{2} R \left(\dfrac{n}{V}\right) T \;.$

Using the equation of state of ideal gas $pV = nRT$, we have

$$\frac{U}{V} = \frac{5}{2} p \;.$$

Because the pressure of the atmosphere does not change at the two temperatures in the problem, neither does the energy density.

1075

A perfect gas may be defined as one whose equation of state is $pV = NkT$ and whose internal energy is only a function of temperature. For a perfect gas show that

(a) $c_p = c_v + k$, where c_p and c_v are the heat capacities (per molecule) at constant pressure and constant volume respectively.

(b) The quantity pV^γ is constant during an adiabatic expansion. (Assume that $\gamma = c_p/c_v$ is constant.)

(MIT)

Solution:

Let C_p and C_v be the principal molar specific heats.

(a) From $pV = NkT$ and $TdS = dU + pdV$, we find

$$C_p - C_v = T\left(\frac{\partial S}{\partial T}\right)_p - T\left(\frac{\partial S}{\partial T}\right)_v = p\left(\frac{\partial V}{\partial T}\right)_p = Nk .$$

Hence $C_p - C_v = k$.

(b) For an adiabatic process, $TdS = 0$ and hence $C_v dT = -pdV$. From $pV = NkT$, we have

$$pdV + Vdp = NkdT = (C_p - C_v)dT ,$$

giving $\gamma pdV + Vdp = 0$, i.e.,

$$pV^\gamma = \text{const.}$$

1076

The difference between the speficif heat at constant pressure and the specific heat at constant volume is nearly equal for all simple gases. What is the approximate numerical value of $c_p - c_v$? What is the physical reason for the difference between c_p and c_v? Calculate the difference for an ideal gas.

(Wisconsin)

Solution:

$$c_p - c_v = \frac{1}{m}\left[T\left(\frac{\partial S}{\partial T}\right)_p - T\left(\frac{\partial S}{\partial T}\right)_v\right]$$

where m is the mass of the gas. From the functional relationship

$$S(T,p) = S(T, V(T,p)) ,$$

we can find

$$\left(\frac{\partial S}{\partial T}\right)_p = \left(\frac{\partial S}{\partial T}\right)_v + \left(\frac{\partial S}{\partial V}\right)_T\left(\frac{\partial V}{\partial T}\right)_p .$$

Utilizing Maxwell's relation $\left(\frac{\partial S}{\partial V}\right)_T = \left(\frac{\partial p}{\partial T}\right)_v$, the above formula becomes

$$c_p - c_v = \frac{T}{m}\left(\frac{\partial p}{\partial T}\right)_v\left(\frac{\partial V}{\partial T}\right)_p = \frac{VT\alpha^2}{mK} , \tag{$*$}$$

where α is the coefficient of thermal expansion, and K is the coefficient of compression. For an ideal gas, $\alpha = \dfrac{1}{T}$ and $K = \dfrac{1}{p}$, thus $c_p - c_v = nR/m = R/M$. (M is the molecular weight of the gas).

The formula (*) relates the difference of two specific heats to the equation of state. For some materials, the specific heat at constant volume or constant pressure is not easily measured in experiments; it can be determined with formula (*) by measuring K and α. For a simple gas, its values of α and K are near to those of an ideal gas. Thus, the difference between the two specific heats is approximately R/M. The reason that $c_p > c_v$ is that the gas expanding at constant pressure has to do work so that more heat is absorbed for this purpose.

1077

A paramagnetic system in an uniform magnetic field H is thermally insulated from the surroundings. It has an induced magnetization $M = aH/T$ and a heat capacity $c_H = \alpha T^2$, $\alpha = a + bH$ at constant H, where a and b are constants and T is the temperature. How will the temperature of the system change when H is quasi-statically reduced to zero? In order to have the final temperature change by a factor of 2 from the initial temperature, how strong should be the initial H?

(*UC, Berkeley*)

Solution:

From the relation $dU = TdS + HdM$, we have $\left(\dfrac{\partial T}{\partial M}\right)_S = \left(\dfrac{\partial H}{\partial S}\right)_M$, so that

$$\frac{\partial(T,S)}{\partial(H,M)} = -1 .$$

Therefore

$$\left(\frac{\partial T}{\partial H}\right)_S = \frac{\partial(T,S)}{\partial(H,S)} = \frac{\partial(H,T)}{\partial(H,S)} \cdot \frac{\partial(H,M)}{\partial(H,T)} \cdot \frac{\partial(T,S)}{\partial(H,M)}$$

$$= -\left(\frac{\partial T}{\partial S}\right)_H \left(\frac{\partial M}{\partial T}\right)_H = -\frac{T}{c_H} \cdot \left(\frac{\partial M}{\partial T}\right)_H ,$$

and $T_i = T_f(1 + bH^2/a)^{a/2b}$. This shows that the temperature of the system will decrease as H is reduced to zero.

If $T_f = T_i/2$, then $H = \sqrt{a(2^{2b/a} - 1)/b}$.

1078

The thermodynamics of a classical paramagnetic system are expressed by the variables: magnetization M, magnetic field B, and absolute temperature T.

The equation of state is

$$M = CB/T, \quad \text{where} \quad C = \text{Curie constant} .$$

The system's internal energy is

$$U = -MB .$$

The increment of work done by the system upon the external environment is $dW = MdB$.

(a) Write an expression for the heat input, dQ, to the system in terms of thermodynamic variables M and B:

$$dQ = (\qquad)dM + (\qquad)dB .$$

(b) Find an expression for the differential of the system entropy:

$$dS = (\qquad)dM + (\qquad)dB .$$

(c) Derive an expression for the entropy: $\qquad S =$

(*Wisconsin*)

Solution:

(a) $dQ = dU + dW = -d(MB) + MdB = -BdM.$

(b) $dS = \dfrac{dQ}{T} = -\dfrac{B}{T}dM = -\dfrac{M}{C}dM.$

(c) $S = S_0 - \dfrac{M^2}{2C}.$

(Note: the internal energy and the work done in the problem have been given new definitions).

1079

The state equation of a new matter is

$$p = AT^3/V ,$$

where p, V and T are the pressure, volume and temperature, respectively, A is a constant. The internal energy of the matter is

$$U = BT^n \ln(V/V_0) + f(T) ,$$

where B, n and V_0 are all constants, $f(T)$ only depends on the temperature. Find B and n.

(CUSPEA)

Solution:

From the first law of thermodynamics, we have

$$dS = \frac{dU + pdV}{T} = \left[\frac{1}{T} \left(\frac{\partial U}{\partial V} \right)_T + \frac{p}{T} \right] dV + \frac{1}{T} \left(\frac{\partial U}{\partial T} \right)_V dT .$$

We substitute in the above the expressions for internal energy U and pressure p and get

$$dS = \frac{BT^{n-1} + AT^2}{V} dV + \left[\frac{f'(T)}{T} + nBT^{n-2} \ln \frac{V}{V_0} \right] dT .$$

From the condition of complete differential, we have

$$\frac{\partial}{\partial T} \left(\frac{BT^{n-1} + AT^2}{V} \right) = \frac{\partial}{\partial V} \left[\frac{f'(T)}{T} + nBT^{n-2} \ln \frac{V}{V_0} \right] ,$$

giving

$$2AT - BT^{n-2} = 0 .$$

Therefore $n = 3, B = 2A$.

1080

The following measurements can be made on an elastic band:

(a) The change in temperature when the elastic band is stretched. (In case you have not tried this, hold the attached band with both hands, test the temperature by touching the band to your lips, stretch the band and check the temperature, relax the band and check the temperature once more).

(b) One end of the band is fixed, the other attached to weight W, and the frequency ν of small vibrations is measured.

(c) With the weight at rest σQ is added, and the equilibrium length L is observed to change by δL.

Derive the equation by which you can predict the result of the last measurement from the results of the first two.

(Princeton)

Solution:

The elastic coefficient of the band is $k = W(2\pi\nu)^2/g$.

When heat δQ is added with the weight at rest, i.e., with the stress kept unchanged, we have $\delta S = \delta Q/T$. Therefore,

$$\delta L = \left(\frac{\partial L}{\partial S}\right)_W \delta S = \frac{\partial(L,W)}{\partial(T,W)} \cdot \frac{\partial(T,W)}{\partial(S,W)} \cdot \frac{\delta Q}{T}$$

$$= \left(\frac{\partial L}{\partial T}\right)_W \cdot \frac{\delta Q}{C_W} .$$

As $L - L_0 = W/k$, we get

$$\left(\frac{\partial L}{\partial T}\right)_W = \frac{dL_0}{dT} - \frac{W}{k^2}\frac{dk}{dT} ,$$

Thus

$$\delta L = \left(\frac{dL_0}{dT} - \frac{W}{k^2}\frac{dk}{dT}\right)\frac{\delta Q}{C_W} ,$$

where

$$k = \frac{4\pi^2 W}{g}\nu^2 .$$

1081

The tension F in an ideal elastic cylinder is given by the equation of state

$$F = aT\left(\frac{L}{L_0(T)} - \frac{L_0^2(T)}{L^2}\right) ,$$

where a is a constant, L_0 is the length at zero tension, and $L(T)$ is a function of temperature T only.

(a) The cylinder is stretched reversibly and isothermally from $L = L_0$ to $L = 2L_0$. Find the heat transferred to the cylinder, Q, in terms of a, T, L_0 and α_0, the thermal expansion coefficient at zero tension, being

$$\alpha_0 = \frac{1}{L_0(T)} \frac{dL_0(T)}{dT} .$$

(b) When the length is changed adiabatically, the temperature of the cylinder changes. Derive an expression for the elastocaloric coefficient, $(\partial T/\partial L)_S$ where S is the entropy, in terms of a, T, L, L_0, α_0, and C_L, the heat capacity at constant length.

(c) Determine whether C_L is a function of T alone, $C_L(T)$, or whether it must also depend on the length, $C_L(T, L)$, for this system.

(MIT)

Solution:

Let Φ be the free energy. From $d\Phi = -SdT + FdL$, we get

$$\left(\frac{\partial \phi}{\partial L}\right)_T = F = aT \left(\frac{L}{L_0(T)} - \frac{L_0^2(T)}{L^2}\right) .$$

Thus

$$\phi(T, L) = aT \left(\frac{L^2}{2L_0} + \frac{L_0^2}{L} - \frac{3}{2}L_0\right) + \phi(T, L_0) ,$$

and

$$S = -\left(\frac{\partial \phi}{\partial T}\right)_L = a \left(\frac{3L_0}{2} - \frac{L_0^2}{L} - \frac{L^2}{2L_0}\right)$$
$$- aTL_0 \left(\frac{3}{2} - \frac{2L_0}{L} + \frac{L^2}{2L_0^2}\right) \alpha_0 + S_0 .$$

Fig. 1.26.

(a) $Q = T[S(T, 2L_0) - S_0] = -aTL_0 \left(1 + \frac{5}{2}T\alpha_0\right)$.

(b) $\left(\dfrac{\partial T}{\partial L}\right)_S = \dfrac{\partial(T, S)}{\partial(L, S)} = -\dfrac{T\left(\dfrac{\partial S}{\partial L}\right)_T}{C_L}$

$\qquad = \dfrac{aTL_0^2}{C_L L^2}\left[-1 + \dfrac{L^3}{L_0^3} + T\left(2 + \dfrac{L^3}{L_0^3}\right)\alpha_0\right]$.

(c) $\left(\dfrac{\partial C_L}{\partial L}\right)_T = T\dfrac{\partial^2 S}{\partial L \partial T}$

$\qquad = -aT\left\{\dfrac{\partial}{\partial T}\left[-\dfrac{L_0^2}{L^2} + \dfrac{L}{L_0} + T\left(2\dfrac{L_0^2}{L^2} + \dfrac{L}{L_0}\right)\alpha_0\right]\right\}_L \neq 0$.

Thus $C_L = C_L(T, L)$.

1082

Information: If a rubber band is stretched adiabatically, its temperature increases.

(a) If the rubber band is stretched isothermally, does its entropy increase, decrease, or stay the same?

(b) If the rubber band is stretched adiabatically, does the internal energy increase, decrease, or stay the same?

(*Wisconsin*)

Solution:

(a) We assume that when the rubber band is stretched by dx the work done on it is

$$dW = kx\,dx ,$$

where k, the elastic coefficient, is greater than 0. From the formula $dF = -SdT + kx\,dx$, we can obtain the Maxwell relation:

$$\left(\frac{\partial S}{\partial x}\right)_T = -\left(k\frac{\partial x}{\partial T}\right)_x = 0 .$$

Hence the entropy of the rubber band stays the same while it is stretched isothermally.

(b) According to the formula $dU = TdS + kx\,dx$, we have $\left(\dfrac{\partial U}{\partial x}\right)_S = kx > 0$, that is, its internal energy increases while it is stretched adiabatically.

1083

The tension of a rubber band in equilibrium is given by

$$t = AT\left(\frac{x}{l_0} - \frac{l_0^2}{x^2}\right) ,$$

where t = tension, T = absolute temperature, x = length of the band, l_0 = length of the band when $t = 0$, A = constant.

When x is the constant length l_0, the thermal capacity $c_x(x, T)$ is observed to be a constant K.

(a) Find as functions of T and x:

(1) $\left(\dfrac{\partial E}{\partial x}\right)_T$ where E = internal energy, (2) $\left(\dfrac{\partial c_x}{\partial x}\right)_T$, (3) $c_x(x, T)$, (4) $E(x, T)$, (5) $S(x, T)$, where S = entropy.

(b) The band is stretched adiabatically from $x = l_0$ to $x = 1.5l_0$. Its initial temperature was T_0. What is its final temperature?

(*CUSPEA*)

Solution:

(a) From the theory of thermodynamics, we know $dE = TdS + tdx$. Then as

$$c_x = T\left(\frac{\partial S}{\partial T}\right)_x ,$$

we have

$$c_x = \left(\frac{\partial E}{\partial T}\right)_x .$$

Generally, $E = E(x, T)$, and we have

$$dE = \left(\frac{\partial E}{\partial T}\right)_x dT + \left(\frac{\partial E}{\partial x}\right)_T dx ,$$

i.e., $dE = c_x dT + \left(\dfrac{\partial E}{\partial x}\right)_T dx.$

On the other hand,

$$dS = \frac{1}{T}(dE - tdx) = \frac{c_x}{T}dT + \frac{1}{T}\left[\left(\frac{\partial E}{\partial x}\right)_T - t\right]dx .$$

From $\dfrac{\partial^2 E}{\partial x \partial T} = \dfrac{\partial^2 E}{\partial T \partial x}$, $\dfrac{\partial^2 S}{\partial x \partial T} = \dfrac{\partial^2 S}{\partial T \partial x}$ we obtain

$$\begin{cases} \dfrac{\partial}{\partial x} c_x = \dfrac{\partial^2 E}{\partial T \partial x}\,, \\[2mm] \dfrac{\partial}{\partial x}\dfrac{c_x}{T} = \dfrac{\partial}{\partial T}\dfrac{1}{T}\left[\left(\dfrac{\partial E}{\partial x}\right)_T - t\right]\,, \end{cases}$$

Thus $(\partial E/\partial x)_T = t - T(\partial t/\partial T)_x$.

Substituting the expression for t, we have $(\partial E/\partial x)_T = 0$. It follows that $(\partial c_x/\partial x)_T = 0$. Integrating, we get

$$c_x(x,T) = c_x(l_0, T) + \int_{l_0}^{x} \frac{\partial c_x(x,T)dx}{\partial x}$$

$$= c_x(l_0, T) + 0 = c_x(l_0, T) = K\,.$$

$$E(x,T) = E(T) = \int_{T_0}^{T} dE + E(T_0) = \int_{T_0}^{T}\frac{dE}{dT}dT + E(T_0)$$

$$= \int_{T_0}^{T} K\,dT + E(T_0) = K(T - T_0) + E(T_0)\,.$$

From

$$dS = \frac{c_x}{T}dT + \frac{1}{T}\left[\left(\frac{\partial E}{\partial x}\right)_T - t\right]dx$$

$$= \frac{K}{T}dT - A\left(\frac{x}{l_0} - \frac{l_0^2}{x^2}\right)dx\,,$$

we find after integration

$$S(x,T) = K\ln T - A\left(\frac{x^2}{2l_0} + \frac{l^2}{x}\right) + \text{const.}$$

(b) For an adiabatic process $dS = 0$ so that

$$\frac{K}{T}dT - A\left(\frac{x}{l_0} - \frac{l_0^2}{x^2}\right)dx = 0\,.$$

After integration we have

$$K\ln\frac{T_f}{T_0} = A\left[\frac{(1.5l_0)^2}{2l_0} + \frac{l_0^2}{1.5l_0} - \frac{l_0^2}{2l_0} - \frac{l_0^2}{l_0}\right]$$

$$= 0.292\,Al_0\,,$$

Hence $T_f = T_0 \exp(0.292\,Al_0/K)$.

1084

Consider a gas which undergoes an adiabatic expansion (throttling process) from a region of constant pressure p_i and initial volume V_i to a region with constant pressure p_f and final volume V_f (initial volume 0).

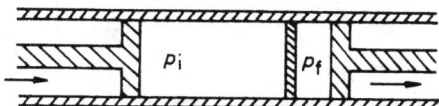

Fig. 1.27.

(a) By considering the work done by the gas in the process, show that the initial and final enthalpies of the gas are equal.

(b) What can be said about the intermediate states of the system?

(c) Show for small pressure differences $\Delta p = p_f - p_i$ that the temperature difference between the two regions is given by $\Delta T = \dfrac{V}{c_p}(T\alpha - 1)\Delta p$, where $\alpha = \dfrac{1}{V}\left(\dfrac{\partial V}{\partial T}\right)_p$ and $c_p = \left(\dfrac{\partial U}{\partial T}\right)_p$.

(d) Using the above result, discuss the possibility of using the process to cool either an ideal gas, or a more realistic gas for which $p = RT/(V - b)$. Explain your result.

(*SUNY, Buffalo*)

Solution:

(a) The work done by the gas in the throttling process is $p_f V_f - p_i V_i$, which is equal to a reduction of the internal energy:

$$U_i - U_f = p_f V_f - p_i V_i .$$

Thus $U_i + p_i V_i = U_f + p_f V_f$, i.e., $H_i = H_f$.

(b) Because the process is quasi-static, the final and initial states can be any two intermediate states. Thus the conclusion is still valid for intermediate states.

(c) From $dH = TdS + Vdp = 0$ and

$$dS = \left(\frac{\partial S}{\partial T}\right)_p dT + \left(\frac{\partial S}{\partial p}\right)_T dp = \frac{c_p}{T}dT - \left(\frac{\partial V}{\partial T}\right)_p dp ,$$

we obtain

$$dT = \frac{1}{c_p} \left[T \left(\frac{\partial V}{\partial T} \right)_p - V \right] dp = \frac{V}{c_p} (T\alpha - 1) dp .$$

Thus for a small pressure difference Δp, we have approximately

$$\Delta T = \frac{V}{c_p} (T\alpha - 1) \Delta p .$$

(d) For an ideal gas, we have $pV = NRT$ and $\alpha = 1/T$. Hence

$$\Delta T = V(T\alpha - 1)\Delta p / c_v = 0 .$$

As $\Delta T = 0$ this process cannot be used to cool ideal gases. For a realistic gas for which $p = RT/(V - b), \alpha = R/Vp$ and $V(\alpha T - 1) = -b$. Hence $\Delta T = -b\Delta p/c_p$. As $\Delta p < 0$ for a throttling process, $\Delta T > 0$, such a gas cannot be cooled by this process either.

1085

(a) Using the equation of state $pV = NRT$ and the specific heat per mole $C_v = 3R/2$ for a monatomic ideal gas, find its Helmholtz free energy F as a function of number of moles N, V, and T.

(b) Consider a cylinder separated into two parts by an adiabatic, impermeable piston. Compartments a and b each contains one mole of a monatomic ideal gas, and their initial volumes are $V_{ai} = 10$ litres and $V_{bi} = 1$ litre, respectively. The cylinder, whose walls allow heat transfer only, is immersed in a large bath at $0°C$. The piston is now moved reversibly so that the final volumes are $V_{af} = 6$ and $V_{bf} = 5$ litres. How much work is delivered by (or to) the system?

(*Princeton*)

Solution:

(a) For an ideal gas, we have $dU = NC_v dT$ and $U = NC_v T + U_0$, where U_0 is the internal energy of the system when $T = 0$. As

$$dS = \frac{NC_v}{T} dT + \frac{p}{T} dV ,$$

$$S = \frac{3NR}{2} \ln T + NR \ln V + S_0' ,$$

where S_0' is a constant. Assuming the entropy of the system is S_0 when $T = T_0$, $V = V_0$, we have

$$S = \frac{3NR}{2} \ln \frac{T}{T_0} + NR \ln \frac{V}{V_0} + S_0 \, ,$$

$$F = U - TS = \frac{3NRT}{2} - \left(\frac{3NRT}{2} \ln \frac{T}{T_0} - NRT \ln \frac{V}{V_0} \right) + F_0 \, .$$

where $F_0 = U_0 - T_0 S_0$.

(b) The process described is isothermal. When $dT = 0, dF = -pdV$. The work delivered by the system is

$$W = \int dW_a + \int dW_b = -\int dF_a - \int dF_b$$

$$= \left(N_a \ln \frac{V_{af}}{V_{ai}} + N_b \ln \frac{V_{bf}}{V_{bi}} \right) RT = 2.6 \times 10^3 \text{ J} \, .$$

1086

A Van der Waal's gas has the equation of state

$$\left(p + \frac{a}{V^2} \right) (V - b) = RT \, .$$

(a) Discuss the physical origin of the parameters a and b. Why is the correction to p inversely proportional to V^2?

(b) The gas undergoes an isothermal expansion from volume V_1 to volume V_2. Calculate the change in the Helmholtz free energy.

(c) From the information given can you calculate the change in internal energy? Discuss your answer.

(*Wisconsin*)

Solution:

(a) On the basis of the equation of state of an ideal gas, we introduce the constant b when considering the volume of a real gas to allow for the finite volumes of the molecules and we introduce the constant a to allow for the mutual attraction between molecules of the gas. Now we discuss why the pressure correction term is inversely proportional to V^2.

Each of the molecules of the gas has a certain interaction region. For the molecules near the center of the volume, the forces on them are isotropic because of the uniform distribution of molecules around them. For the molecules near the walls (the distances from which are smaller than the interaction distance of molecules), they will have a net attractive force directing inwards because the distribution of molecules there is not uniform. Thus the pressure on the wall must have a correction Δp. If Δk denotes the decrease of a molecule's momentum perpendicular to the wall due to the net inward attractive force, these $\Delta p =$ (The number of molecules colliding with unit area of the wall in unit time)$\times 2\Delta k$. As k is obviously proportional to the attractive force, the force is proportional to the number of molecules in unit volume, n, i.e., $\Delta k \propto n$, and the number of molecules colliding with unit area of the wall in unit time is proportional to n too, we have

$$\Delta p \propto n^2 \propto 1/V^2 .$$

(b) The equation of state can be written as

$$p = \frac{kT}{V - b} - \frac{a}{V^2} .$$

In the isothermal process, the change of the Helmholtz free energy is

$$\Delta F = -\int_{V_1}^{V_2} p dV = -\int_{V_1}^{V_2} \frac{kT}{V - b} - \frac{a}{V^2} dV$$

$$= -kT \ln \left(\frac{V_2 - b}{V_1 - b} \right) + a \left(\frac{1}{V_1} - \frac{1}{V_2} \right) .$$

(c) We can calculate the change of internal energy in the terms of T and V:

$$dU = \left(\frac{\partial U}{\partial T} \right)_V dT + \left(\frac{\partial U}{\partial V} \right)_T dV .$$

For the isothermal process, we have

$$dU = \left(\frac{\partial U}{\partial V} \right)_T dV .$$

The theory of thermodynamics gives

$$\left(\frac{\partial U}{\partial V} \right)_T = T \left(\frac{\partial p}{\partial T} \right)_V - p .$$

Use of the equation of state then gives

$$dU = \frac{a}{V^2} dV \ .$$

Integrating, we find

$$\Delta U = \int_{V_1}^{V_2} \frac{a}{V^2} dV = a \left(\frac{1}{V_1} - \frac{1}{V_2} \right) \ .$$

1087

A 100-ohm resistor is held at a constant temperature of 300 K. A current of 10 amperes is passed through the resistor for 300 sec.

(a) What is the change in the entropy of the resistor?

(b) What is the change in the entropy of the universe?

(c) What is the change in the internal energy of the universe?

(d) What is the change in the Helmholtz free-energy of the universe?

(*Wisconsin*)

Solution:

(a) As the temperature of the resistor is constant, its state does not change. The entropy is a function of state. Hence the change in the entropy of the resistor is zero: $\Delta S_1 = 0$.

(b) The heat that flows from the resistor to the external world (a heat source of constant temperature) is

$$I^2 R t = 3 \times 10^6 \text{ J} \ .$$

The increase of entropy of the heat source is $\Delta S_2 = 3 \times 10^6 / 300 = 10^4$ J/K. Thus the total change of entropy is $\Delta S = \Delta S_1 + \Delta S_2 = 10^4$ J/K.

(c) The increase of the internal energy of the universe is

$$\Delta U = 3 \times 10^6 \text{ J} \ .$$

(d) The increase of the free energy of the universe is

$$\Delta F = \Delta U - T \Delta S = 0 \ .$$

1088

Blackbody radiation.

(a) Derive the Maxwell relation

$$(\partial S/\partial V)_T = (\partial p/\partial T)_V .$$

(b) From his electromagnetic theory Maxwell found that the pressure p from an isotropic radiation field is equal to $\frac{1}{3}$ the energy density $u(T)$:
$p = \frac{1}{3}u(T) = \frac{U(T)}{3V}$, where V is the volume of the cavity. Using the first and second laws of thermodynamics together with the result obtained in part (a) show that u obeys the equation

$$u = \frac{1}{3}T\frac{du}{dT} - \frac{1}{3}u .$$

(c) Solve this equation and obtain Stefan's law relating u and T.

(Wisconsin)

Solution:

(a) From the equation of thermodynamics $dF = -SdT - pdV$, we know

$$\left(\frac{\partial F}{\partial T}\right)_V = -S , \quad \left(\frac{\partial F}{\partial V}\right)_T = -p .$$

Noting $\frac{\partial^2 F}{\partial V \partial T} = \frac{\partial^2 F}{\partial T \partial V}$, we get $\left(\frac{\partial S}{\partial V}\right)_T = \left(\frac{\partial p}{\partial T}\right)_V .$

(b) The total energy of the radiation field is $U(T,V) = u(T)V$. Substituting it into the second law of thermodynamics:

$$\left(\frac{\partial U}{\partial V}\right)_T = T\left(\frac{\partial S}{\partial V}\right)_T - p = T\left(\frac{\partial p}{\partial T}\right)_V - p ,$$

we find $u = \frac{T}{3}\frac{du}{dT} - \frac{1}{3}u.$

(c) The above formula can be rewritten as $T\frac{du}{dT} = 4u$, whose solution is $u = aT^4$, where a is the constant of integration. This is the famous Stefan's law of radiation for a black body.

1089

A magnetic system of spins is at thermodynamic equilibrium at temperature T. Let μ be the magnetic moment of each spin; and let M be the mean magnetization per spin, so $-\mu < M < \mu$. The free energy per spin, for specified magnetization M, is $F(M)$.

(1) Compute the magnetization M as a function of external magnetic field strength B, given that

$$F(M) = \lambda \begin{cases} 0, & |M/\mu| \leq 1/2 \ , \\ (|M/\mu| - 1/2)^2 \ , & 1 \geq |M/\mu| \geq 1/2 \ , \end{cases}$$

where λ is a constant.

(2) Suppose, instead, that someone gives you

$$F(M) = \lambda[(M/\mu)^4 - (M/\mu)^2] \ ,$$

you should respond that this is unacceptable – this expression violates a fundamental convexity principle of thermodynamics. (a) State the principle. (b) Check it against the above expression. (c) Discuss, by at least one example, what would go wrong with thermodynamics if the principle is not satisfied.

(*Princeton*)

Solution:

(1) From $dF = -SdT + HdM$, we have

$$H = \left(\frac{\partial F}{\partial M}\right)_T = \begin{cases} 0, & |\frac{M}{\mu}| \leq \frac{1}{2} \ , \\ \frac{2}{\mu}\left(\frac{M}{\mu} - \frac{1}{2}\right) \ , & \frac{1}{2} \leq \frac{M}{\mu} \leq 1 \ , \\ \frac{2}{\mu}\left(\frac{M}{\mu} + \frac{1}{2}\right) \ , & -1 \leq \frac{M}{\mu} \leq -\frac{1}{2} \ . \end{cases}$$

Hence

$$B = H + M = \begin{cases} M & |\frac{M}{\mu}| \leq \frac{1}{2} \ , \\ M + \frac{2}{\mu}\left(\frac{M}{\mu} - \frac{1}{2}\right) \ , & \frac{1}{2} \leq \frac{M}{\mu} \leq 1 \ , \\ M + \frac{2}{\mu}\left(\frac{M}{\mu} + \frac{1}{2}\right) \ , & -1 \leq \frac{M}{\mu} \leq -\frac{1}{2} \ . \end{cases}$$

(2) (a) The convexity principle of free energy says that free energy is a concave function of T while it is a convex function of M, and if $\left(\dfrac{\partial^2 F}{\partial M^2}\right)_T$ exists then $\left(\dfrac{\partial^2 F}{\partial M^2}\right)_T \geq 0$.

(b) Supposing $F(M) = \lambda\left[\left(\dfrac{M}{\mu}\right)^4 - \left(\dfrac{M}{\mu}\right)^2\right]$, we have

$$\left(\frac{\partial^2 F}{\partial M^2}\right)_T = \frac{2\lambda}{\mu^2}\left(\frac{6M^2}{\mu^2} - 1\right) .$$

When $\left|\dfrac{M}{\mu}\right| < \sqrt{\dfrac{1}{6}}$, $\left(\dfrac{\partial^2 F}{\partial M^2}\right)_T < 0$, i.e., F is not convex.

(c) If the convexity principle is untenable, for example if

$$\left(\frac{\partial^2 F}{\partial M^2}\right)_T = 1/\chi_T < 0 ,$$

that is, $\left(\dfrac{\partial M}{\partial H}\right)_T < 0$, then the entropy of the equilibrium state is a minimum and the equilibrium state will be unstable.

1090

A certain system is found to have a Gibbs free energy given by

$$G(p, T) = RT \ln\left[\frac{ap}{(RT)^{5/2}}\right]$$

where a and R are constants. Find the specific heat at constant pressure, C_p.

(*MIT*)

Solution:

The entropy is given by

$$S = -\left(\frac{\partial G}{\partial T}\right)_p = \frac{5}{2}R - R\ln\left[\frac{ap}{(RT)^{5/2}}\right] .$$

The specific heat at constant pressure is

$$C_p = T\left(\frac{\partial S}{\partial T}\right)_p = \frac{5}{2}R .$$

1091

Consider a substance held under a pressure p and at a temperature T. Show that $(\partial \,(\text{heat emitted})/\partial p)_T = T(\partial V/\partial T)_p$.

(Wisconsin)

Solution:

From Maxwell's relation

$$-\left(\frac{\partial S}{\partial p}\right)_T = \left(\frac{\partial V}{\partial T}\right)_p \,,$$

we find

$$-\left(T\frac{\partial S}{\partial p}\right)_T = T\left(\frac{\partial V}{\partial T}\right)_p \,,$$

i.e., $\left(\dfrac{\partial(\text{heat emitted})}{\partial p}\right)_T = T\left(\dfrac{\partial V}{\partial T}\right)_p .$

1092

A given type of fuel cell produces electrical energy W by the interaction of O_2 fed into one electrode and H_2 to the other. These gases are fed in at 1 atmosphere pressure and 298 K, and react isothermally and isobarically to form water. Assuming that the reaction occurs reversibly and that the internal resistance of the cell is negligible, calculate the e.m.f. of the cell. Given: one Faraday = 96,500 coulombs/g·mole.

Enthalpies in joules/g·mole at 1 atmospheric and 298 K for oxygen, hydrogen, and water are respectively 17,200, 8,100 and −269,300.

Entropies in joules/mole·K at 1 atmosphere and 298 K for oxygen, hydrogen, and water are respectively 201, 128 and 66.7.

(Wisconsin)

Solution:

The chemical equation is

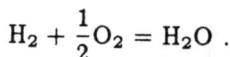

$$H_2 + \frac{1}{2}O_2 = H_2O \ .$$

In the reversible process at constant temperature and pressure, the decrease of Gibbs function of the system is equal to the difference between the total external work and the work the system does because of the change of volume. Thus

$$-\Delta g = \varepsilon \Delta q \ ,$$

or

$$-\sum_{i}(\Delta h_i - T\Delta S_i) = \varepsilon \Delta q .$$

If 1 mole of water forms, there must have been electric charges of 2F flowing in the circuit, i.e., $\Delta g = 2F$. Thus the e.m.f. is

$$\varepsilon = \frac{1}{2F}\left[TS_W - h_W - T\left(S_H + \frac{1}{2}S_0\right) + \left(h_H + \frac{1}{2}h_0\right)\right] .$$

As given , $S_0 = 201$ J/mol·K, $S_H = 128$ J/mol K ,
$S_W = 66.7$ J/mol·K, $h_0 = 17200$ J/mol
$h_H = 8100$ J/mol, $h_W = -269300$ J/mol , and $T = 298$ K,
We have $\varepsilon = 1.23$ V.

1093

It is found for a simple magnetic system that if the temperature T is held constant and the magnetic field H is changed to $H + \Delta H$, the entropy S changes by an amount ΔS,

$$\Delta S = -\frac{CH\Delta H}{T^2}$$

where C is a constant characteristic of the system. From this information determine how the magnetization M depends on the temperature and sketch a plot of M versus T for small H.

(*Wisconsin*)

Solution:
We are given that
$$\left(\frac{\partial S}{\partial H}\right)_T = -\frac{CH}{T^2} .$$

From $dG = -SdT - MdH$, we have

$$\left(\frac{\partial M}{\partial T}\right)_H = \left(\frac{\partial S}{\partial H}\right)_T .$$

Thus $\left(\dfrac{\partial M}{\partial T}\right)_H = -\dfrac{CH}{T^2}$, that is $M = \dfrac{CH}{T}$.

The diagram of M vs T is shown in Fig. 1.28.

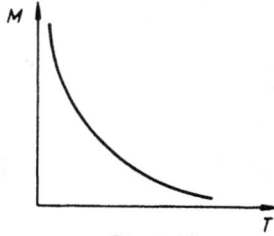

Fig. 1.28.

1094

A certain magnetic salt is found to obey Curie's law, and to have a heat capacity per unit volume (at constant magnetic field) inversely proportional to the square of the absolute temperature, i.e., $\chi = b/T, c_H = \alpha V/T^2$, where $\alpha = b + aH^2$, a and b being constants, and χ is the susceptibility. A sample of this salt at temperature T_i is placed in a magnetic field of strength H. The sample is adiabatically demagnetised by slowly reducing the strength of the field to zero. What is the final temperature, T, of the salt?

(Columbia)

Solution:

This process can be taken as reversible adiabatic. Then

$$dS = \left(\frac{\partial S}{\partial T}\right)_H dT + \left(\frac{\partial S}{\partial H}\right)_T dH = 0 \ .$$

From $c_H = T\left(\frac{\partial S}{\partial T}\right)_H$ and $dG = -S\,dT - \mu_0 MV\,dH$, we can write

$$\left(\frac{\partial S}{\partial H}\right)_T = \mu_0 V\left(\frac{\partial M}{\partial T}\right)_H \ .$$

As $M = \chi H$, we have $\left(\frac{\partial S}{\partial H}\right)_T = \mu_0 V H\left(\frac{\partial \chi}{\partial T}\right)_H = \mu_0 V H\frac{d\chi}{dT}$. Therefore, for the above adiabatic process, we have

$$\frac{dT}{dH} = -\left(\frac{\partial S}{\partial H}\right)_T \bigg/ \left(\frac{\partial S}{\partial T}\right)_H$$

$$= -\frac{\mu_0 V H T}{c_H}\frac{d\chi}{dT}$$

The final temperature is obtained by integration to be

$$T_f = \left[\frac{b}{b + aH^2} \right]^{\frac{\mu_0 b}{2a}} T_i \; .$$

1095

Explain the principles of cooling by adiabatic demagnetization. What factors limit the temperature obtained with this method?

(*Wisconsin*)

Solution:

The fundamental equation of the thermodynamics of a magnetic medium is $dU = TdS + HdM$. The Gibbs function is $G = U - TS - HM$, giving $dG = -SdT - MdH$. From the condition of complete differential

$$\left(\frac{\partial S}{\partial H} \right)_T = \left(\frac{\partial M}{\partial T} \right)_H$$

and

$$\left(\frac{\partial S}{\partial H} \right)_T = -\left(\frac{\partial S}{\partial T} \right)_H \left(\frac{\partial T}{\partial H} \right)_S \; ,$$

and the definition of specific heat, $C_H = T \left(\dfrac{\partial S}{\partial T} \right)_H$, we obtain

$$\left(\frac{\partial T}{\partial H} \right)_S = -\frac{T}{C_H} \left(\frac{\partial M}{\partial T} \right)_H \; .$$

If we assume the magnetic medium satisfies Curie's law

$$M = \frac{CV}{T} H \; ,$$

and substitute it into the above formula, we have

$$\left(\frac{\partial T}{\partial H} \right)_S = \frac{CV}{C_H T} H \; .$$

We can see that if the magnetic field is decreased adiabatically, the temperature of the magnetic medium will decrease also. This is the principle of cooling by adiabatic demagnetization.

Adiabatic demagnetization can produce temperatures as low as 1 K to 10^{-3} K; but when the temperature is of the order of magnitude of 10^{-3} K, the interactions between the paramagnetic ions cannot be neglected. The interactions are equivalent to a magnetic field. It thus limits the lowest temperature obtainable with this method.

1096

A flask of conical shape (see figure) contains raw milk. The pressure is measured inside the flask at the bottom. After a sufficiently long time, the cream rises to the top and the milk settles to the bottom. [You may assume that the total volume of liquid remains the same.] Does the pressure increase, decrease, or remain the same? Explain.

(*MIT*)

Solution:

Let the volume of the cream be V_1, its thickness be H_1, and its density be ρ_1; and let the volume of the milk be V_2, the thickness be H_2 and the density be $\rho_2 \cdot \rho_0$ stands for the density of raw milk.

Fig. 1.29.

As $\rho(V_1 + V_2) = \rho_1 V_1 + \rho_2 V_2$, i.e., $(\rho_2 - \rho)V_2 = (\rho - \rho_1)V_1$, we have $V_2/V_1 = (\rho - \rho_1)/(\rho_2 - \rho)$.

Then as $V_2/V_1 > H_2/H_1, (\rho - \rho_1)/(\rho_2 - \rho) > H_2/H_1$, or $\rho_1 H_1 + \rho_2 H_2 < \rho(H_1 + H_2)$, which means the pressure decreases.

1097

Assume the atmosphere to be an ideal gas of constant specific heat ratio $\gamma = C_p/C_v$. Also assume the acceleration due to gravity, g, to be constant over the range of the atmosphere. Let $z = 0$ at sea level, T_0, p_0, ρ_0 be the absolute temperature, pressure, and density of the gas at $z = 0$.

(a) Assuming that the thermodynamic variables of the gas are related in the same way they would be for an adiabatic process, find $p(z)$ and $\rho(z)$.

(b) Show that for this case no atmosphere exists above a z_{max} given by $z_{max} = \dfrac{\gamma}{\gamma - 1}\left(\dfrac{RT_0}{g}\right)$, where R is the universal gas constant per gram.

(SUNY, Buffalo)

Solution:

(a) When equilibrium is reached, we have

$$g\rho(z) = -\frac{dp(z)}{dz} .$$

By using the adiabatic relation $p\rho^{-\gamma} = p_0\rho_0^{-\gamma}$, we obtain,

$$\rho^{\gamma-2}(z)d\rho(z) = -\frac{g\rho_0^{\gamma}}{\gamma p_0}dz .$$

With the help of the equation of state $p = \rho RT$, we find

$$\rho(z) = \rho_0\left[1 - \frac{\gamma - 1}{\gamma}\frac{gz}{RT_0}\right]^{1/(\gamma-1)} ,$$

and

$$p(z) = p_0\left[1 - \frac{\gamma - 1}{\gamma}\frac{gz}{RT_0}\right]^{\gamma/(\gamma-1)} .$$

(b) In the region where no atmosphere exists, $\rho(z_{max}) = 0$. Thus

$$z_{max} = \frac{\gamma}{\gamma - 1} \cdot \frac{RT_0}{g} .$$

1098

Consider simple models for the earth's atmosphere. Neglect winds, convection, etc, and neglect variation in gravity.

(a) Assume that the atmosphere is isothermal (at $0°C$). Calculate an expression for the distribution of molecules with height. Estimate roughly the height below which half the molecules lie.

(b) Assume that the atmosphere is perfectly adiabatic. Show that the temperature then decreases linearly with height. Estimate this rate of temperature decrease (the so-called adiabatic lapse rate) for the earth.

(CUSPEA)

Solution:

(a) The molecular number density at height h is denoted by $n(h)$. From the condition of mechanical equilibrium $dp = -nmgdh$ and the equation of state $p = nkT$, we find

$$\frac{1}{p}dp = -\frac{mg}{kT}dh .$$

Thus $n(h) = n_0 \exp(-mgh/kT)$. Let $\int_0^H n(h)dh / \int_0^\infty n(h)dh = \frac{1}{2}$, then

$$H = \frac{kT}{mg}\ln 2 = \frac{RT}{N_0 mg}\ln 2 .$$

The average molecular weight of the atmosphere is 30. We have

$$H = \frac{8.31 \times 10^7 \times 273}{30 \times 980} \times \ln 2 \approx 8 \times 10^5 \text{ cm} = 8 \text{ km} .$$

(b) $\frac{1}{p}dp = -\frac{mg}{kT}dh$ is still correct and the adiabatic process follows

$$p^{(1-\gamma)/\gamma}T = \text{const}$$

where $\gamma = \frac{c_p}{c_v} \approx 7/2$ (for diatomic molecules). Therefore $\frac{dT}{T}\frac{\gamma}{\gamma-1} = -\frac{mg}{kT}dh$. Integrating we get

$$T - T_0 = -(\gamma - 1)mg(h - h_0)/\gamma k .$$

Furthermore,

$$\frac{dT}{dh} = -\frac{\gamma - 1}{\gamma}\frac{mg}{k} \approx -0.1 \text{ K/m} .$$

1099

The atmosphere is often in a convective steady state at constant entropy, not constant temperature. In such equilibrium pV^γ is independent of altitude, where $\gamma = C_p/C_v$. Use the condition of hydrostatic equilibrium in a uniform gravitational field to find an expression for dT/dz, where z is the altitude.

(UC, Berkeley)

Solution:

In the atmosphere, when the gas moves, pressure equilibrium can be quickly established with the new surroundings, whereas the establishment of temperature equilibrium is much slower. Thus, the process of formation of gas bulk can be regarded as adiabatic. Resulting from many times of mixing by convection, the temperature distribution of the atmosphere can be considered such that there is no temperature difference between the compressed or expanded gas bulk and its new surroundings. This is the so called "convective steady state at constant entropy". From $dp/dz = -nmg$ (where n is the molecular number density and z the altitude) and the equation of state of an ideal gas $p = nkT$, we get

$$\frac{dp}{dz} = -\frac{p}{kT}mg \ .$$

Together with the equation of adiabatic process

$$T^{\gamma} = \text{const.} p^{\gamma - 1} \ ,$$

we find

$$\frac{dT}{dz} = -\frac{\gamma - 1}{\gamma} \cdot \frac{mg}{k} \ .$$

It can be seen that the temperature decreases linearly. The temperature drops $\approx 1°\text{C}$ when the height increases by 100 metres.

1100

The gas group that is slowly and adiabatically arising and unrestricted near the ground cannot continuously rise; neither can it fall (the atmosphere almost does not convect). If the height z is small, the pressure and temperature of the atmosphere are respectively $p = p_0(1 - \alpha z)$ and $T = T_0(1 - \beta z)$, where p_0 and T_0 are respectively the pressure and temperature near the surface. Find α and β as functions of the temperature T_0, gravitational acceleration near the surface, g, and the molecular weight M. Suppose that air consists of $\frac{4}{5}N_2$ and $\frac{1}{5}O_2$, and that T_0 is low enough so that the molecule oscillations cannot be excited, but is high enough so that the molecule rotation can be treated by the classical theory.

(CUSPEA)

Solution:

Near the ground, we have

$$dp/dz = -\alpha p_0 .$$

Dynamic considerations give $dp/dz = -\rho g$.

Thus $\alpha = \rho_0 g/p_0$, where ρ_0 is the density of air near the ground. Treating air as an ideal gas, we have

$$p_0 = RT_0/V_0 = RT_0\rho_0/M ,$$

where R is the gas constant, V_0 is the volume and M the molecular weight $\left(\dfrac{4}{5} \cdot 28 + \dfrac{1}{5} \cdot 32 = 29\right)$. Thus we have $\alpha = Mg/RT_0$.

The slow rising of the gas group can be taken as a quasi-static process. It has the same p and ρ as the atmosphere surrounding it. Thus the same is also true of the temperature T. In the adiabtic process,

$$T^\gamma p^{1-\gamma} = \text{const} ,$$

with

$$\gamma = C_p/C_v = (C_v + R)/C_v = 7/5 .$$

Differentiating we have

$$\frac{dT}{T} = \frac{\gamma - 1}{\gamma} \frac{dp}{p} .$$

On the ground, $dT/T = -\beta dz$ and $dp/p = -\alpha dz$. We substitute them into above formula and obtain

$$\beta = \frac{\gamma - 1}{\gamma}\alpha = \frac{2}{7}\alpha .$$

1101

Suppose that the earth's atmosphere is an ideal gas with molecular weight μ and that the gravitational field near the surface is uniform and produces an acceleration g.

(a) Show that the pressure p varies as

$$\frac{1}{p}dp = -\frac{\mu g}{RT} dz$$

where z is the height above the surface, T is the temperature, and R is the gas constant.

(b) Suppose that the pressure decrease with height is due to adiabatic expansion. Show that

$$\frac{dp}{p} = \frac{\gamma}{\gamma - 1} \frac{dT}{T}, \quad \gamma = \frac{C_p}{C_v} .$$

(c) Evaluate dT/dz for a pure N_2 atmosphere with $\gamma = 1.4$.

(d) Suppose the atmosphere is isothermal with temperature T. Find $p(z)$ in terms of T and p_0, the sea level pressure.

(e) Suppose that at sea level, $p = p_0$ and $T = T_0$. Find $p(z)$ for an adiabatic atmosphere.

(*Columbia*)

Solution:

(a) Mechanical equilibrium gives $dp = -n\mu g dz$, where n is the mole number of unit volume. Thus using the equation of state of an ideal gas $p = nRT$, we find

$$dp = -\frac{p}{RT}\mu g dz ,$$

or

$$\frac{dp}{p} = -\frac{\mu g}{RT} dz .$$

(b) The adiabatic process satisfies $T^{\gamma/(1-\gamma)}p = $ const. Thus

$$\frac{dp}{p} = \frac{\gamma}{\gamma - 1} \frac{dT}{T} .$$

(c) Comparing the result of (b) with that of (a), we deduce

$$\frac{dT}{dz} = \left(\frac{1}{\gamma} - 1\right) \frac{\mu g}{R} .$$

For $N_2, \gamma = 1.4$, we get $dT/dz \approx -4.7$ K/km.

(d) From (a) we find

$$p(z) = p_0 \exp\left(-\frac{\mu g z}{RT}\right) .$$

(e) From (a) and (b) we have

$$\frac{dp}{p} = -\frac{\mu g}{RT} dz \ , \qquad pT^{\gamma/(1-\gamma)} = p_0 T_0^{\gamma/(1-\gamma)} \ .$$

Thus

$$p^{-1/\gamma} dp = -\frac{\mu g}{RT_0} p_0^{1-\frac{1}{\gamma}} dz \ ,$$

Integrating, we get

$$p(z) = p_0 \left[1 - \frac{\mu g}{RT_0} \left(1 - \frac{1}{\gamma} \right) z \right]^{\gamma/(\gamma-1)} \ .$$

This is, of course, valid only if

$$z < \frac{RT_0 \gamma}{\mu g(\gamma - 1)} \ .$$

1102

A fully ionized gas containing a single species of ion with charge $Z|e|$ and atomic weight A is in equilibrium in a uniform gravitational field g. The gas is isothermal with temperature T and there is thermal equilibrium between the ions and the electrons. The gas has a low enough density that local interactions between the particles can be neglected.

(a) Show that to avoid charge separation there must be a uniform electric field E given by

$$E = -\frac{(Am_{\rm p} - m_{\rm e})}{(1 + Z)|e|} g \ ,$$

where $m_{\rm p}$ and $m_{\rm e}$ are the proton and electron masses respectively.

(b) Show that the above equation is also valid if the plasma is not isothermal. (Hint: Treat each component i as an ideal gas subject to the equation of hydrostatic equilibrium

$$\frac{dp_i}{dx} = n_i F_{ix} \ ,$$

where p_i is the partial pressure of the ith component, n_i is its number density, and F_{ix} is the total force per particle in the x direction.)

(c) The equation in (a) is also valid throughout the sun where **E** and **g** are now directed radially. Show that the charge on the sun is given approximately by

$$Q = \frac{A}{1+Z} \frac{GMm_p}{|e|} ,$$

where M is the mass of the sun.

(d) For the sun $M = 2 \times 10^{33}$ grams. If the composition of the sun were pure hydrogen, what would be Q in coulombs? Given this value of Q, is the approximation that there is no charge separation a good one?

(*MIT*)

Solution:
(a) Take an arbitrary point in the gravitational field as the zero potential point. The number density at this point is n and the height is taken opposite to the direction of **g**. Suppose there exists an uniform electric field **E** in the direction opposite to **g**. The electron and ion distributions as functions of height are respectively

$$n_e(h) = n_{oe} \exp[-(m_e g h + E|e|h)/kT] ,$$
$$n_I(h) = n_{oI} \exp[-(A m_p g h - E Z|e|h)/kT] .$$

To avoid charge separation, the following condition must be satisfied:

$$n_I(h)/n_e(h) = n_{oI}/n_{oe} .$$

This gives

$$A m_p g - E Z|e| = m_e g + E|e| ,$$

from which we get

$$E = \frac{A m_p - m_e}{(Z+1)|e|} g , \quad \text{or} \quad \mathbf{E} = -\frac{A m_p - m_e}{(1+Z)|e|} \mathbf{g} .$$

(b) $\dfrac{dp_I}{dh} = n_I(-A m_p g + Z|e|E),$

$\dfrac{dp_e}{dh} = n_e(-m_e g - |e|E).$

At equilibrium, the partial pressure for each type of particles (at the same height) should be the same. Thus

$$\frac{1}{n_I} \frac{dp_I}{dh} = \frac{1}{n_e} \frac{dp_e}{dh} ,$$

i.e.,

$$-Am_p g + Z|e|E = -m_e g - |e|E .$$

Hence $E = \dfrac{Am_p - m_e}{(1+z)|e|} g.$

(c) As $\mathbf{E} = \dfrac{Q}{r^2}\dfrac{\mathbf{r}}{r}, \quad \mathbf{g} = -\dfrac{GM}{r^2}\dfrac{\mathbf{r}}{r}$, we have

$$Q/r^2 = \frac{GM}{r^2}\frac{Am_p - m_e}{(1+Z)|e|} \approx \frac{GM}{r^2}\frac{Am_p}{(1+Z)|e|} .$$

Hence $Q = \dfrac{GMAm_p}{(1+Z)|e|}.$

(d) For hydrogen, one has $A = 1, Z = 1$, giving

$$Q \approx \frac{GMm_p}{2|e|} \approx 1.5 \times 10^3 \ {}^\circ C .$$

1103

Consider a thermally isolated system consisting of two volumes, V and $2V$ of an ideal gas separated by a thermally conducting and movable partition.

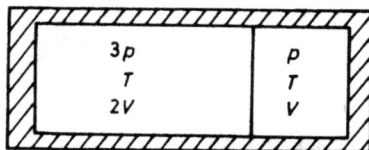

Fig. 1.30.

The temperatures and pressures are as shown. The partition is now allowed to move without the gases mixing.

When equilibrium is established what is the change in the total internal energy? The total entropy?

What is the equilibrium temperature? Pressure?

(SUNY, Buffalo)

Solution:

Let the molar numbers of the gas in the two sides be n_1 and n_2 respectively. From the equations $6pV = n_1 RT$ and $pV = n_2 RT$, we obtain $n_1 = 6n_2$. As this is an isolated system of ideal gas, the final temperature is $T_f = T$ since both the initial temperatures are equal to T. The final pressure p_f is

$$p_f = (n_1 + n_2)RT/3V = \frac{p}{3}\left(1 + \frac{n_1}{n_2}\right) = \frac{7}{3}p \ .$$

We calculate the change of the state function S by designing a quasi-static isothermal process. Then

$$\Delta S = \frac{1}{T}\left(\int p_1 dV_1 + p_2 dV_2\right)$$

$$= n_1 R \int_{2V}^{V_1} \frac{dV}{V} + n_2 R \int_{V}^{V_2} \frac{dV}{V}$$

$$= n_1 R \ln \frac{V_1}{2V} + n_2 R \ln \frac{V_2}{V} \ .$$

Since $V_1 + V_2 = 3V$ and $\dfrac{V_1}{V_2} = \dfrac{n_1}{n_2} = 6, V_1 = 6V_2 = \dfrac{18}{7}V$. Hence

$$\Delta S = n_1 R \ln \frac{9}{7} + n_2 R \ln \frac{3}{7} \approx \frac{pV}{T} \ .$$

1104

A thermally insulated cylinder, closed at both ends, is fitted with a frictionless heat-conducting piston which divides the cylinder into two parts. Initially, the piston is clamped in the center, with 1 litre of air at 200 K and 2 atm pressure on one side and 1 litre of air at 300 K and 1 atm on the other side. The piston is released and the system reaches equilibrium in pressure and temperature, with the piston at a new position.

(a) Compute the final pressure and temperature.
(b) Compute the total increase in entropy.
Be sure to give all your reasoning.

(SUNY, Buffalo)

Solution:

(a) The particle numbers of the two parts do not change. Let these be N_1 and N_2, the final pressure be p, and the final temperature be T. Taking air as an ideal gas, we have

$$p_1 V_0 = N_1 k T_1 , \quad p_2 V_0 = N_2 k T_2 ,$$

where $p_1 = 2$ atm, $T_1 = 200$ K, $p_2 = 1$ atm, $T_2 = 300$ K, $V_0 = 1l$.

The piston does not consume internal energy of the gas as it is frictionless, so that the total internal energy of the gas is conserved in view of the cylinder being adiabatical. Thus

$$\mu N_1 k T_1 + \mu N_2 k T = \mu (N_1 + N_2) k T ,$$

where μ is the degree of freedom of motion of an air molecule. Hence

$$T = \frac{T_1 + \frac{N_2}{N_1} T_2}{\frac{N_2}{N_1} + 1} = \frac{1 + \frac{p_2}{p_1}}{1 + \frac{p_2}{p_1} \cdot \frac{T_1}{T_2}} T_1 = 225 \text{ K} .$$

By $V_1 + V_2 = 2V_0$, we find

$$N_1 k T / p + N_2 k T / p = 2 V_0 ,$$

and hence

$$p = \frac{(N_1 + N_2)}{2 V_0} k T = \frac{T}{2} \left(\frac{p_1}{T_1} + \frac{p_2}{T_2} \right) = 1.5 \text{ atm} .$$

(b) Entropy is a state function independent of the process. To calculate the change of entropy by designing a quasi-static process, we denote the entropies of the two parts by S_1 and S_2. Then

$$\Delta S = \Delta S_1 + \Delta S_2 = \int_{(T_1, V_0)}^{(T, V_1)} dS_1 + \int_{(T_2, V_0)}^{(T, V_2)} dS_2 ,$$

$$\int_{(T_1, V_0)}^{(T, V_1)} dS_1 = \int_{T_1}^{T} \frac{dU_1}{T} + \int_{V_0}^{V_1} \frac{p}{T} dV$$

$$= n_1 \left(c_v \ln \frac{T}{T_1} + R \ln \frac{V_1}{V_0} \right) ,$$

$$\int_{(T_2, V_0)}^{(T, V_2)} dS_2 = n_2 \left(c_v \ln \frac{T}{T_2} + R \ln \frac{V_2}{V_0} \right) ,$$

where n_1 and n_2 are the molar numbers of the particles in the two parts, c_v is the molar specific heat at constant volume, and R is the gas constant. Thus

$$\Delta S = \frac{p_1 V_0}{T_1}\left(\frac{c_v}{R}\ln\frac{T}{T_1} + \ln\frac{p_1 T}{pT_1}\right)$$
$$+ \frac{p_2 V_0}{T_2}\left(\frac{c_v}{R}\ln\frac{T}{T_2} + \ln\frac{p_2 T}{pT_2}\right) .$$

Taking $c_v = \frac{3}{2}R$ as the temperature of the system is not high, we have $\Delta S = 0.4$ J/cal.

1105

A cylindrical container is initially separated by a clamped piston into two compartments of equal volume. The left compartment is filled with one mole of neon gas at a pressure of 4 atmospheres and the right with argon gas at one atmosphere. The gases may be considered as ideal. The whole system is initially at temperature $T = 300$ K, and is thermally insulated from the outside world. The heat capacity of the cylinder-piston system is C (a constant).

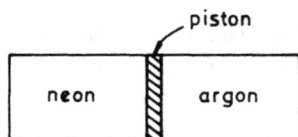

Fig. 1.31.

The piston is now unclamped and released to move freely without friction. Eventually, due to slight dissipation, it comes to rest in an equilibrium position. Calculate:

(a) The new temperature of the system (the piston is thermally conductive).

(b) The ratio of final neon to argon volumes.

(c) The total entropy change of the system.

(d) The additional entropy change which would be produced if the piston were removed.

(e) If, in the initial state, the gas in the left compartment were a mole of argon instead of a mole of neon, which, if any, of the answers to (a), (b) and (c) would be different?

(*UC, Berkeley*)

Solution:

(a) The internal energy of an ideal gas is a function dependent only on temperature, so the internal energy of the total system does not change. Neither does the temperature. The new equilibrium temperature is 300 K.

(b) The volume ratio is the ratio of molecular numbers, and is also the ratio of initial pressures. Thus

$$V_{Ne} : V_{Ar} = 4 : 1 = 1 : n .$$

where $n = 1/4$ is the mole number of the argon gas.

(c) The increase of entropy of the system is

$$\Delta S = N_{Ne} R \ln \left(\frac{V_2}{V_1} \right)_{Ne} + N_{Ar} R \ln \left(\frac{V_2}{V_1} \right)_{Ar}$$

$$= R \ln \frac{\frac{4}{5}}{\frac{1}{2}} + \frac{R}{4} \ln \frac{\frac{1}{4}}{\frac{1}{2}} = 2.0 \text{ J/K} .$$

(d) The additional entropy change is

$$\Delta S' = R \ln(1 + n) + n R \ln \left(\frac{1+n}{n} \right) = 5.2 \text{ J/K} .$$

(e) If initially the gas on the left is a mole of argon, the answers to (a), (b) and (c) will not change. As for (d), we now have $\Delta S' = 0$.

4. CHANGE OF PHASE AND PHASE EQUILIBRIUM (1106-1147)

1106

Is the melting point of tungsten 350, 3500, 35,000, or 350,000°C?

(*Columbia*)

Solution:

The answer is 3500°C.

1107

Assuming that 1/20 eV is required to liberate a molecule from the surface of a certain liquid when $T = 300$ K, what is the heat of vaporization in ergs/mole?

[1 eV $= 1.6 \times 10^{-12}$ erg]

(*Wisconsin*)

Solution:

The heat of vaporization is

$$L_{\text{vapor}} = \frac{1}{20} \times 1.6 \times 10^{-12} \times 6.023 \times 10^{23}$$
$$= 4.8 \times 10^{10} \text{ ergs/mol.}$$

1108

Twenty grams of ice at 0°C are dropped into a beaker containing 120 grams of water initially at 70°C. Find the final temperature of the mixture neglecting the heat capacity of the beaker. Heat of fusion of ice is 80 cal/g.

(*Wisconsin*)

Solution:

We assume the temperature of equilibrium to be T after mixing. Thus

$$M_1 L_{\text{fusion}} + M_1 C_{p,\text{water}} T = M_2 C_{p,\text{water}} (T_0 - T) .$$

We substitute $M_1 = 20$ g, $M_2 = 120$ g, $T_0 = 70°$C, $L_{\text{fusion}} = 80$ cal/g and $C_{p,\text{water}} = 1$ cal/g, and obtain the final temperature $T = 48.57°$C.

1109

The entropy of water at atmospheric pressure and 100°C is 0.31 cal/g·deg, and the entropy of steam at the same temperature and pressure is 1.76 cal/g·deg.

(a) What is the heat of vaporization at this temperature?

(b) The enthalpy $(H = U + PV)$ of steam under these conditions is 640 cal/g. Calculate the enthalpy of water under these conditions.

(c) Calculate the Gibbs functions $(G = H - TS)$ of water and steam under these conditions.

(d) Prove that the Gibbs function does not change in a reversible isothermal isobaric process.

(UC, Berkeley)

Solution:

(a) Heat of vaporization is
$$L = T\Delta S = 540 \text{ cal/g}.$$

(b) From $dH = TdS + Vdp$, we get
$$H_{\text{water}} = H_{\text{steam}} - T\Delta S = 100 \text{ cal/g}.$$

(c) Since $G = H - TS$,
$$G_{\text{water}} = H_{\text{water}} - TS_{\text{water}} = -16 \text{ cal/g },$$
$$G_{\text{steam}} = H_{\text{steam}} - TS_{\text{steam}} = -16 \text{ cal/g },$$

(d) From $dG = -SdT + Vdp$, we see that in a reversible isothermal isobaric process, G does not change.

1110

Given 1.0 kg of water at 100°C and a very large block of ice at 0°C. A reversible heat engine absorbs heat from the water and expels heat to the ice until work can no longer be extracted from the system. At the completion of the process:

(a) What is the temperature of the water?

(b) How much ice has been melted? (The heat of fusion of ice is 80 cal/g)

(c) How much work has been done by the engine?

(Wisconsin)

Solution:

(a) Because the block of ice is very large, we can assume its temperature to be a constant. In the process the temperature of the water gradually decreases. When work can no longer be extracted from the system, the efficiency of the cycle is zero:

$$\eta = 1 - T_{\text{ice}}/T = 0, \text{ or } T = T_{\text{ice}} = 0° \text{ C }.$$

Therefore, the final temperature of the water is 0°C.

(b) The heat absorbed by the ice block is

$$Q_2 = \int [1 - \eta(t)] dQ = mC_v \int_{273}^{373} \frac{273}{T} dT = 8.5 \times 10^4 \text{ cal }.$$

This heat can melt ice to the amount of

$$M = \frac{Q_2}{L_{\text{fusion}}} = \frac{8.5 \times 10^4}{80}$$
$$= 1.06 \text{ kg }.$$

(c) The work done by the engine is

$$W = Q_1 - Q_2 = 1000 \times 100 \times 1 - 8.5 \times 10^4 = 1.5 \times 10^4 \text{ cal }.$$

1111

What is the smallest possible time necessary to freeze 2 kg of water at 0°C if a 50 watt motor is available and the outside air (hot reservoir) is at 27°C?

(Wisconsin)

Solution:

When 2 kg of water at 0°C becomes ice, the heat released is

$$Q_2 = 1.44 \times 2 \times 10^3 / 18 = 1.6 \times 10^2 \text{ kcal }.$$

The highest efficiency of the motor is

$$\varepsilon = \frac{T_2}{T_1 - T_2} = \frac{Q_2}{W_{\text{min}}} .$$

Thus,

$$W_{\text{min}} = Q_2 \frac{T_1 - T_2}{T_2} .$$

If we use the motor of $P = 50$ W, the smallest necessary time is

$$\tau = \frac{W_{\text{min}}}{P} = \frac{Q_2}{P} \cdot \frac{T_1 - T_2}{T_2} .$$

With $T_1 = 300$ K, $T_2 = 273$ K, we find

$$\tau = 1.3 \times 10^3 \text{ s} .$$

1112

Compute the theoretical minimum amount of mechanical work needed to freeze 1 kilogram of water, if the water and surroundings are initially at a temperature $T_0 = 25°C$. The surroundings comprise the only large heat reservoir available.

$$(L_{\text{ice}} = 80 \text{ cal}/g, \quad C_p = 1 \text{ cal}/g \cdot °\text{ C}) .$$

(*UC, Berkeley*)

Solution:

The minimum work can be divided into two parts W_1 and W_2: W_1 is used to lower the water temperature from $25°C$ to $0°C$, and W_2 to transform water to ice. We find

$$W_1 = -\int_{T_0}^{T_t} (T_0 - T)MC_p dT/T$$
$$= MC_p T_0 \ln(T_0/T_t) - MC_p(T_0 - T_t)$$
$$= 1.1 \times 10^3 \text{ cal} ,$$
$$W_2 = (T_0 - T_t)LM/T_t = 7.3 \times 10^3 \text{ cal} ,$$
$$W = W_1 + W_2 = 8.4 \times 10^3 \text{ cal} = 3.5 \times 10^3 \text{ J} .$$

1113

An ideal Carnot refrigerator (heat pump) freezes ice cubes at the rate of 5 g/s starting with water at the freezing point. Energy is given off to the room at $30°C$. If the fusion energy of ice is 320 joules/gram,

(a) At what rate is energy expelled to the room?

(b) At what rate in kilowatts must electrical energy be supplied?

(c) What is the coefficient of performance of this heat pump?

(*Wisconsin*)

Solution:

(a) The rate that the refrigerator extracts heat from water is

$$Q_2 = 5 \times 320 = 1.6 \times 10^3 \text{ J/s} .$$

The rate that the energy is expelled to the room is

$$Q_1 = \frac{T_1}{T_2} Q_2 = (303/273) \times 1.6 \times 10^3$$
$$= 1.78 \times 10^3 \text{ J/s} .$$

(b) The necessary power supplied is

$$W = Q_1 - Q_2 = 0.18 \text{ kW} .$$

(c) The coefficient of performance is

$$\varepsilon = \frac{T_2}{T_1 - T_2} = \frac{273}{30} = 9.1 .$$

1114

A Carnot cycle is operated with liquid-gas interface. The vapor pressure is p_v, temperature T, volume V. The cycle is operated according to the following $p - V$ diagram.

The cycle goes isothermally from 1 to 2, evaporating n moles of liquid. This is followed by reversible cooling from 2 to 3, then there is an isothermal contraction from 3 to 4, recondensing n moles of liquid, and finally a reversible heating from 4 to 1, completes the cycle.

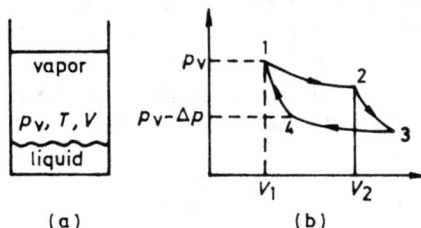

Fig. 1.32.

(a) Observe that $V_2 - V_1 = V_g - V_\ell$ where V_g = volume of n moles of gas, V_ℓ = volume of n moles of liquid. Calculate the efficiency in terms of

$\Delta p, V_g - V_\ell$, and L_v = latent heat vaporization of a mole of liquid. Treat Δp and ΔT as small.

(b) Recognizing that any two Carnot engines operating between T and $T - \Delta T$ must have the same efficiency (why?) and that this efficiency is a function of T and T alone, use the result of part (a) to obtain an expression for dp_v/dT in terms of $V_g - V_\ell, n, L_v$ and T.

(*CUSPEA*)

Solution:

(a) The temperature T in the process from 1 to 2 is constant. Because the total volume does not change, $V_2 - V_1 = V_g - V_\ell$. The engine does work $\Delta p(V_2 - V_1)$ on the outside world in the cyclic process. The heat it absorbs is nL_v. Therefore, the efficiency is

$$\eta = \Delta p(V_2 - V_1)/(nL_v) .$$

(b) The efficiency of a reversible Carnot engine working between T and $T - \Delta T$ is

$$\eta = \frac{\Delta T}{T} = \frac{\Delta p(V_g - V_\ell)}{L_v n} ,$$

Thus $\dfrac{dp_V}{dT} = \dfrac{nL_v}{T(V_g - V_\ell)} .$

1115

Many results based on the second law of thermodynamics may be obtained without use of the concepts of entropy or such functions. The method is to consider a (reversible) Carnot cycle involving heat absorption Q at $(T + dT)$ and release at T such that external work $(W + dW)$ is done externally at $(T + dT)$ and $-W$ is done at T. Then $Q = \Delta U + W$, where ΔU is the increase in the internal energy of the system. One must go around the cycle so positive net work dW is performed externally, where $dW/dT = Q/T$. In the following problems devise such a cycle and prove the indicated relations.

(a) A liquid or solid has vapor pressure p in equilibrium with its vapor. For 1 mole of vapor treated as a perfect gas, V (vapor) $\gg V$ (solid or liquid), let l be the 1 mole heat of vaporization. Show that

$$d \ln p/dT = l/RT^2 .$$

(b) A liquid has surface energy density u and surface tension τ.

i) Show that $u = \tau - T\dfrac{d\tau}{dT}$.

ii) If $\dfrac{d\tau}{dT} < 0$, and $\dfrac{d^2\tau}{dT^2} > 0$, will T increase or decrease for an adiabatic increase in area?

(*Columbia*)

Solution:

(a) Consider the following cycle: 1 mole of a liquid vaporizes at temperature $T + dT$, pressure $p + dp$, the vapor expands adiabatically to T, p and then condenses at T, p and finally it arrives adiabatically at its initial state. Thus we have $Q = l$, $dW \approx (p + dp)V - pV = V\,dp$, where V is the molar volume of the vapor, and

$$\frac{V\,dp}{dT} = \frac{l}{T} \ .$$

From the equation of state of an ideal gas $V = RT/p$, we have

$$\frac{d\ln p}{dT} = \frac{l}{RT^2} \ .$$

(b)(i) Consider the following cycle: A surface expands by one unit area at $T + dT$, and then expands adiabatically to T, it contracts at T, and comes back adiabatically to its initial state. For this cycle:

$$Q = u - \tau \ ,$$

$$dW = -\tau(T + dT) + \tau(T) = -\frac{d\tau}{dT}dT \ .$$

Thus

$$\frac{dW}{dt} = -\frac{d\tau}{dT} = \frac{u - \tau}{T} \ ,$$

or

$$u = \tau - T\frac{d\tau}{dT} \ .$$

(ii) From conservation of energy, we have

$$d(Au) = dQ + \tau(T)dA \ ,$$

where A is the surface area. As $dQ = 0$ in the adiabatic process,

$$(u - \tau)dA + Adu = 0 \ ,$$

or

$$\left(\frac{dT}{dA}\right)_{\text{adia}} = -\frac{u - T}{A\left(\dfrac{du}{dT}\right)} \ .$$

From (i) we have

$$\frac{du}{dT} = -T\left(\frac{d^2\tau}{dT^2}\right) \ .$$

With $d\tau/dT < 0$ and $d^2\tau/dT^2 > 0$, the above equations give

$$\left(\frac{dT}{dA}\right)_{\text{adia}} > 0 \ .$$

Hence when the surface area increases adiabatically, its temperature increases also.

1116

The heat of melting of ice at 1 atmosphere pressure and 0°C is 1.4363 kcal/mol. The density of ice under these conditions is 0.917 g/cm^3 and the density of water is 0.9998 g/cm^3. If 1 mole of ice is melted under these conditions, what will be

(a) the work done?
(b) the change in internal energy?
(c) the change in entropy?

(Wisconsin)

Solution:

(a) The work done is

$$W = p(V_2 - V_1)$$
$$= 1.013 \times 10^5 \times \left[\left(\frac{18}{0.9998}\right) - \left(\frac{18}{0.917}\right)\right]$$
$$= -0.1657 \text{ J} = -0.034 \text{ cal} \ .$$

(b) The heat absorbed by the 1 mole of ice is equal to its heat of fusion:

$$Q = 1.4363 \times 10^3 \text{ cal} \ .$$

Thus the change in internal energy is

$$\Delta U = Q - W \approx Q = 1.4363 \times 10^3 \text{ cal} \ .$$

(c) The change in entropy is

$$\Delta S = \frac{Q}{T} = \frac{1.4363 \times 10^3}{273} \approx 5.26 \text{ cal/K} .$$

1117

10 kg of water at 20°C is converted to ice at $-10°$C by being put in contact with a reservoir at $-10°$C. This process takes place at constant pressure and the heat capacities at constant pressure of water and ice are 4180 and 2090 J/kg·deg respectively. The heat of fusion of ice is 3.34×10^5 J/kg. Calculate the change in entropy of the universe.

(*Wisconsin*)

Solution:

The conversion of water at 20°C to ice at $-10°$C consists of the following processes. Water at 20° C $\xrightarrow{\text{a}}$ water at 0°C $\xrightarrow{\text{b}}$ ice at 0°C $\xrightarrow{\text{c}}$ ice at $-10°$C, where a and c are processes giving out heat with decreases of entropy and b is the process of condensation of water giving off the latent heat with a decrease of entropy also. As the processes take place at constant pressure, the changes of entropy are

$$\Delta S_1 = \int_{293}^{273} \frac{MC_p}{T} dT = MC_p \ln\left(\frac{273}{293}\right) = -2955 \text{ J/K} ,$$

$$\Delta S_2 = -\frac{|Q|}{T_0} = -\frac{10 \times 3.34 \times 10^5}{273} = -1.2234 \times 10^4 \text{ J/K} ,$$

$$\Delta S_3 = \int_{273}^{263} \frac{MC_p}{T} dT = MC_p \ln\frac{263}{273} = -757 \text{ J/K} .$$

In the processes, the increase of entropy of the reservoir due to the absorbed heat is

$$\Delta S_e = \frac{10 \times (4180 \times 20 + 3.34 \times 10^5 + 2090 \times 10)}{263}$$

$$= 16673 \text{ J/K} .$$

Thus, the total change of entropy of the whole system is

$$\Delta S = \Delta S_1 + \Delta S_2 + \Delta S_3 + \Delta S_e = 727 \text{ J/K} .$$

1118

Estimate the surface tension of a liquid whose heat of vaporization is 10^{10} ergs/g (250 cal/g).

(Columbia)

Solution:

The surface tension is the free energy of surface of unit area; therefore the surface tension is $\sigma = Qr\rho$, where Q is the heat of vaporization, r is the thickness of the surface ($r = 10^{-8}$ cm) and ρ is the liquid density ($\rho = 1$ g/cm^3). Thus

$$\sigma = 10^{10} \times 10^{-8} \times 1 = 100 \text{ dyn/cm} \ .$$

1119

Put letters from a to h on your answer sheet. After each put a T or an F to denote whether the correspondingly numbered statement which follows is true or false.

(a) The liquid phase can exist at absolute zero.

(b) The solid phase can exist at temperatures above the critical temperature.

(c) Oxygen boils at a higher temperature than nitrogen.

(d) The maximum inversion temperature of He is less than 20 K.

(e) γ of a gas is always greater than one.

(f) A compressor will get hotter when compressing a diatomic gas than when compressing a monatomic gas at the same rate.

(g) The coefficient of performance of a refrigerator can be greater than one.

(h) A slightly roughened ball is thrown from north to south. As one looks down from above, the ball is seen to be spinning counterclockwise. The ball is seen to curve toward east.

(Wisconsin)

Solution:

   (a) F;   (b) F;   (c) T;   (d) F;   (e) T;   (f) F;   (g) T;   (h) T .

1120

One gram each of ice, water, and water vapor are in equilibrium together in a closed container. The pressure is 4.58 mm of Hg, the temperature is 0.01°C. Sixty calories of heat are added to the system. The total volume is kept constant. Calculate to within 2% the masses of ice, water, and water vapor now present in the container. Justify your answers.

(Hint: For water at 0.01°C, the latent heat of fusion is 80 cal/g, the latent heat of vaporization is 596 cal/g, and the latent heat of sublimation is 676 cal/g. Also note that the volume of the vapor is much larger than the volume of the water or the volume of the ice.)

(*Wisconsin*)

Solution:

It is assumed that the original volume of water vapor is V, it volume is also V after heating, and the masses of ice, water, and water vapor are respectively x, y and z at the new equilibrium. We have

$$x + y + z = 3 , \tag{1}$$

$$(1 - x)L_{\text{sub}} + (1 - y)L_{\text{vap}} = Q = 60 , \tag{2}$$

$$\frac{1 - x}{\rho_{\text{ice}}} + \frac{(1 - y)}{\rho_{\text{water}}} + V_0 = V . \tag{3}$$

$$V_0 = \frac{RT}{\mu p} . \tag{4}$$

$$V = \frac{z}{\mu p} RT \tag{5}$$

where $\mu = 18$ g/mole, $p = 4.58$ mmHg, $T = 273.16$ K, $R = 8.2 \times 10^{-5}$ m$^3 \cdot$ atm/mol $\cdot$ K, $\rho_{\text{ice}} = \rho_{\text{water}} = 1$ g/cm^3, $L_{\text{sub}} = 676$ cal/g, and $L_{\text{vap}} = 596$ cal/g. Solving the equations we find

$$x = 0.25 \text{ g} , \quad y = 1.75 \text{ g} , \quad z = 1.00 \text{ g} .$$

That is, the heat of 60 cal is nearly all used to melt the ice.

1121

Define (a) critical point and (b) triple point in phase transformation.

Helium boils at 4.2 K under the atmospheric pressure $p = 760$ mm of mercury. What will be the boilding temperature of helium if p is reduced to 1 mm of mercury?

(UC, Berkely)

Solution:

Critical point is the terminal point of the vaporization line. It satisfies equations

$$\left(\frac{\partial p}{\partial V}\right)_T = 0 , \quad \left(\frac{\partial^2 p}{\partial V^2}\right)_T = 0 .$$

Triple point is the coexistence point for solid, liquid, and gas. When $p' = 1$ mmHg, the boilding temperature is 2.4 K.

1122

(a) State Van der Waals' equation of state for a real gas.

(b) Give a physical interpretation of the equation.

(c) Express the constants in terms of the critical data T_c, V_c, and p_c.

(Wisconsin)

Solution:

(a) Van der Waal's equation of state for a real gas is

$$\left(p + \frac{a}{V^2}\right)(V - b) = nRT .$$

(b) On the basis of the state equation for an ideal gas, we account for the intrinsic volumes of real gas molecules by introducing a constant b, and for the attractive forces among the molecules by introducing a pressure correction a/V^2.

(c) From $\left(p + \frac{a}{V^2}\right)(V - b) = nRT$, we have

$$p = \frac{nRT}{V - b} - \frac{a}{V^2}$$

so that

$$\left(\frac{\partial p}{\partial V}\right)_T = -\frac{nRT}{(V - b)^2} + \frac{2a}{V^3} ;$$

$$\left(\frac{\partial^2 p}{\partial V^2}\right)_T = \frac{2nRT}{(V - b)^3} - \frac{6a}{V^4} .$$

At the critical point, we have $\left(\frac{\partial p}{\partial V}\right)_T = 0$, $\left(\frac{\partial^2 p}{\partial V^2}\right)_T = 0$, so that

$$V_c = 3b, \quad p_c = \frac{a}{27b^2}, \quad nRT_c = \frac{8a}{27b}.$$

namely, $a = 3p_c V_c^2, b = V_c/3$.

1123

The Van der Waals equation of state for one mole of an imperfect gas reads

$$\left(p + \frac{a}{V^2}\right)(V - b) = RT.$$

[Note: part (d) of this problem can be done independently of part (a) to (c).]

(a) Sketch several isotherms of the Van der Waals gas in the p-V plane (V along the horizontal axis, p along the vertical axis). Identify the critical point.

(b) Evaluate the dimensionless ratio pV/RT at the critical point.

(c) In a portion of the p-V plane below the critical point the liquid and gas phases can coexist. In this region the isotherms given by the Van der Waals equation are unphysical and must be modified. The physically correct isotherms in this region are lines of constant pressure, $p_0(T)$. Maxwell proposed that $p_0(T)$ should be chosen so that the area under the modified isotherm should equal the area under the original Van der Waals isotherm. Draw a modified isotherm and explain the idea behind Maxwell's construction.

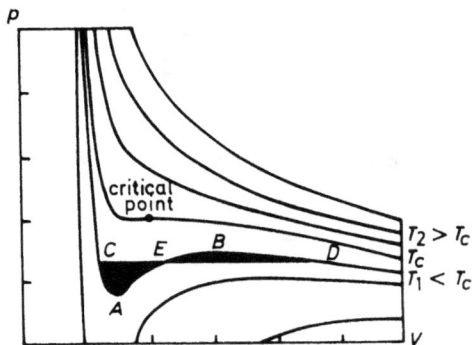

Fig. 1.33.

(d) Show that the heat capacity at constant volume of a Van der Waals gas is a function of temperature alone (i.e., independent of V).

$$(MIT)$$

Solution:

(a) As shown in Fig. 1.33, from $(\partial p/\partial V)_{T=T_c} = 0$ and $(\partial^2 p/\partial V^2)_{T=T_c} = 0$, we get

$$T_c = \frac{3a}{V_c^4} \frac{(V_c - b)^3}{R} \ .$$

So

$$V_c = 3b, p_c = \frac{a}{27b^2}, \quad T_c = \frac{8a}{27bR} \ .$$

(b) $p_c V_c/RT_c = 3/8$.

(c) In Fig. 1.33, the horizontal line CD is the modified isotherm. The area of CAE is equal to that of EBD. The idea is that the common points, i.e., C and D of the Van der Waals isotherm and the physical isotherm have the same Gibbs free energy. Because of $G = G(T, p)$, the equality of T's and p's respectively will naturally cause the equality of G. In this way,

$$\int_C^D dG = \int_C^D V\,dp = \int_C^A V\,dp + \int_A^E V\,dp + \int_E^B V\,dp + \int_B^D V\,dp = 0 \ .$$

That is,

$$\int_A^E V\,dp - \int_A^C V\,dp = \int_D^B V\,dp - \int_E^B V\,dp, \quad \text{or} \quad \Delta S_{CAE} = \Delta S_{EBD} \ .$$

(d) $$\left(\frac{\partial C_v}{\partial V}\right)_T = T\frac{\partial^2 S}{\partial V\,\partial T} = -T\frac{\partial^2}{\partial T^2}\left(\frac{\partial F}{\partial V}\right)$$

$$= T\left(\frac{\partial^2 p}{\partial T^2}\right)_V \ .$$

For a Van der Waals gas, the equation of state gives

$$\left(\frac{\partial^2 p}{\partial T^2}\right)_V = 0$$

so that

$$\left(\frac{\partial C_v}{\partial V}\right)_T = 0 \ .$$

1124

Determine the ratio (pV/RT) at the critical point for a gas which obeys the equation of state (Dieterici's equation)

$$p(V - b) = RT \exp(-a/RTV) .$$

Give the numerical answer accurately to two significant figures.

(UC, Berkeley)

Solution:

The critical point satisfies

$$\left(\frac{\partial p}{\partial V}\right)_T = 0 , \quad \left(\frac{\partial^2 p}{\partial V^2}\right)_T = 0 .$$

From the equation of state, we get

$$\left(\frac{\partial p}{\partial V}\right)_T = RT \frac{\left[\dfrac{a(V - b)}{RTV^2} - 1\right]}{(V - b)^2} e^{-\frac{a}{RTV}} ,$$

Consequently, $\dfrac{a(V - b)}{RTV^2} - 1 = 0.$

Using this result, we get

$$\left(\frac{\partial^2 p}{\partial V^2}\right)_T = \frac{a}{(V - b)V^3} \left(\frac{a}{RTV} - 2\right) e^{-\frac{a}{RTV}} .$$

Thus, $\dfrac{a}{RTV} - 2 = 0.$ Then, $V = 2b, RT = \dfrac{a}{4b}.$

Substituting these back in the equation of state, we find $\dfrac{pV}{RT} = 0.27.$

1125

Find the relation between the equilibrium radius r, the potential ϕ, and the excess of ambient pressure over internal pressure Δp of a charged soap bubble, assuming that surface tension can be neglected.

(Wisconsin)

Solution:

We assume that the air inside the bubble is in α-phase, the air outside the bubble is in β-phase, and the soap bubble itself is in γ-phase. We can solve this problem using the principle of minimum free energy. If the temperature is constant, we have

$$\delta F^\alpha = -p^\alpha \delta V^\alpha, \quad \delta F^\beta = -p^\beta \delta V^\beta, \quad \text{and} \quad \delta F^\gamma = q(\partial \phi/\partial r)\delta r \ ,$$

where $V^\alpha = \dfrac{4}{3}\pi r^3, \delta V^\alpha = 4\pi r^2 \delta r, \delta V^\beta = -\delta V^\alpha$.

The condition of minimum free energy demands

$$-p^\alpha \delta V^\alpha - p^\beta \delta V^\beta + q\frac{\partial \phi}{\partial r}\delta r = 0 \ .$$

Thus we have

$$\left(p^\beta - p^\alpha + \frac{q}{4\pi r^2}\frac{\partial \phi}{\partial r}\right)4\pi r^2 dr = 0 \ .$$

It follows that

$$\Delta p = p^\beta - p^\alpha = -\frac{q}{4\pi r^2}\frac{\partial \phi}{\partial r} \ .$$

With $\phi = q/r$, we have $\Delta p = \dfrac{\phi^2}{4\pi r^2}$.

1126

Consider a spherical soap bubble made from a soap film of constant surface tension, σ, and filled with air (assumed to be a perfect gas). Denote the ambient external pressure by p_0 and temperature by T.

(a) Find a relation between the equilibrium radius r of the soap bubble and the mass of air inside it.

(b) Solve the relation of part (a) for the radius r in the limit that the bubble is "large". Define precisely what is meant by "large".

(MIT)

Solution:

(a) Let $d\tau$ be an infinitesimal area of soap bubble surface, p_1 and p_0 be the pressures inside and outside the soap bubble, and μ_1, μ_2 be their chemical potentials. We have $dU = TdS - p_1 dV_1 - p_0 dV_2 + \sigma d\tau + \mu_1 dN_1 + \mu_2 dN_2$.

From the condition of equilibrium: $dU = 0, dS = 0, \mu_1 = \mu_2, dV_1 = -dV_2$ and $d(N_1 + N_2) = 0$, we get $(p_1 - p_0)dV_1 = \sigma d\tau$, or $p_1 - p_0 = \sigma d\tau / dV_1$, where $\dfrac{d\tau}{dV_1} = \dfrac{2}{r}$. Hence $p_1 - p_0 = 2\sigma / r$.

Since $p_1 V_1 = \dfrac{m}{M} RT$, where m is the mass of air inside the bubble, M is the molecular weight of air, we have

$$m = \frac{4\pi}{3} \frac{M}{RT} r^3 \left(p_0 + \frac{2\sigma}{r} \right) .$$

(b) When $p_0 \gg 2\sigma/r$, i.e., $r \gg 2\sigma/p_0$, we have $m = \dfrac{4\pi M p_0 r^3}{3RT}$.

1127

Derive the vapor pressure equation (Clausius-Clapeyron equation): $dp/dT = ?$

(*UC, Berkeley*)

Solution:

Conservation of energy gives

$$d\mu_1 = -S_1 dT + V_1 dp , \quad d\mu_2 = -S_2 dT + V_2 dp ,$$

where V_1 is the volume of the vapor, and V_2 is the volume of the liquid. In phase transition from liquid to vapor, chemical potential is invariant, i.e., $\mu_1 = \mu_2$, so that one has the vapor pressure equation:

$$\frac{dp}{dT} = \frac{S_1 - S_2}{V_1 - V_2} = \frac{L}{T(V_1 - V_2)} ,$$

where L is the latent heat of vaporization.

Usually $V_2 \ll V_1$, and this equation can be simplified to

$$\frac{dp}{dT} = \frac{L}{TV_1} .$$

1128

(a) By equating the Gibbs free energy or chemical potential on the two sides of the liquid-vapor coexistence curve derive the Clausius-Clapeyron equation: $\dfrac{dp}{dT} = \dfrac{q}{T(V_V - V_L)}$, where q is the heat of vaporization per particle and V_L is the volume per particle in the liquid and V_V is the volume per particle in the vapor.

(b) Assuming the vapor follows the ideal gas law and has a density which is much less than that of the liquid, show that $p \sim \exp(-q/kT)$, when the heat of vaporization is independent of T.

(Wisconsin)

Solution:

(a) From the first law of thermodynamics

$$d\mu = -S\,dT + V\,dp$$

and the condition that the chemical potential of the liquid is equal to that of the vapor at equilibrium, we obtain

$$-S_L\,dT + V_L\,dp = -S_V\,dT + V_V\,dp \ .$$

It follows that

$$\frac{dp}{dT} = \frac{S_V - S_L}{V_V - V_L} \ .$$

With $q = T(S_V - S_L)$, we have

$$\frac{dp}{dT} = \frac{q}{T(V_V - V_L)}$$

which is the Clausius-Clapeyron equation.

(b) If the vapor is regarded as an ideal gas, we have

$$pV_V = kT \ .$$

Because the density of vapor is much smaller than that of liquid, we can neglect V_L in the Clausius-Clapeyron equation and write

$$\frac{1}{p}\frac{dp}{dT} = \frac{q}{kT^2} \ ,$$

The solution is $p \sim \exp(-q/kT)$.

1129

A gram of liquid and vapor with heat of vaporization L is carried around the very flat reversible cycle shown in Fig. 1.34. Beginning at point A, a volume V_1 of liquid in equilibrium with a negligible amount of its saturated vapor is raised in temperature by ΔT and in pressure by Δp so as to maintain the liquid state. Then heat is applied at constant pressure and the volume increases to V_2 leaving a negligible amount of liquid. Then the pressure is lowered by Δp and the temperature decreased by ΔT so that essentially all the material remains in the vapor state. Finally, heat is removed, condensing essentially all the vapor back into the liquid state at point A.

Consider such a Carnot cycle and write the change of boiling point with pressure, dT/dp, for the liquid in terms of the heat of vaporization and other quantities.

(*Wisconsin*)

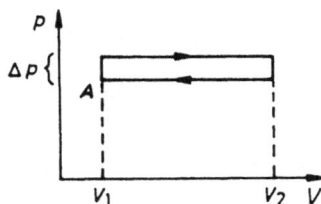

Fig. 1.34.

Solution:

In this cycle, the process at constant pressure is isothermal. We assume the net heat absorbed by the system is Q. Then its efficiency is $\eta = Q/L$. For the reversible Carnot cycle, the efficiency is $\eta = \dfrac{\Delta T}{T}$, giving $Q = \dfrac{\Delta T}{T} L$. Q must be equal to the external work W of the system in the cycle, $W = \Delta p(V_2 - V_1)$, so that

$$\frac{\Delta T}{T} L = \Delta p(V_2 - V_1) .$$

Therefore,

$$dT/dp = \lim_{\Delta p \to 0} \frac{\Delta T}{\Delta p} = \frac{T(V_2 - V_1)}{L} .$$

1130

(a) Deduce from the 1st and 2nd laws of thermodynamics that, if a substance such as H_2O expands by 0.091 cm^3/g when it freezes, its freezing temperature must decrease with increasing pressure.

(b) In an ice-skating rink, skating becomes unpleasant (i.e., falling frequently) if the temperature is too cold so that the ice becomes too hard. Estimate the lowest temperature of the ice on a skating rink for which ice skating for a person of normal weight would be possible and enjoyable. (The latent heat of ice is 80 cal/g).

(SUNY, Buffalo)

Solution:

Denote the liquid and solid phases by 1 and 2 respectively.

(a) The condition for coexistence of the two phases is

$$\mu_2 = \mu_1, \quad \text{so that } d\mu_2 = d\mu_1 ,$$

giving

$$V_1 dp_1 - S_1 dT_1 = V_2 dp_2 - S_2 dT_2 .$$

As $p_2 = p_1 = p$ and $T_2 = T_1 = T$ on the coexistence line, we have

$$\left(\frac{dp}{dT}\right)_{\text{phase line}} = \frac{S_2 - S_1}{V_2 - V_1} .$$

For regions whose temperatures are higher than those of phase transformation we have $\mu_1 < \mu_2$, and for the regions whose temperatures are lower than those of phase transformation we have $\mu_1 > \mu_2$. This means that

$$\left(\frac{\partial \mu_1}{\partial T}\right)_p < \left(\frac{\partial \mu_2}{\partial T}\right)_p$$

i.e., for any temperature, $S_1 > S_2$.

For substances such as water, $V_2 > V_1$, so $\left(\dfrac{dp}{dT}\right)_{\text{phase line}} < 0.$

(b) The lowest temperature permitted for enjoyable skating is the temperature at which the pressure on the coexistence line is equal to the pressure exerted by the skater on ice. The triple point of water is at $T_0 = 273.16$ K, $p_0 = 1$ atm. For a skater of normal weight $\bar{p} \sim 10$ atm, so that

$$(\bar{p} - p_0)/(T_{\text{min}} - T_0) = -h/T_{\text{min}} \Delta V .$$

With $h = 80$ cal/g, $\Delta V = 0.091$ cm^3/g, we have

$$T_{\min} = \frac{T_0}{1 + \dfrac{(\bar{p} - p_0)\Delta V}{h}} = (1 - 2.5 \times 10^{-3})T_0 = -0.06°C \ .$$

1131

The following data apply to the triple point of H_2O.

Temperature: 0.01°C; Pressure: 4.6 mmHg

Specific volume of solid: 1.12 cm^3/g

Specific volume of liquid: 1.00 cm^3/g

Heat of melting: 80 cal/g

Heat of vaporization: 600 cal/g.

(a) Sketch a $p - T$ diagram for H_2O which need not be to scale but which should be qualitatively correct. Label the various phases and critical points.

(b) The pressure inside a container enclosing H_2O (which is maintained at $T = -1.0°C$) is slowly reduced from an initial value of 10^5 mmHg. Describe what happens and calculate the pressure at which the phase changes occur. Assume the vapor phase behaves like an ideal gas.

(c) Calculate the change in specific latent heat with temperature dL/dT at a point (p, T) along a phase equilibrium line. Express your result in terms of L and the specific heat C_p, coefficient of expansion α, and specific volume V of each phase at the original temperature T and pressure p.

(d) If the specific latent heat at 1 atm pressure on the vaporization curve is 540 cal/g, estimate the change in latent heat 10°C higher than the curve. Assume the vapor can be treated as an ideal gas with rotational degrees of freedom.

(*MIT*)

Solution:

(a) The $p - T$ diagram of H_2O is shown in Fig. 1.35.

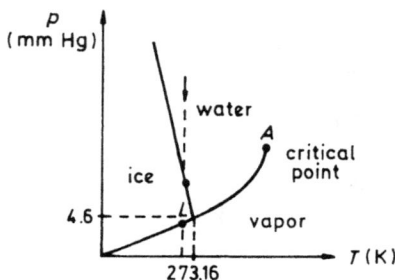

Fig. 1.35.

(b) The Clausius-Clapeyron equation gives

$$\left(\frac{dp}{dT}\right)_{\text{ice-water}} = \frac{L}{T(V_{\text{water}} - V_{\text{ice}})} = -2.4 \text{ cal/cm}^3 \cdot \text{K} .$$

$$\left(\frac{dp}{dT}\right)_{\text{water-vapor}} > 0 .$$

When the pressure, which is slowly reduced, reaches the solid-liquid phase line, heat is released by the water while the pressure remains unchanged until all the water is changed into ice. Then at the vapor-solid line, the ice absorbs heat until it is completely changed into vapor. Afterwards the pressure begins to decrease while the vapor phase is maintained. The pressure at which water is converted to ice is given by

$$p_{\text{water-ice}} = p_0 + \frac{L}{V_{\text{water}} - V_{\text{ice}}} \cdot \frac{T - T_0}{T_0} = 6.3 \times 10^3 \text{cmHg}$$

where we have used the values $T = 272.15$ K, $T_0 = 273.16$ K and $p_0 = 4.6$ mmHg. As $V_{\text{vapor}} = \dfrac{kT}{pm} \gg V_{\text{ice}}$, we have

$$\frac{dp}{dT} \approx \frac{L}{TV_{\text{vapor}}} = \frac{mLp}{T^2 k} .$$

The pressure at which ice is converted to vapor is

$$p_{\text{ice-vapor}} \approx p_0 \exp\left[\frac{mL}{k}\left(\frac{1}{T_0} - \frac{1}{T}\right)\right] = 4.4 \text{ mmHg}$$

where m is the molecular mass of water.

(c) From $L = T(S_1 - S_2)$, we have

$$\frac{dL}{dT} = \frac{L}{T} + T\left(\frac{dS_1}{dT} - \frac{dS_2}{dT}\right) .$$

As $dS_1 = \dfrac{C_{p_1}}{T}dT - \alpha_1 V_1 dp$, where $\alpha_1 = \dfrac{1}{V_1}\left(\dfrac{\partial V_1}{\partial T}\right)_p$, we have

$$\frac{dL}{dT} = \frac{L}{T} + (C_{p_1} - C_{p_2}) - (\alpha_1 V_1 - \alpha_2 V_2)T\frac{dp}{dT} .$$

Using

$$\frac{dp}{dT} = \frac{L}{T(V_1 - V_2)} ,$$

we obtain

$$\frac{dL}{dT} = \frac{L}{T} + (C_{p_1} - C_{p_2}) - (\alpha_1 V_1 - \alpha_2 V_2)\frac{L}{V_1 - V_2} .$$

(d) Let 1 and 2 stand for water and vapor respectively. From $V_2 \gg V_1$, we know

$$\frac{dL}{dT} \approx \frac{L}{T} + (C_{p_1} - C_{p_2}) - \alpha_2 L ,$$

where $\alpha_2 = 1/T$, so $\Delta L = (C_{p_1} - C_{p_2})\Delta T$.

Letting $C_{p_1} = 1$ cal/g $\cdot$ °C, $C_{p_2} = \dfrac{2}{9}R$ cal/g·°C, $\Delta T = 10$°C, we get $\Delta L = 6$ cal/g.

1132

(a) Derive an expression for the dependence of the equilibrium vapor pressure of a material on the toal pressure (i.e., how does the equilibrium partial pressure of a material depend on the addition of an overpressure of some inert gas?).

(b) Use this result to discuss qualitatively the difference between the triple point and the ice point of water.

(Wisconsin)

Solution:

(a) We assume the mole concentration of the solute in the solution is $x \ll 1$. Thus the mole chemical potential of the solution is

$$\mu_1(p, T) = \mu_1^0(p, T) - xRT ,$$

where $\mu_1^0(p, T)$ is the mole chemical potential of the pure solvent. If the mole chemical potential of the vapor phase is $\mu_2^0(p, T)$, the equilibrium vapor pressure of the solvent, p_0, is determined by

$$\mu_1^0(p_0, T_0) = \mu_2^0(p_0, T_0) .$$

When the external pressure (the total pressure) is p, the condition of equilibrium of vapor and liquid is

$$\mu_1^0(p, T) - xRT = \mu_2^0(p, T) .$$

Making use of Taylor's theorem, we have from the above two equations

$$\frac{\partial \mu_1^0}{\partial p}(p - p_0) + \frac{\partial \mu_1^0}{\partial T}(T - T_0) - xRT$$

$$= \frac{\partial \mu_2^0}{\partial p}(p - p_0) + \frac{\partial \mu_2^0}{\partial T}(T - T_0) .$$

Using the thermodynamic relation $d\mu = -SdT + Vdp$, we can write the above as

$$p - p_0 = [(S_2 - S_1)(T - T_0) - xRT]/(V_2 - V_1) ,$$

or

$$p_0 = p - [L(T - T_0)/T - xRT]/(V_2 - V_1) ,$$

where V is the mole volume, S is the mole entropy, and L is the latent heat, $L = T(S_2 - S_1)$.

(b) The triple point of water is the temperature T_0 at which ice, water and vapor are in equilibrium. The ice point is the temperature T at which pure ice and air-saturated water are in equilibrium at 1 atm. Utilizing the result in (a) we have

$$T - T_0 = T(V_2 - V_1)(p - p_0)/L + xRT^2/L ,$$

where V_2 and V_1 are respectively the mole volumes of ice and water. From $V_2 > V_1$ and $L < 0$, we know the ice point is lower than the triple point.

The first term of the above formula comes from the change of pressure, the second term appears because water is not pure. The quantitative result of the first term is -0.0075 K, of the second term is -0.0023 K.

1133

Some researchers at the Modford Institute of Taxidermy claim to have measured the following pressure-temperature phase diagram of a new substance, which they call "embalmium". Their results show that along the phase lines near the triple point

$$0 < \left(\frac{dp}{dT}\right)_{\text{sublimation}} < \left(-\frac{dp}{dT}\right)_{\text{fusion}} < \left(\frac{dp}{dT}\right)_{\text{vaporization}}$$

as indicated in the diagram. If these results are correct, "embalmium" has one rather unusual property and one property which violates the laws of thermodynamics. What are the two properties?

(MIT)

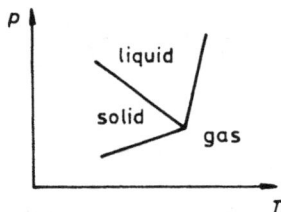

Fig. 1.36.

Solution:

The property $\left(\dfrac{dp}{dT}\right)_{\text{fusion}} < 0$ is unusual as only a few substances like water behaves in this way. The Clausius-Clapeyron equation gives

$$\left(\frac{dp}{dT}\right)_{\text{vaporization}} = \frac{(S_{\text{gas}} - S_{\text{liquid}})}{V_{\text{gas}}}$$

$$\left(\frac{dp}{dT}\right)_{\text{sublimation}} = \frac{(S_{\text{gas}} - S_{\text{solid}})}{V_{\text{gas}}} .$$

$\left(\dfrac{dp}{dT}\right)_{\text{vaporization}} > \left(\dfrac{dp}{dT}\right)_{\text{sublimiation}}$ means $S_{\text{solid}} > S_{\text{liquid}}$, i.e., the mole entropy of the solid phase is greater than that of the liquid phase, which violates the second law of thermodynamics, since a substance absorbs heat to transform from solid to liquid and the process should be entropy increasing.

1134

The latent heat of vaporization of water is about 2.44×10^6 J/kg and the vapor density is 0.598 kg/m^3 at 100°C. Find the rate of change of the boiling temperature with altitude near sea level in °C per km. Assume the temperature of the air is 300 K.

(Density of air at 0°C and 1 atm is 1.29 kg/m^3).

(Wisconsin)

Solution:

The Boltzmann distribution gives the pressure change with height:

$$p(z) = p(0) \exp - \frac{mgz}{kT_0} \, ,$$

where $p(0)$ is the pressure at sea level $z = 0$, m is the molecular weight of air, and $T_0 = 300$ K is the temperature of the atmosphere. The Clausius-Clapeyron equation can be written as

$$\frac{dp}{dT} = \frac{L}{T(V_2 - V_1)} = \frac{L}{TM \left(\frac{1}{\rho_2} - \frac{1}{\rho_1} \right)} = \frac{\alpha}{T} \, .$$

with $\rho_1 = 1000$ kg/m^3, $\rho_2 = 0.598$ kg/m^3 and $L/M = 2.44 \times 10^6$ J/kg, we have

$$\alpha = \frac{L\rho_1\rho_2}{M(\rho_1 - \rho_2)} = 1.40 \times 10^6 \text{ J/m}^3 \, .$$

So the rate of change of the boiling point with height is

$$\frac{dT}{dz} = \frac{dT}{dp} \cdot \frac{dp}{dz} = \frac{T}{\alpha} \cdot \left(\frac{-mg}{kT_0} \right) p(z) \, .$$

Using the equation of state for ideal gas $p = \rho kT_0/m$, we have near the sea level

$$\frac{dT}{dz} = -\rho g T(0)/\alpha \, ,$$

where $\rho = 1.29$ kg/m^3 is the density of air, $g = 9.8$ m/s^2 and $T(0) = 100$°C.

Thus $\dfrac{dT}{dz} = -0.87$°C/km.

1135

A long vertical cylindrical column of a substance is at temperature T in a gravitational field g. Below a certain point along the column the substance is found to be a solid; above that point it is a liquid. When the temperature is lowered by ΔT, the position of the solid-liquid interface is observed to move upwards a distance l. Neglecting the thermal expansion of the solid, find an expression for the density ρ_1 of the liquid in terms of the density ρ_s of the solid, the latent heat L of the solid-liquid phase transition, g and the absolute temperature T and ΔT.

Assume that $\Delta T/T \ll 1$.

(*Princeton*)

Solution:

The Clausius-Clapeyron equation gives

$$\frac{dp}{dT} = \frac{L}{T(V_1 - V_s)} = \frac{L}{T\left(\dfrac{1}{\rho_1} - \dfrac{1}{\rho_s}\right)} .$$

In the problem, $dT = -\Delta T, dp = -gl\rho_s$. Hence

$$\rho_1 = \rho_s \frac{1}{1 + \dfrac{\Delta T}{T}\dfrac{L}{lg}} .$$

1136

(a) Use simple thermodynamic considerations to obtain a relation between $\dfrac{1}{T_m}\dfrac{dT_m}{dp}$, the logarithmic rate of variation of melting point with change of pressure, the densities of the solid and liquid phases of the substance in question and the latent heat of melting. (You may find it convenient to relate the latent heat to the entropy change.)

(b) Use simple hydrostatic considerations to relate the pressure gradient within the earth to the earth's density and the acceleration of gravity. (Assume that the region in question is not at great depth below the surface.)

(c) Combine the foregoing to calculate the rate of variation of the melting point of silicate rock with increasing depth below the earth's surface in a region where the average melting point of the rock is 1300°C. Assume a density ratio

$$\rho_{\text{liquid}} / \rho_{\text{solid}} \approx 0.9$$

and a latent heat of melting of 100 cal/g. Give your answer in degrees C per kilometer.

(UC, Berkeley)

Solution:

(a) During the phase transtion, $\mu_l = \mu_s$, where l and s represent liquid phase and solid phase respectively. By thermodynamic relation

$$d\mu = -S\,dT + V\,dP \ ,$$

we have $(S_l - S_s)dT_m = (V_l - V_s)dP$, so

$$\frac{1}{T_m}\frac{dT_m}{dp} = \frac{V_l - V_s}{T_m(S_l - S_s)} = \frac{V_l - V_s}{L} \ .$$

Substituting $V = 1/\rho$ into the equation above, we get

$$\frac{dT_m}{T_m\,dp} = \frac{1}{L\rho_l}\left(1 - \frac{\rho_l}{\rho_s}\right) \ .$$

(b) Denote the depth as z, we have $\dfrac{dp}{dz} = \rho g$.

(c) From the above results, we have

$$\frac{dT_m}{dz} = \frac{T_m \rho g}{L\rho_l}\left(1 - \frac{\rho_l}{\rho_s}\right) \approx \frac{T_m g}{L}\left(1 - \frac{\rho_l}{\rho_s}\right)$$

$$= 37 \times 10^{-6}\ {}^\circ\text{C/cm} = 3.7^\circ\text{C/km} \ .$$

1137

The vapor pressure, in mm of Hg, of solid ammonia is given by the relation: $\ln p = 23.03 - 3754/T$ where $T = $ absolute temperature.

The vapor pressure, in mm of Hg, of liquid ammonia is given by the relation: $\ln p = 19.49 - 3063/T$.

(a) What is the temperature of the triple point?

(b) Compute the latent heat of vaporization (boiling) at the triple point. Express your answer in cal/mole. (You may approximate the behavior of the vapor by treating it as an ideal gas, and may use the fact that the density of the vapor is negligibly small compared to that of the liquid.)

(c) The latent heat of sublimation at the triple point is 7508 cal/mole. What is the latent heat of melting at the triple point?

(UC, Berkeley)

Solution:

(a) The temperature T of the triple point satisfies the equation $23.03 - 3754/T_0 = 19.49 - 3063/T_0$, which gives $T_0 = 195$ K.

(b) From the relation between the vapor pressure and temperature of liquid ammonia

$$\ln p = C - 3063/T ,$$

we get $dp/dT = 3063p/T^2$.

The Clausius-Clapeyron equation $\dfrac{dp}{dT} = \dfrac{L}{TV}$ then gives

$$L = 3063pV/T = 3063R = 2.54 \times 10^4 \text{ J/mol}$$
$$= 6037 \text{ cal/mol} .$$

(c) Denote S_g, S_l and S_s as the entropy for vapor, liquid and solid at triple point. Then the latent heat of vaporization is $T_0(S_g - S_l)$, that of sublimation is $T_0(S_g - S_s)$, and that of melting is

$$T(S_l - S_s) = T(S_g - S_l) - T(S_g - S_l)$$
$$= 7508 - 6037 = 1471 \text{ cal/mol} .$$

1138

The high temperature behavior of iron can be summarized as follows.

(a) Below 900°C and above 1400°C α-iron is the stable phase.

(b) Between these temperatures γ-iron is stable.

(c) The specific heat of each phase may be taken as constant: $C_\alpha = 0.775\,\text{J/g} \cdot \text{K}; C_\gamma = 0.690$ J/g $\cdot$ K.

What is the latent heat at each transition?

(UC, Berkeley)

Solution:

Referring to Fig. 1.37, we regard the whole process as isobaric.

Choose the entropy at T_1 as zero for the α-phase. Since $T\dfrac{dS}{dT} = C$, one has

$$S = C \ln T + \text{const.} : \quad S_\alpha = C_\alpha \ln \left(\frac{T}{T_1} \right), \quad S_\gamma = S_1 + C_\gamma \ln \left(\frac{T}{T_1} \right) .$$

The changes in chemical potential are

$$\Delta\mu^\alpha = \mu^\alpha(T_2) - \mu^\alpha(T_1) = -\int_{T_1}^{T_2} S_\alpha dT = -C_\alpha T_2 \ln \left(\frac{T_2}{T_1} \right) + C_\alpha(T_2 - T_1) ,$$

$$\Delta\mu^\gamma = \mu^\gamma(T_2) - \mu^\gamma(T_1) = -\int_{T_1}^{T_2} S_\gamma dT = -C_\gamma T_2 \ln \left(\frac{T_2}{T_1} \right)$$
$$+ (C_\gamma - S_1)(T_2 - T_1) .$$

Since $\Delta\mu^\alpha = \Delta\mu^\gamma$, we have

$$S_1 = \left(\frac{T_2}{T_2 - T_1} \ln \frac{T_2}{T_1} - 1 \right) (C_\alpha - C_\gamma)$$
$$= 1.60 \times 10^{-2} \text{ J/g·K} .$$

Therefore

$$L_1 = S_1 T_1 = 18.8 \text{ J/g} .$$
$$L_2 = T_2(S_3 - S_2)$$
$$= (C_\alpha - C_\gamma)T_2 \ln \left(\frac{T_2}{T_1} \right) - S_1 T_2$$
$$= 23.7 \text{ J/g} .$$

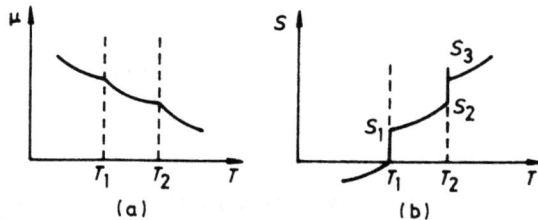

Fig. 1.37.

1139

Liquid helium-4 has a normal boiling point of 4.2 K. However, at a pressure of 1 mm of mercury, it boils at 1.2 K. Estimate the average latent heat of vaporization of helium in this temperature range.

(*UC, Berkeley*)

Solution:

According to the Clausius-Clapeyron equation and the equation of state for ideal gas

$$\frac{dp}{dT} = \frac{L}{T(V_g - V_l)} \approx \frac{L}{TV_g}, \quad pV_g = RT ,$$

and assuming L to be constant, we get

$$L = R \ln \frac{p}{p_0} \bigg/ \left(\frac{1}{T_0} - \frac{1}{T} \right) .$$

Therefore $L = 93$ J/mol.

1140

(a) The pressure-volume diagram shows two neighbouring isotherms in the region of a liquid-gas phase transition. By considering a Carnot cycle between temperatures T and $T + dT$ in the region shown shaded in the diagram, derive the Clausius-Clapeyron equation relating vapor pressure and temperature, $dp/dT = L/(T\Delta V)$, where L is the latent heat of vaporization per mole and ΔV is the volume change between gas and liquid per mole.

(b) Liquid helium boils at temperature $T_0 = 4.2$ K when its vapor pressure is equal to $p_0 = 1$ atm. We now pump on the vapor and reduce the pressure to a much smaller value p. Assuming that the latent heat L is approximately independent of temperature and that the helium vapor density is much smaller than that of the liquid, calculate the approximate temperature T_m of the liquid in equilibrium with its vapor at pressure p. Express your answer in terms of L, T_0, p_0, p_m, and any other required constants.

(*CUSPEA*)

Solution:

(a) From the $p - V$ diagram, we can see that the work done by the working material on the outside world is $dW = dp\Delta V$ in this infinitesimal Carnot cyce. The heat absorbed in the process is $Q = L$. The formula for the efficiency of a Carnot engine gives $\dfrac{dW}{L} = \dfrac{dT}{T}$.

Thus $\dfrac{dp}{dT} = \dfrac{L}{T\Delta V}$.

(b) Since

$$\Delta V = V_{\text{gas}} - V_{\text{liquid}} \approx V_{\text{gas}} = RT/p ,$$
$$\frac{dp}{dT} = \frac{Lp}{RT^2} .$$

Hence

$$\ln \frac{p_0}{p_m} = \frac{L}{R}\left(\frac{1}{T_m} - \frac{1}{T_0}\right) .$$

Therefore

$$T_m = \frac{T_0}{\left(1 + \dfrac{RT_0}{L}\ln \dfrac{p_0}{p_m}\right)} .$$

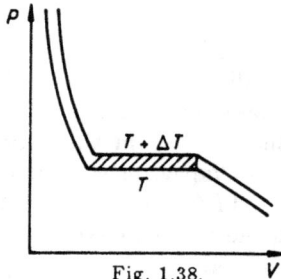

Fig. 1.38.

1141

When He^3 melts the volume increases. The accompanying plot is a sketch of the He^3 melting curve from 0.02 to 1.2 K. Make a sketch to show the change in entropy which accompanies melting in this temperature range.

(*Wisconsin*)

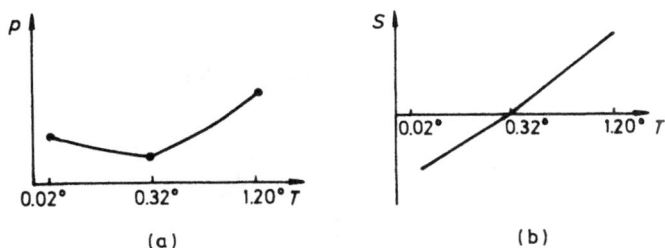

Fig. 1.39.

Solution:

From the Clausius-Clapeyron equation, we have

$$\frac{dp}{dT} = \frac{\Delta S}{\Delta V}, \quad \text{and so} \quad \Delta S = \Delta V \frac{dp}{dT} .$$

When He^3 melts, the volume increases, i.e., $\Delta V > 0$.

When 0.02 K$< T < 0.32$ K, because $\frac{dp}{dT} < 0, \Delta S < 0$.

When 0.32 K $< T < 1.2$ K, because $\frac{dp}{dT} > 0, \Delta S > 0$.

When $T = 0.32$ K, $\Delta S = 0$. The results are shown in Fig. 1.39(b).

1142

The phase transition between the aromatic (a) and fragrant (f) phases of the liquid mythological-mercaptan is second order in the Ehrenfest scheme, that is, ΔV and ΔS are zero at all points along the transition line $p_{a-f}(T)$.

Use the fact that $\Delta V = V_a(T,p) - V_f(T,p) = 0$, where V_a and V_f are the molar volumes in phase a and phase f respectively, to derive the slope of the transition line, $dp_{a-f}(T)/dT$, in terms of changes in the thermal expansion coefficient, α, and the isothermal compressibility, k_T at the transition.

(*MIT*)

Solution:

Along the transition line, one has

$$V_a(T,p) = V_f(T,p) .$$

Thus $dV_a(T,p) = dV_f(T,p)$.

Since

$$dV(T,p) = \left(\frac{\partial V}{\partial T}\right)_p dT + \left(\frac{\partial V}{\partial p}\right)_T dp = V(\alpha dT - k_T dp) ,$$

we have

$$(\alpha_a - \alpha_f)dT = (k_{Ta} - k_{Tf})dp$$

or

$$dp_{a-f}(T)/dT = (\alpha_a - \alpha_f)/(k_{Ta} - k_{Tf}) \ .$$

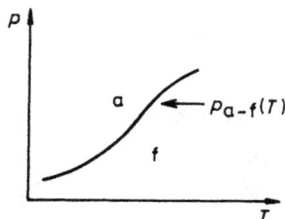

Fig. 1.40.

1143

State Curie's law for the magnetization of a paramagnetic gas. Why does the magnetization depend on temperature? What modification of the law is necessary as $T \to 0$?

(Wisconsin)

Solution:

Curie's law states that the magnetization of a paramagnetic substance in a magnetic field is inversely proportional to the absolute temperature: $M = CH/T$, where C is the Curie constant. As the temperature changes, so does the distribution of the directions of spins of the atoms and ions; thus the magnetization is dependent on T.

At low temperatures the paramagnetic phase changes into the ferromagnetic phase. At this time, the external magnetic field B_a produces a certain magnetization M, which in turn produces an exchange magnetic field $B_E = \lambda M$ (λ is a constant). From $M = \chi(B_a + B_E) = \chi(B_a + \lambda M)$ and $\chi = C/T$ (Curie's law), we have

$$\chi = \frac{M}{B_a} = \frac{C}{T - T_C} \ ,$$

where $T_C = C\lambda$ is the Curie temperature.

1144

A substance is found to have two phases, N and S. In the normal state, the N phase, the magnetization M is negligible. At a fixed temperature $T < T_c$, as the external magnetic field H is lowered below the critical field

$$H_c(T) = H_0 \left[1 - \left(\frac{T}{T_c} \right)^2 \right] ,$$

the normal state undergoes a phase transition to a new state, the S phase. In the S state, it is found that $B = 0$ inside the material. The phase diagram is shown below.

(a) Show that the difference in Gibbs free energies (in cgs units) between the two phases at temperature $T \le T_c$ is given by

$$G_S(T, H) - G_N(T, H) = \frac{1}{8\pi} [H^2 - H_c^2(T)] .$$

(You may express your answer in another system of units. The Gibbs free energy in a magnetic field is given by $G = U - TS - HM$.)

(b) At $H \le H_0$, compute the latent heat of transition L from the N to the S phase. (Hint: one approach is to consider a "Clausius-Clapeyron" type of analysis.)

(c) At $H = 0$, compute the discontinuity in the specific heat as the material transforms from the N to the S phase.

(d) Is the phase transition first or second order at $H = 0$?

(*UC, Berkeley*)

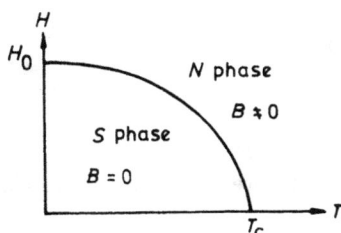

Fig. 1.41.

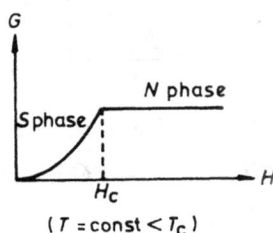

Fig. 1.42.

Solution:

(a) Differentiating the expression for Gibbs free energy, we find $dG = -SdT - MdH$, where $B = H + 4\pi M$ in cgs units. Referring to Fig. 1.42, we have

N phase: $M = 0$, $G_N = G_0(T)$,

S phase: $B = 0$, $M = -H/4\pi$.

Integrating $dG = -M\,dH$, we obtain

$$G_S = H^2/8\pi + \text{const.}$$

Noting that $G_S\,(\text{H}_c,\,T) = G_0(T)$ at the transition point, we have

$$G_S = G_0(T) + \frac{1}{8\pi}(H^2 - H_c^2)\ .$$

It follows that

$$G_S - G_N = \frac{1}{8\pi}(H^2 - H_c^2)\ .$$

(b) Since

$$S = -\left(\frac{\partial G}{\partial T}\right)_H\ ,$$

we have

$$L = T(S_N - S_S) = T\left[\frac{\partial(G_S - G_N)}{\partial T}\right]_H = -\frac{TH_c}{4\pi}\left(\frac{\partial H_c}{\partial T}\right)$$

$$= \frac{H_0^2}{2\pi}\left(\frac{T}{T_c}\right)^2\left[1 - \left(\frac{T}{T_c}\right)^2\right]\ .$$

(c) $C_S - C_N = T\left[\frac{\partial(S_S - S_N)}{\partial T}\right]_H$

$$= \frac{T}{4\pi}\left[\left(\frac{\partial H_c}{\partial T}\right)^2 + H_c\left(\frac{\partial^2 H_c}{\partial T^2}\right)\right]$$

$$= \frac{H_0^2}{2\pi}\frac{T}{T_c^2}\left[3\left(\frac{T}{T_c}\right)^2 - 1\right]\ .$$

When $H = 0, C_S - C_N = H_0^2/\pi T_c$

(d) At $H = 0, L = 0, C_S - C_N \neq 0$, therefore the phase transition is second order.

1145

The phase boundary between the superconducting and normal phases of a metal in the $H_e - T$ plane (H_e = magnitude of applied external field) is given by Fig. 1.43.

The relevant thermodynamic parameters are T, p, and H_e. Phase equilibrium requires the generalized Gibbs potential G (including magnetic paramters) to be equal on either side of the curve. Consider state A in the normal phase and A' in the superconducting phase; each lies on the phase boundary curve and has the same T, p and H_e but different entropies and magnetizations. Consider two other states B and B' arbitrarily close to A and A'; as indicated by $p_A = p_B$.

(a) Use this information to derive a Clapeyron-Clausius relation (that is, a relation between the latent heat of transition and the slope dH_e/dT of the curve). What is the latent heat at either end of the curve? (For a long rod-shaped superconducting sample with volume V oriented parallel to the field, the induced magnetic moment is given by $M'_H = -V H_e/4\pi$; in the normal state, set $M_H = 0$.)

(b) What is the difference in specific heats at constant field and pressure (C_{p,H_e}) for the two phases? What is the discontinuity in C_{p,H_e} at $H_e = 0, T = T_c$? At $T = 0, H_e = H_c$?

(*Princeton*)

Solution:

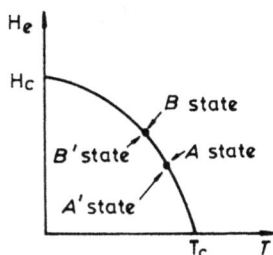

Fig. 1.43.

(a) $dG = -S \, dT + V \, dp - M_H \, dH_e$.
The condition of phase equilibrium is

$$G_A = G'_A, \quad G_B = G_{B'} .$$

Thus $dG = dG'$.
With $dp = 0$, one obtains for the superconducting sample

$$\frac{dH_e}{dT} = -\frac{S' - S}{M'_H - M_H} = \frac{L}{T(M'_H - M_H)} = -\frac{4\pi L}{V H_e T} .$$

where $L = T(S - S')$ is the latent heat of phase transition. At the two ends of the curve: $H_e(T_c) = 0$ at $T = T_c$ gives $L = 0$; $dH_e/dT = 0$ at $T = 0$ gives $L = 0$ also.

(b) From the above equation, we have

$$S' - S = \frac{V H_e}{4\pi} \cdot \frac{dH_e}{dT} .$$

As $C = T(dS/dT)$

$$\Delta C = C'_{p,H_e} - C_{p,H_e} = \frac{VT}{4\pi} \left[H_e \frac{d^2 H_e}{dT^2} + \left(\frac{dH_e}{dT} \right)^2 \right] .$$

At $T = T_c$; $H_e = 0$, we have

$$\Delta C = \frac{VT_c}{4\pi} \left[\frac{dH_e}{dT} \right]^2_{T_c} ,$$

At $T = 0$, $H_e = H_c$, we have

$$\Delta C = \frac{VT}{4\pi} H_c \left[\frac{d^2 H_e}{dT^2} \right]_{T=0} = 0 .$$

1146

A simple theory of the thermodynamics of a ferromagnet uses the free energy F written as a function of the magnetization M in the following form: $F = -HM + F_0 + A(T - T_c)M^2 + BM^4$, where H is the magnetic field, F_0, A, B are positive constants, T is the temperature and T_c is the critical temperature.

(a) What condition on the free energy F determines the thermodynamically most probable value of the magnetization M in equilibrium?

(b) Determine the equilibrium value of M for $T > T_c$ and sketch a graph of M versus T for small constant H.

(c) Comment on the physical significance of the temperature dependence of M as T gets close to T_c for small H in case (b).

(Wisconsin)

Solution:

According to the problem F denotes the Gibbs function.

(a) $F = $ minimum is the condition to determine the most probable value of M in equilibrium. Thus M is determined from $(\partial F/\partial M)_{T,H} = 0$.

(b) $(\partial F/\partial M)_{T,H} = -H + 2A(T - T_c)M + 4BM^3 = 0.$ (*)
If $2A(T - T_c)M \gg 4BM^3$, that is, if T is far from T_c, we have

$$M = \frac{H}{2A(T - T_c)} \cdot$$

This is the Curie-Weiss law. The change of M with T is shown in Fig. 1.44.

(c) If $H = 0$, the equation (*) has solutions

$$M = 0 , \quad M = \pm\sqrt{A(T_c - T)/2B} .$$

For stability consider

$$\left(\frac{\partial^2 F}{\partial M^2}\right)_{T,H} = 2A(T - T_c) + 12BM^2 .$$

When $T > T_c$, the only real solution, $M = 0$, is stable;

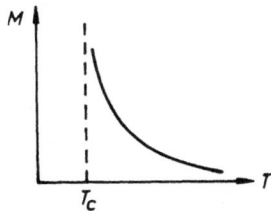

Fig. 1.44.

when $T < T_c$, the $M = 0$ solution is unstable, while the
$M = \pm\sqrt{A(T_c - T)/2B}$ solution is stable;
when $T = T_c, M = 0$, T_c is the point of phase transition of the second order. (If $T > T_c$, the substance is paramagnetic; if $T < T_c$, the substance is ferromagnetic.)

If $H \neq 0$, (*) requires $M \neq 0$. Then as long as $M^2 > A(T_c - T)/6B$, the system is stable. When $T \to T_c, 2A(T - T_c)M \ll 4BM^2$, and (*) has the solution $M = (H/4B)^{\frac{1}{3}}$. Thus T_c is the point of first-order phase transition.

1147

In the absence of external magnetic fields a certain substance is super-conducting for temperatures $T < T_0$. In the presence of a uniform field B and for $T < T_0'$, the system can exist in two thermodynamic phases:

For $B < B_c(T)$, it is in the superconducting phase and in this phase the magnetization per unit volume is

$$\text{(Superconducting phase)} \quad M = -B/4\pi.$$

For $B > B_c(T)$, the system is in the normal phase and here (Normal phase) $M = 0$.

The two phases can coexist in equilibrium along the curve $B = B_c(T)$ in the $B - T$ plane.

Evidently there is a discontinuity in magnetization across the coexistence curve. There is also a discontinuity in entropy. Let $S_N(T)$ and $S_s(T)$ be the entropies per unit volume respectively for the normal and superconducting phases along the coexistence curve. Given that $B_c(T) = B_0 \left(1 - \dfrac{T^2}{T_0^2} \right)$, compute $\Delta S = S_N(T) - S_s(T)$ as a function of T and the other parameters.

$$(CUSPEA)$$

Solution:

Comparing this magnetic system with a $p-V$ system, we have $-B \to P$ and $M \to V$. From the Clausius-Clapeyron equation of the $p - V$ system,

$$\frac{dp}{dT} = \frac{\Delta S}{\Delta V} \, ,$$

we have for the magnetic system, on the line of two-phase coexistence,

$$\frac{dB}{dT} = \frac{-\Delta S}{\Delta M} \, .$$

where $\Delta S = S_N - S_s, \Delta M = M_N - M_s = B/4\pi$.

Therefore

$$\Delta S = -\Delta M \cdot \frac{dB}{dT} = -\frac{B}{4\pi} \frac{dB}{dT}$$

$$= \frac{B_0^2}{2\pi} \cdot \frac{T}{T_0^2} \left(1 - \frac{T^2}{T_0^2} \right) \, .$$

5. NONEQUILIBRIUM THERMODYNAMICS (1148-1159)

1148

A tube of length L contains a solution with sugar concentration at time $t = 0$ given by

$$n(x,0) = n_0 + n_1 \left\{ \cos \frac{\pi x}{L} + \frac{1}{9} \cos \frac{3\pi x}{L} + \frac{1}{25} \cos \frac{5\pi x}{L} \right\}.$$

Assume that $n(x,t)$ obeys a one-dimensional diffusion equation with diffusion constant D.

(a) Write down the diffusion equation for $n(x,t)$.

(b) Calculate $n(x,t)$ for $t > 0$.

(MIT)

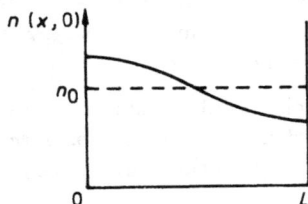

Fig. 1.45.

Solution:

(a) The diffusion equation is

$$\frac{\partial n(x,t)}{\partial t} = D \frac{\partial^2 n(x,t)}{\partial x^2},$$

and the condition for existence of solutions are

$$\left(\frac{\partial n}{\partial x} \right)_{x=0} = \left(\frac{\partial n}{\partial x} \right)_{x=L} = 0,$$

(b) Let $n(x,t) = X(x)T(t)$. We then have

$$X''(x) + \lambda X(x) = 0,$$
$$T'(t) + D\lambda T(t) = 0, \quad \text{with } \lambda \neq 0 \quad \text{and} \quad X'(0) = X'(L) = 0.$$

The conditions require $\lambda = (k\pi/L)^2, k = 1, 2, 3, \ldots$. The general solution is

$$n(x,t) = n_0 + \sum_{k=1}^{\infty} c_k e^{-D(k\pi/L)^2 t} \cos \frac{k\pi}{L} x.$$

The coefficients c_k are obtained from the given concentration at $t = 0, n(x, 0)$. Hence

$$n(x, t) = n_0 + n_1 e^{-D\left(\frac{\pi}{L}\right)^2 t} \left[\cos \frac{\pi x}{L} + \frac{1}{9} e^{-8D\left(\frac{\pi}{L}\right)^2 t} \cos \frac{3\pi x}{L} \right.$$

$$\left. + \frac{1}{25} e^{-24D\left(\frac{\pi}{L}\right)^2 t} \cos \frac{5\pi x}{L} + \dots \right] .$$

1149

(a) With neglect of viscosity and heat conductivity, small disturbances in a fluid propagate as undamped sound waves. Given the relation $p = p(\rho, S)$, where p is pressure, ρ is the mass density, S is the entropy, derive an expression for the sound wave speed v.

(b) As an example of such a fluid, consider a system of identical, noninteracting spin 1/2 particles of mass m at the absolute zero of temperature. The number density is n. Compute the sound speed v in such a system.

(*Princeton*)

Solution:

(a) The equations of continuity and momentum in a fluid are respectively

$$\frac{\partial \rho}{\partial t} + \nabla \cdot (\rho \mathbf{v}) = 0 ,$$

$$\frac{\partial}{\partial t}(\rho \mathbf{v}) + (\mathbf{v} \cdot \nabla)(\rho \mathbf{v}) + \nabla p = 0 .$$

For a fluid at rest, $\mathbf{v} = 0, \rho = \rho_0, p = p_0$. Consider small disturbances, the corresponding quantities are $\mathbf{v} = \mathbf{v}', \rho = \rho_0 + \rho', p = p_0 + p'$. We substitute them into the equations above, taking into consideration only first-order terms, and obtain

$$\frac{\partial \rho'}{\partial t} + \rho_0 \nabla \cdot \mathbf{v}' = 0 ,$$

$$\rho_0 \frac{\partial \mathbf{v}'}{\partial t} + \nabla p' = 0 .$$

Hence

$$\frac{\partial^2 \rho'}{\partial t^2} = \nabla^2 p' = \nabla^2 \left[\left(\frac{\partial p}{\partial \rho} \right)_S \rho' \right] = \left(\frac{\partial p}{\partial \rho} \right)_S \cdot \nabla^2 \rho' .$$

Compare it with wave equation $\dfrac{\partial^2 \rho'}{\partial t^2} = v^2 \nabla^2 \rho'$, we have $v^2 = \left(\dfrac{\partial p}{\partial \rho}\right)_S$

(Note: An assumption has been made here that the pressing of the fluid created by the disturbances is adiabatic for which $S = $ const. Generally speaking, such approximation is reasonable as the heat conductivity is negligible.)

(b) At $T = 0$ K, for a system of spin 1/2 Fermion gas we have

$$p = \frac{2}{5}\frac{N\mu_0}{V}$$

$$= \frac{h^2}{5m^2}\left(\frac{3}{8\pi m}\right)^{2/3} \cdot \rho^{5/3} .$$

Hence $v = \dfrac{1}{\sqrt{3}}\dfrac{h}{m}\left(\dfrac{3N}{8\pi V}\right)^{1/3} .$

1150

Gas, in equilibrium at pressure p_0 and mass density ρ_0, is confined to a cylinder of length L and cross sectional area A. The right hand end of the cylinder is closed and fixed. At the left hand end there is a frictionless and massless movable piston. In equilibrium the external force that must be exerted on the piston is of course $f_0 = p_0 A$. However, suppose a small additional force is supplied by an external agency: the harmonic force $f(t) = f_0 \cos(\omega t)$. This produces small motions of the piston and thus small amplitude disturbances in the gas. Let c be the speed of sound in the gas; neglect viscosity. Let $v(t)$ be the velocity of the piston. Compute $v(t)$.

(*CUSPEA*)

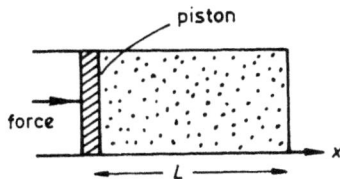

Fig. 1.46.

Solution:

Consider the gas as an ideal fluid. We choose a coordinate system whose origin is the equilibrium point (as shown in Fig. 1.46). Let the velocity of the macroscope motion of the gas be $v(x, t)$ and the pressure of the gas be $p(x, t)$. Because the displacement of the piston is very small, we can solve $v(x, t)$ and $p(x, t)$ approximately in the region $0 \leq x \leq L$ and consider $v(0, t)$. The boundary conditions are $p(0, t) = f(t)/A$ and $v(L, t) = 0$. As $f(t)$ is a sinusoidal function of t and the frequency is ω, the resulting $v(x, t)$ and $p(x, t)$ must be waves of frequency ω and wave vector $k = \omega/c$. In fact, $v(x, t)$ and $p(x, t)$ both satisfy the wave equation with propagating velocity c. We can write

$$f(t) = \mathrm{Re} f_0 \exp(i\omega t) ,$$
$$p = \mathrm{Re} \tilde{p}(x) \exp(i\omega t) ,$$
$$v = \mathrm{Re} \tilde{v}(x) \exp(i\omega t) .$$

Thus, to satisfy the boundary condition of p, we have

$$\tilde{p}(x) = \frac{f_0}{A} \cos(kx) + \lambda \sin(kx) ,$$

where λ is to be determined.

On the other hand, the macroscope motion of fluid satisfies the Euler equation

$$\rho_0 \frac{\partial v}{\partial t} = -\frac{\partial p}{\partial x} ,$$

where ρ_0 is the average density, v is the velocity and p is the pressure. Then $\tilde{v}(x) = \frac{ik}{\omega \rho_0}(-p_0 \sin kx + \lambda \cos kx)$, where $p_0 = \frac{f_0}{A}$.

Using the boundary condition $v(L) = 0$, we have

$$\lambda = p_0 \tan(kL) .$$

Thus

$$\tilde{v}(x = 0) = \frac{ik}{\omega \rho_0} p_0 \tan kL = \frac{i}{c} \frac{p_0}{\rho_0} \tan \frac{\omega L}{c} .$$
$$v(t) = \mathrm{Re}(\tilde{v}(0)e^{i\omega t}) = -\left(\frac{p_0}{c\rho_0} \tan \frac{\omega L}{c} \right) \sin \omega t .$$

1151

Under normal conditions the temperature of the atmosphere decreases steadily with altitude to a height of about 12 km (tropopause), above which the temperature rises steadily (stratosphere) to about 50 km.

(a) What causes the temperature rise in the stratosphere?

(b) The warm stratosphere completely surrounds the earth, above the cooler tropopause, maintained as a permanent state. Explain.

(c) Sound waves emitted by a plane in the tropopause region will travel to great distances at these altitudes, with intensity decreasing, approximately, only as $1/R$. Explain

(*Columbia*)

Solution:

(a) The concentration of ozone in the stratosphere formed by the action of the sun's ultraviolet radiation on the oxygen of the air increases with altitutde. The ozone absorbs the sun's ultraviolet radiation and raises the temperature of surrounding air.

(b) In the stratosphere, the ozone absorbs the ultraviolet radiation of the sun while the carbon dioxide CO_2 there radiates infrared radiation, resulting in an equilibrium of energy.

(c) Sound waves tend to deflect towards the region of lower velocity of propagation, i.e., of lower temperature. In the tropopause, temperature increases for both higher and lower altitudes. Hence the sound waves there are confined to the top layer of the troposphere, spreading only laterally in fan-shape propagation so that the intensity decreases approximately as $\dfrac{1}{R}$ instead of $\dfrac{1}{R^2}$.

1152

Since variations of day and night in temperature are significantly damped at a depth of around 10 cm in granite, the thermal conductivity of granite is $5 \times (10^{-3}, 10^{-1}, 10^2, 10^5)$ cal/s·cm°C.

(*Columbia*)

Solution:

Assume that the temperature at the depth of 10 cm below the surface of granite is constant at T_0°C. When the temperature is the highest in a day, the temperature of the ground surface is assumed to be $T_1 \approx T_0 + 10°C$. The intensity of the solar radiation on the ground is

$$Q = 1400 \text{ W/m}^2 \approx 3.3 \times 10^{-2} \text{ cal/s} \cdot \text{cm}^2 .$$

Q is completely absorbed by the earth within the first 10 cm below surface. Then from the Fourier law of heat conduction, we obtain an estimate of the thermal conductivity of granite:

$$K = Q \cdot \frac{\Delta x}{\Delta T} = Q \cdot \frac{\Delta x}{T_1 - T_0}$$
$$= 3.3 \times 10^{-2} \times (10/10) = 3.3 \times 10^{-2} \text{ cal/s} \cdot \text{cm} \cdot °C ,$$

If we take into account reflection of the radiation from the earth's surface, the value of K will be smaller than the above estimate. Therefore we must choose the answer 5×10^{-3} cal/s · cm · °C.

1153

The heat transferred to and from a vertical surface, such as a window pane, by convection in the surrounding air has been found to be equal to $0.4 \times 10^{-4} (\Delta t)^{5/4}$ cal/sec·cm^2, where Δt is the temperature difference between the surface and the air. If the air temperature is 25°C on the inside of a room and −15°C on the outside, what is the temperature of the inner surface of a window pane in the room? The window pane has a thickness of 2 mm and a thermal conductivity of 2×10^{-3} cal/sec· cm·°C. Heat transfer by radiation can be neglected.

(*Wisconsin*)

Solution:

We consider an area of 1 cm^2, and assume the temperatures of the inner and outer surfaces to be respectively $t_1°C$ and $t_2°C$. Thus we have

$$0.4 \times 10^{-4}(t_2 + 15)^{5/4} = 2 \times 10^{-3} \times \frac{1}{0.2}(t_1 - t_2)$$
$$= 0.4 \times 10^{-4}(25 - t_1)^{5/4} .$$

The solution is $t_1 = 5°C$.

1154

The water at the surface of a lake and the air above it are in thermal equilibrium just above the freezing point. The air temperature suddenly drops by ΔT degrees. Find the thickness of the ice on the lake as a function of time in terms of the latent heat per unit volume L/V and the thermal conductivity Λ of the ice. Assume that ΔT is small enough that the specific heat of the ice may be neglected.

(*MIT*)

Solution:

Consider an arbitrary area ΔS on the surface of water and let $h(t)$ be the thickness of ice. The water of volume $\Delta S dh$ under the ice gives out heat $L\Delta S dh/V$ to the air during time dt and changes into ice. So we have

$$\Delta S dh \cdot \frac{L}{V} = \Lambda \frac{\Delta T}{h} \Delta S dt \,,$$

that is

$$h dh = \frac{\Lambda \Delta T}{(L/V)} dt \,.$$

Hence $h(t) = \left[\dfrac{2\Lambda \Delta T t}{(L/V)}\right]^{1/2} \,.$

1155

A sheet of ice 1 cm thick has frozen over a pond. The upper surface of the ice is at $-20°\,$C.

(a) At what rate is the thickness of the sheet of ice increasing?
(b) How long will it take for the sheet's thickness to double?

The thermal conductivity of ice κ is 5×10^{-3} cal/cm·sec·°C. The latent heat of ice L is 80 cal/g. The mass density of water ρ is 1 g/cm³

(*SUNY, Buffalo*)

Solution:

(a) Let the rate at which the thickness of the sheet of ice increases be η, a point on the surface of ice be the origin of z-axis, and the thickness of ice be z.

The heat current density propagating through the ice sheet is $j = -\kappa \dfrac{T - T_0}{z}$ and the heat released by water per unit time per unit area is $\rho L \dfrac{dz}{dt}$. Hence we obtain the equation $\rho L \dfrac{dz}{dt} = -j$, giving $\eta = \dfrac{dz}{dt} = -j/\rho L = \kappa(T - T_0)/\rho Lz$.

(b) The above expression can be written as

$$dt = \frac{\rho L}{\kappa(T - T_0)} z \, dz \ .$$

$$t = \rho L(z_2^2 - z_1^2)/2\kappa(T - T_0) \ .$$

If we take $z_1 = 1$ cm and $z_2 = 2$ cm, then $\Delta t = 1.2 \times 10^3$ s $= 20$ min.

1156

Consider a spherical black asteroid (made of rock) which has been ejected from the solar system, so that the radiation from the sun no longer has a significant effect on the temperature of the asteroid. Radioactive elements produce heat uniformly inside the asteroid at a rate of $\dot{q} = 3 \times 10^{-14}$ cal/g·sec. The density of the rock is $\rho = 3.5$ g/cm^3, and the thermal conductivity is $k = 5 \times 10^{-3}$ cal/deg·cm·sec. The radius of the asteroid is $R = 100$ km. Determine the central temperature T_c and the surface temperature T_s, of the asteroid assuming that a steady state has been achieved.

(*UC, Berkeley*)

Solution:

The surface temperature satisfies

$$4\pi R^2 \sigma T_s^4 = Q = \frac{4\pi R^3}{3}\rho\dot{q} \ ,$$

so

$$T_s = \left(\frac{R\rho\dot{q}}{3\sigma} \right)^{\frac{1}{4}} = 22.5 \text{ K} \ .$$

The equation of heat conduction inside the asteroid is

$$\nabla \cdot (-k\nabla T) = \dot{q}\rho \ .$$

Using spherical coordinates, we have

$$\frac{d}{dr}\left(r^2 \frac{dT}{dr} \right) = -\frac{\rho\dot{q}r^2}{k}$$

and so

$$T = -\frac{\dot{q}\rho}{6k}(r^2 - R^2) + T_s \ .$$

The central temperature is

$$T_c = \frac{\dot{q}\rho}{6k}R^2 + T_s = 372 \text{ K} \ .$$

1157

Let H be the flow of heat per unit time per unit area normal to the isothermal surface through a point P of the body. Assume the experimental fact

$$\mathbf{H} = -k\nabla T \ ,$$

where T is the temperature and k is the coefficient of thermal conductivity. Finally the thermal energy absorbed per unit volume is given by $c\rho T$, where c is the specific heat and ρ is the density.

(a) Make an analogy between the thermal quantities H, k, T, c, ρ and the corresponding quantities $\mathbf{E}, \mathbf{J}, V, \rho$ of steady currents.

(b) Using the results of (a) find the heat conduction equation.

(c) A pipe of inner radius r_1, outer radius r_2 and constant thermal conductivity k is maintained at an inner temperature T_1 and outer temperature T_2. For a length of pipe L find the rate the heat is lost and the temperature between r_1 and r_2 (steady state).

(SUNY, Buffalo)

Solution:

(a) By comparison with Ohm's law $\mathbf{J} = \sigma \mathbf{E} = -\sigma \text{ grad } V$ (V is voltage) and conservation law of charge $\partial \rho / \partial t = -\nabla \cdot \mathbf{J}$, we obtain the analogy $c\rho T \Longleftrightarrow \rho; \mathbf{H} \Longleftrightarrow \mathbf{J}; \text{grad } T \Longleftrightarrow \text{grad } V; k \Longleftrightarrow \sigma$.

(b) By the above analogy and charge conservation law, we have

$$c\rho \frac{\partial T}{\partial t} = -\text{grad} \cdot (-k \text{ grad } T) = k\nabla^2 T \ .$$

Then the heat conduction equation is

$$\frac{\partial T}{\partial t} - \frac{k}{\rho c}\nabla^2 T = 0 \ .$$

(c) When equilibrium is reached, $\partial T/\partial t = 0$; hence $\nabla^2 T = 0$.
The boundary conditions are $T(r_1) = T_1$ and $T(r_2) = T_2$.

Choosing the cylindrical coordinate system and solving the Laplace equation, we obtain the temperature between r_1 and r_2:

$$T(r) = \frac{1}{\ln \dfrac{r_1}{r_2}} \left[T_1 \ln \frac{r}{r_2} - T_2 \ln \frac{r}{r_1} \right] \ .$$

By

$$\mathbf{H} = -k\nabla T = -k\frac{\partial T}{\partial \mathrm{r}} = \frac{k(T_1 - T_2)}{r \ln(r_2/r_1)} \mathbf{r}^0 \ ,$$

we obtain the rate at which the heat is lost:

$$\dot{q} = 2\pi r L H = 2\pi k(T_1 - T_2)L/\ln\frac{r_2}{r_1} \ .$$

1158

A uniform non-metallic annular cylinder of inner radius r_1, outer radius r_2, length l_0 is maintained with its inner surface at 100°C and its outer surface at 0°C.

(a) What is the temperature distribution inside?

(b) If it is then placed in a thermally insulated chamber of negligible heat capacity and allowed to come to temperature equilibrium, will its entropy increase, decrease or remain the same? Justify your answer.

(*Wisconsin*)

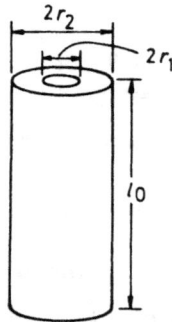

Fig. 1.47.

Solution:

(a) Because the material is uniform, we can assume the heat conductivity is uniform too. According to the formulas $dQ = -k(dT/dr)s\,dt$ and $s = 2\pi l_0 r$, we have

$$dQ/dt = -2\pi l_0 rk\,dT/dr \;.$$

Since dQ/dt is independent of r, we require $dT/dr = A/r$, where A is a constant. Then $T(r) = A\ln r + B$.

From the boundary conditions, we have

$$A = \frac{T_2 - T_1}{\ln\dfrac{r_2}{r_1}}\;, \qquad B = \frac{T_1 \ln r_2 - T_2 \ln r_1}{\ln\dfrac{r_2}{r_1}}\;,$$

where $T_1 = 373$ K and $T_2 = 273$ K, so that

$$T(r) = \frac{1}{\ln r_1 - \ln r_2}\left[(T_1 - T_2)\ln r + T_2 \ln r_1 - T_1 \ln r_2\right]\;.$$

(b) This is an irreversible adiabatic process, so that the entropy increases.

1159

When there is heat flow in a heat conducting material, there is an increase in entropy. Find the local rate of entropy generation per unit volume in a heat conductor of given heat conductivity and given temperature gradient.

(UC, Berkeley)

Solution:

If we neglect volume expansion inside the heat conducting material, then $du = T\,dS$. The heat conduction equation is

$$du/dt + \nabla \cdot \mathbf{q} = 0 \;.$$

Hence

$$dS/dt = -\nabla \cdot \mathbf{q}/T = -\nabla \cdot (\mathbf{q}/T) + \mathbf{q} \cdot \nabla(1/T)\;,$$

where $\mathbf{q}/T$ is the entropy flow, and $\mathbf{q} \cdot \nabla\left(\dfrac{1}{T}\right)$ is the irreversible entropy increase due to the inhomogeneous temperature distribution. Thus, the local rate of entropy generation per unit volume is

$$\dot{S} = \mathbf{q} \cdot \nabla\left(\frac{1}{T}\right) = -\frac{\nabla T}{T^2} \cdot \mathbf{q}\;.$$

According to Fourier's heat conduction law, $\mathbf{q} = -k\nabla T$, the above gives

$$\dot{S} = k \left(\frac{\nabla T}{T} \right)^2 .$$

PART II

STATISTICAL PHYSICS

1. PROBABILITY AND STATISTICAL ENTROPY (2001-2013)

2001

A classical harmonic oscillator of mass m and spring constant k is known to have a total energy of E, but its starting time is completely unknown. Find the probability density function, $p(x)$, where $p(x)dx$ is the probability that the mass would be found in the interval dx at x.

(MIT)

Solution:

From energy conservation, we have

$$E = \frac{k}{2}l^2 = \frac{k}{2}x^2 + \frac{m}{2}\dot{x}^2 ,$$

where l is the oscillating amplitude. So the period is

$$T = 2\int_{-l}^{l} \frac{dx}{\sqrt{\dfrac{2E - kx^2}{m}}} = 2\pi\sqrt{\frac{m}{k}} .$$

Therefore we have

$$p(x)dx = \frac{2dt}{T} = \frac{2}{T}\left(\frac{m}{2E - kx^2}\right)^{\frac{1}{2}} dx ,$$

$$p(x) = \frac{1}{\pi}\left(\frac{k}{2E - kx^2}\right)^{\frac{1}{2}} .$$

2002

Suppose there are two kinds of E. coli (bacteria), "red" ones and "green" ones. Each reproduces faithfully (no sex) by splitting into half, red→red+red or green→green+green, with a reproduction time of 1 hour. Other than the markers "red" and "green", there are no differences between them. A colony of 5,000 "red" and 5,000 "green" E. coli is allowed to eat and reproduce. In order to keep the colony size down, a predator is introduced which keeps the colony size at 10,000 by eating (at random) bacteria.

(a) After a very long time, what is the probability distribution of the number of red bacteria?

(b) About how long must one wait for this answer to be true?

(c) What would be the effect of a 1% preference of the predator for eating red bacteria on (a) and (b)?

(Princeton)

Solution:

(a) After a sufficiently long time, the bacteria will amount to a huge number $N \gg 10,000$ without the existence of a predator. That the predator eats bacteria at random is mathematically equivalent to selecting $n = 10,000$ bacteria out of N bacteria as survivors. $N \gg n$ means that in every selection the probabilities of surviving "red" and "green" E. coli are the same. There are 2^n ways of selection, and there are C_m^n ways to survive m "red" ones. Therefore the probability distribution of the number of "red" E. coli is

$$\frac{1}{2^n} C_m^n = \frac{1}{2^n} \cdot \frac{n!}{m!(n-m)!}, \quad m = 0, 1, \ldots, n.$$

(b) We require $N \gg n$. In practice it suffices to have $N/n \approx 10^2$. As $N = 2^t n, t = 6$ to 7 hours would be sufficient.

(c) If the probability of eating red bacteria is $\left(\frac{1}{2} + p\right)$, and that of eating green is $\left(\frac{1}{2} - p\right)$, the result in (a) becomes

$$C_m^n \left(\frac{1}{2} + p\right)^m \left(\frac{1}{2} - p\right)^{n-m}$$

$$= \frac{n!}{m!(n-m)!} \left(\frac{1}{2} + p\right)^m \cdot \left(\frac{1}{2} - p\right)^{n-m}.$$

The result in (b) is unchanged.

2003

(a) What are the reduced density matrices in position and momentum spaces?

(b) Let us denote the reduced density matrix in momentum space by $\phi(\mathbf{p_1}, \mathbf{p_2})$. Show that if ϕ is diagonal, that is,

$$\phi(\mathbf{p_1}, \mathbf{p_2}) = f(\mathbf{p_1}) \delta_{\mathbf{p_1}, \mathbf{p_2}},$$

then the diagonal elements of the position density matrix are constant.

(*SUNY, Buffalo*)

Solution:

(a) The reduced density matrices are matrix expressions of density operator $\hat{\rho}(t)$ in an orthogonal complete set of singlet states, where the density operator $\hat{\rho}(t)$ is defined such that the expectation value of an arbitrary operator $\hat{0}$ is $\langle\hat{0}\rangle = \text{tr}[\hat{0}\hat{\rho}(t)]$. We know that an orthogonal complete set of singlet states in position space is $\{|\mathbf{r}\rangle\}$, from which we can obtain the reduced density matrix in position space $\langle\mathbf{r}'|\hat{\rho}(t)|\mathbf{r}\rangle$. Similarly, the reduced density matrix in momentum space is $\langle\mathbf{p}'|\hat{\rho}(t)|\mathbf{p}\rangle$, where $\{|\mathbf{p}\rangle\}$ is an orthogonal complete set of singlet states in momentum space.

(b)
$$\langle\mathbf{r}'|\hat{\rho}(t)|\mathbf{r}\rangle = \sum_{\mathbf{p}'\mathbf{p}}\langle\mathbf{r}'|\mathbf{p}'\rangle\langle\mathbf{p}'|\hat{\rho}(t)|\mathbf{p}\rangle\langle\mathbf{p}|\mathbf{r}\rangle$$
$$= \frac{1}{V}\sum_{\mathbf{p}'\mathbf{p}}\phi(\mathbf{p}',\mathbf{p})\exp(i(\mathbf{r}'\cdot\mathbf{p}' - \mathbf{r}\cdot\mathbf{p}))$$
$$= \frac{1}{V}\sum_{\mathbf{p}'\mathbf{p}}f(\mathbf{p})\delta_{\mathbf{p}',\mathbf{p}}\exp(i(\mathbf{r}' - \mathbf{r})\cdot\mathbf{p})$$
$$= \frac{1}{V}\sum_{\mathbf{p}}f(\mathbf{p})\exp(i(\mathbf{r}' - \mathbf{r})\cdot\mathbf{p})$$

Then the diagonal elements $\langle\mathbf{r}|\hat{\rho}(t)|\mathbf{r}\rangle = \frac{1}{V}\Sigma_{\mathbf{p}}f(\mathbf{p})$ are obviously constant.

2004

(a) Consider a large number of N localized particles in an external magnetic field $\mathbf{H}$. Each particle has spin $1/2$. Find the number of states accessible to the system as a function of M_s, the z-component of the total spin of the system. Determine the value of M_s for which the number of states is maximum.

(b) Define the absolute zero of the thermodynamic temperature. Explain the meaning of negative absolute temperature, and give a concrete example to show how the negative absolute temperature can be reached.

(*SUNY, Buffalo*)

Solution:

(a) The spin of a particle has two possible orientations $1/2$ and $-1/2$. Let the number of particles with spin $1/2$ whose direction is along $\mathbf{H}$ be

$N_\uparrow$ and the number of particles with spin $-1/2$ whose direction is opposite to $\mathbf{H}$ be $N_\downarrow$; then the component of the total spin in the direction of $\mathbf{H}$ is $M_s = \frac{1}{2}(N_\uparrow - N_\downarrow)$. By $N_\uparrow + N_\downarrow = N$, we can obtain $N_\uparrow = \frac{N}{2} + M_s$ and $N_\downarrow = \frac{N}{2} - M_s$. The number of states of the system is

$$Q = \frac{N!}{N_\uparrow! N_\downarrow!} \frac{N!}{\left[\frac{N}{2} + M_s\right]! \left[\frac{N}{2} - M_s\right]!} .$$

Using Stirling's formula, one obtains

$$
\begin{aligned}
\ln Q &= \ln \frac{N!}{N_\uparrow! N_\downarrow!} \\
&= N \ln N - N_\uparrow \ln N_\uparrow - N_\downarrow \ln N_\downarrow \\
&= N \ln N - N_\uparrow \ln N_\uparrow - (N - N_\uparrow) \ln(N - N_\uparrow) .
\end{aligned}
$$

By

$$\frac{\partial \ln Q}{\partial N_\uparrow} = -\ln N_\uparrow + \ln(N - N_\uparrow) = 0 ,$$

we get $N_\uparrow = \frac{N}{2}$, i.e., $M_s = 0$ when the number of states of the system is maximum.

(b) See Question 2009.

2005

There is an one-dimensional lattice with lattice constant a as shown in Fig. 2.1. An atom transits from a site to a nearest-neighbor site every τ seconds. The probabilities of transiting to the right and left are p and $q = 1 - p$ respectively.

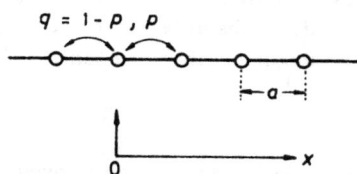

Fig. 2.1.

(a) Calculate the average position $\bar{x}$ of the atom at the time $t = N\tau$, where $N \gg 1$;

(b) Calculate the mean-square value $\overline{(x - \bar{x})^2}$ at the time t.

(MIT)

Solution:

(a) Choose the initial position of the atom as the origin $x = 0$, with the x-axis directing to the right. We have

$$\bar{x} = \sum_{n=0}^{N} \frac{N!}{n!(N-n)!}(2n - N)ap^n q^{N-n}$$

$$= 2ap\frac{\partial}{\partial p}\left(\sum_{n=0}^{N} \frac{N!}{n!(N-n)!}p^n q^{N-n}\right) - Na$$

$$= 2ap\frac{\partial}{\partial p}(p + q)^N - Na = Na(p - q) \ .$$

(b) $$\overline{x^2} = \sum_{n=0}^{N} \frac{N!}{n!(N-n)!}(2n - N)^2 a^2 p^n q^{N-n}$$

$$= 4a^2 p^2 \frac{\partial^2}{\partial p^2}(p + q)^N - 4(N - 1)a^2 p\frac{\partial}{\partial p}(p + q)^N + N^2 a^2$$

$$= Na^2[(N - 1)(p - q)^2 + 1] \ ,$$

$$\overline{(x - \bar{x})^2} = \overline{x^2} - \bar{x}^2 = 4Na^2 pq \ .$$

2006

(a) Give the definition of entropy in statistical physics.

(b) Give a general argument to explain why and under what circumstances the entropy of an isolated system A will remain constant, or increase. For convenience you may assume that A can be divided into subsystems B and C which are in weak contact with each other, but which themselves remain in internal thermodynamic equilibrium.

(UC, Berkeley)

Solution:

(a) $S = k \ln \Omega$, where k is Boltzmann's constant and Ω is the total number of microscopic states of the given macroscopic state.

(b) Assume that the temperatures of the two subsystems are T_B and T_C respectively, and that $T_B \geq T_C$. According to the definition of entropy, if there is a small energy exchange $\Delta > 0$ between them (from B to C), then

$$\Delta S_B = -\frac{\Delta}{T_B} , \quad \Delta S_C = \frac{\Delta}{T_C} ,$$

$$\Delta S = \Delta S_B + \Delta S_C = \frac{(T_B - T_C)}{T_B T_C} \Delta \geq 0 .$$

When $T_B > T_C$, there is no thermal equilibrium between the subsystems, and $\Delta S > 0$;
When $T_B = T_C$, i.e., the two subsystems are in equilibrium, $\Delta S = 0$.

2007

Give Boltzmann's statistical definition of entropy and present its physical meaning briefly but clearly. A two-level system of $N = n_1 + n_2$ particles is distributed among two eigenstates 1 and 2 with eigenenergies E_1 and E_2 respectively. The system is in contact with a heat reservoir at temperature T. If a single quantum emission into the reservoir occurs, population changes $n_2 \to n_2 - 1$ and $n_1 \to n_1 + 1$ take place in the system. For $n_1 \gg 1$ and $n_2 \gg 1$, obtain the expression for the entropy change of
(a) the two level system, and of
(b) the reservoir, and finally
(c) from (a) and (b) derive the Boltzmann relation for the ratio n_1/n_2.
(UC, Berkeley)

Solution:

$S = k \ln \Omega$, where Ω is the number of microscopic states of the system. Physically entropy is a measurement of the disorder of a system.

(a) The entropy change of the two-level system is

$$\Delta S_1 = k \ln \frac{N!}{(n_2 - 1)!(n_1 + 1)!} - k \ln \frac{N!}{n_1! n_2!}$$

$$= k \ln \frac{n_2}{n_1 + 1} \cong k \ln \frac{n_2}{n_1} .$$

(b) The entropy change of the reservoir is

$$\Delta S_2 = \frac{E_2 - E_1}{T} .$$

(c) From $\Delta S_1 + \Delta S_2 = 0$, we have

$$\frac{n_2}{n_1} = \exp\left(-\frac{E_2 - E_1}{kT}\right).$$

2008

Consider a system composed of a very large number N of distinguishable atoms, non-moving and mutually non-interacting, each of which has only two (non-degenerate) energy levels: $0, \varepsilon > 0$. Let E/N be the mean energy per atom in the limit $N \to \infty$.

(a) What is the maximum possible value of E/N if the system is not necessarily in thermodynamic equilibrium? What is the maximum attainable value of E/N if the system is in equilibrium (at positive temperature, of course)?

(b) For thermodynamic equilibrium, compute the entropy per atom, S/N, as a function of E/N.

(*Princeton*)

Solution:

(a) If the system is not necesssarily in thermodynamic equilibrium, the maximum possible value of E/N is ε; and if the system is in equilibrium (at positive temperature), the maximum possible value of E/N is $\varepsilon/2$ corresponding to $T \to \infty$.

(b) When the mean energy per atom is E/N, E/ε particles are on the level of energy ε and the microscopic state number is

$$Q = \frac{N!}{\left(\dfrac{E}{\varepsilon}\right)!\left(N - \dfrac{E}{\varepsilon}\right)!}.$$

So the entropy of the system is

$$S = k \ln \frac{N!}{\left(\dfrac{E}{\varepsilon}\right)!\left(N - \dfrac{E}{\varepsilon}\right)!}.$$

If $E/\varepsilon \gg 1, N - E/\varepsilon \gg 1$, we have

$$\frac{S}{N} = k \left[\ln N - \frac{E/\varepsilon}{N} \ln \frac{E}{\varepsilon} - \left(1 - \frac{E/\varepsilon}{N}\right) \ln \left(N - \frac{E}{\varepsilon}\right) \right]$$

$$= k \left[\frac{E}{\varepsilon N} \ln \frac{\varepsilon N}{E} + \left(1 - \frac{E}{N\varepsilon}\right) \ln \frac{1}{1 - \frac{E}{\varepsilon N}} \right] .$$

2009

Consider a system of N non-interacting particles, each fixed in position and carrying a magnetic moment μ, which is immersed in a magnetic field H. Each particle may then exist in one of the two energy states $E = 0$ or $E = 2\mu H$. Treat the particles as distinguishable.

(a) The entropy, S, of the system can be written in the form $S = k \ln \Omega(E)$, where k is the Boltzmann constant and E is the total system energy. Explain the meaning of $\Omega(E)$.

(b) Write a formula for $S(n)$, where n is the number of particles in the upper state. Crudely sketch $S(n)$.

(c) Derive Stirling's approximation for large n:

$$\ln n! = n \ln n - n$$

by approximating $\ln n!$ by an integral.

(d) Rewrite the result of (b) using the result of (c). Find the value of n for which $S(n)$ is maximum.

(e) Treating E as continuous, show that this system can have negative absolute temperature.

(f) Why is negative temperature possible here but not for a gas in a box?

(*CUSPEA*)

Solution:

(a) $\Omega(E)$ is the number of all the possible microscopic states of the system when its energy is E, where

$$0 \le E \le N\varepsilon, \quad \varepsilon = 2\mu H .$$

(b) As the particles are distinguishable,

$$Q = \frac{N!}{n!(N-n)!} \ .$$

Hence $S = k \ln \dfrac{N!}{n!(N-n)!} = S(n)$.

We note that $S(n = 0) = S(n = N) = 0$, and we expect $S_{\max}$ to appear at $n = N/2$ (to be proved in (d) below). The graph of $S(n)$ is shown in Fig. 2.2.

(c) $\ln n! = \displaystyle\sum_{m=1}^{n} \ln m \approx \int_{1}^{n} \ln x\, dx = n \ln n - n + 1 \approx n \ln n - n$, (for large n).

(d) $\dfrac{S}{k} \approx N \ln \dfrac{N}{N-n} - n \ln \dfrac{n}{N-n}$.

$\dfrac{dS}{dn} = 0$ gives

$$\frac{N}{N-n} - 1 - \ln n - \frac{n}{N-n} + \ln(N-n) = 0 \ .$$

Therefore, $S = S_{\max}$ when $n = N/2$.

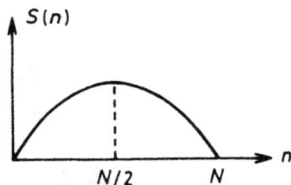

Fig. 2.2.

(e) As $E = n\varepsilon$, $S = S_{\max}$ when $E = \dfrac{1}{2}N\varepsilon$. When $E > \dfrac{1}{2}N\varepsilon$, $\dfrac{\partial S}{\partial E} < 0$ (see Fig. 2.2). Because $\dfrac{1}{T} = \dfrac{dS}{dE}$, we have $T < 0$ when $E > N\varepsilon/2$.

(f) The reason is that here the energy level of a single particle has an upper limit. For a gas system, the energy level of a single particle does not have an upper limit, and the entropy is an increasing function of E; hence negative temperature cannot occur.

From the point of view of energy, we can say that a system with negative temperature is "hotter" than any system with a positive temperature.

2010

A solid contains N magnetic atoms having spin $1/2$. At sufficiently high temperatures each spin is completely randomly oriented. At sufficiently low temperatures all the spins become oriented along the same direction (i.e., Ferromagnetic). Let us approximate the heat capacity as a function of temperature T by

$$C(T) = \begin{cases} c_1 \left(\dfrac{2T}{T_1} - 1 \right) & \text{if } T_1/2 < T < T_1 \\ 0 & \text{otherwise}, \end{cases}$$

where T_1 is a constant. Find the maximum value c_1 of the specific heat (use entropy considerations).

(*UC, Berkeley*)

Solution:

From $C = T \dfrac{dS}{dT}$, we have

$$S(\infty) - S(0) = \int_0^\infty \frac{C}{T} dT = c_1 (1 - \ln 2) .$$

On the other hand, we have from the definition of entropy $S(0) = 0, S(\infty) = Nk \ln 2$, hence

$$c_1 = \frac{Nk \ln 2}{1 - \ln 2} .$$

2011

The elasticity of a rubber band can be described in terms of a one-dimensional model of polymer involving N molecules linked together end-to-end. The angle between successive links is equally likely to be $0°$ or $180°$.

(a) Show that the number of arrangements that give an overall length of $L = 2md$ is given by

$$g(N, m) = \frac{2N!}{\left(\dfrac{N}{2} + m \right)! \left(\dfrac{N}{2} - m \right)!}, \quad \text{where } m \text{ is positive} .$$

Indicate clearly the reasoning you used to get this result.

(b) For $m \ll N$, this expression becomes

$$g(N, m) \approx g(N, 0) \exp(-2m^2/N) \ .$$

Find the entropy of the system as a function of L for $N \gg 1, L \ll Nd$.

(c) Find the force required to maintain the length L for $L \ll Nd$.

(d) Find the relationship between the force and the length, without using the condition in (c), i.e., for any possible value of L, but $N \gg 1$.

(UC, Berkeley)

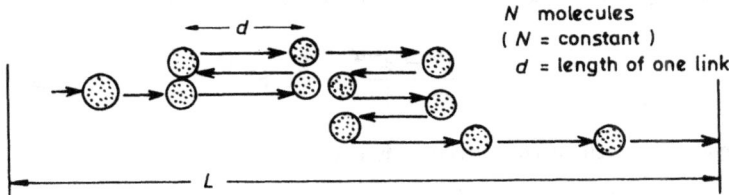

Fig. 2.3.

Solution:

(a) Assume that there are N_+ links of $0°$ angle and N_- links of $180°$ angle then

$$N_+ - N_- = 2m \ , \quad N_+ + N_- = N \ .$$

Therefore
$$N_+ = \frac{N}{2} + m \ , \quad N_- = \frac{N}{2} - m \ .$$

This corresponds to $N!/(N_+!N_-!)$ arrangements. Note that for every arrangement if the angles are reversed, we still get the overall length of $2md$. Thus

$$g = \frac{2N!}{\left(\dfrac{N}{2} + m\right)! \left(\dfrac{N}{2} - m\right)!} \ .$$

(b) When $m \ll N, g(N, m) \approx g(N, 0) \exp(-2m^2/N)$, the entropy of the system becomes

$$S = k \ln g(N, m) = k \ln g(N, 0) - \frac{kL^2}{2Nd^2} \ .$$

(c) From the thermodynamic relations $dU = TdS + fdL$ and $F =$

$U - TS$ we obtain $dF = -SdT + fdL$. Therefore

$$\left(\frac{\partial f}{\partial T}\right)_L = -\left(\frac{\partial S}{\partial L}\right)_T = \frac{kL}{Nd^2} ,$$

$$f = \frac{kTL}{Nd^2} + C .$$

As $f = 0$ when $L = 0$,

$$f = \frac{kTL}{Nd^2} .$$

(d) Consider only one link. When an external force f is exerted, the probability that the angle is $0°$ or $180°$ is proportional to e^{α} or $e^{-\alpha}$ respectively, where $\alpha = fd/kT$. The average length per link is therefore

$$\bar{l} = d\frac{e^{\alpha} - e^{-\alpha}}{e^{\alpha} + e^{-\alpha}} = d\tanh\alpha .$$

The overall length of the polymer is then

$$L = N\bar{l} = Nd\tanh(fd/kT) .$$

2012

Consider a one-dimensional chain consisting of $n \gg 1$ segments as illustrated in the figure. Let the length of each segment be a when the long dimension of the segment is parallel to the chain and zero when the segment is vertical (i.e., long dimension normal to the chain direction). Each segment has just two states, a horizontal orientation and a vertical orientation, and each of these states is not degenerate. The distance between the chain ends is nx.

(a) Find the entropy of the chain as a function of x.

(b) Obtain a relation between the temperature T of the chain and the tension F which is necessary to maintain the distance nx, assuming the joints turn freely.

(c) Under which conditions does your answer lead to Hook's law?

(*Princeton*)

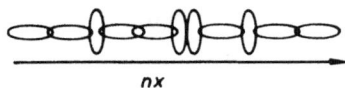

nx

Fig. 2.4.

Solution:

(a) When the length of the chain is nx, there are $m = nx/a$ segments parallel to the chain; so the microscopic state number is

$$\Omega = C_m^n = \frac{n!}{m!(n-m)!} \; .$$

We have

$$S = k \ln \Omega$$

$$= k \ln \frac{n!}{\left(\frac{x}{a}n\right)! \left(n - \frac{x}{a}n\right)!} \; .$$

(b) Under the action of stress F, the energy difference between the vertical and parallel states of a segment is Fa. The mean length of a segment is

$$l = \frac{ae^{Fa/kT}}{1 + e^{Fa/kT}} \; ,$$

so that

$$nx = nl = \frac{nae^{Fa/kT}}{1 + e^{Fa/kT}} \; .$$

(c) At high temperatures,

$$L = nx = na \left(\frac{1}{2} + \frac{1}{2}\frac{Fa}{kT}\right) \; ,$$

which is Hooke's Law.

2013

Consider an idealization of a crystal which has N lattice points and the same number of interstitial positions (places between the lattice points where atoms can reside). Let E be the energy necessary to remove an atom from a lattice site to an interstitial position and let n be the number of atoms occupying interstitial sites in equilibrium.

(a) What is the internal energy of the system?

(b) What is the entropy S? Give an asymptotic formula valid when $n \gg 1$?

(c) In equilibrium at temperature T, how many such defects are there in the solid, i.e., what is n? (Assume $n \gg 1$.)

(Princeton)

Solution:

(a) Let U_0 be the internal energy when no atom occupies the interstitial sites. When n interstitial positions are occupied, the internal energy is then

$$U = U_0 + nE .$$

(b) There are C_n^N ways of selecting n atoms from N lattice sites, and C_n^N ways to place them to N interstitial sites; so the microscopic state number is $\Omega = (C_n^N)^2$. Hence

$$S = k \ln \Omega = 2k \ln \frac{N!}{n!(N-n)!} .$$

When $n \gg 1$ and $(N-n) \gg 1$, we have $\ln(n!) = n \ln n - n$, so that

$$S = 2k[N \ln N - n \ln n - (N-n) \ln(N-n)] .$$

(c) With fixed temperature and volume, free energy is minimized at equilibrium.

From $F = U_0 + nE - TS$ and $\partial F/\partial n = 0$, we have

$$n = \frac{N}{e^{E/2kT} + 1} .$$

2. MAXWELL-BOLTZMANN STATISTICS (2014-2062)

2014

(a) Explain Boltzmann statistics, Fermi statistics and Bose statistics, especially about their differences. How are they related to the indistinguishability of identical particles?

(b) Give as physical a discussion as you can, on why the distinction between the above three types of statistics becomes unimportant in the limit of high temperature (how high is high?). Do not merely quote formulas.

(c) In what temperature range will quantum statistics have to be applied to a collection of neutrons spread out in a two-dimensional plane with the number of neutrons per unit area being $\sim 10^{12}/cm^2$?

(SUNY, Buffalo)

Solution:

(a) *Boltzmann statistics.* For a localized system, the particles are distinguishable and the number of particles occupying a singlet quantum state is not limited. The average number of particles occupying energy level ε_l is

$$a_l = w_l \exp(-\alpha - \beta \varepsilon_l) \ ,$$

where w_l is the degeneracy of l-th energy level.

Fermi statistics. For a system composed of fermions, the particles are indistinguishable and obey Pauli's exclusion principle. The average number of particles occupying energy level ε_l is

$$a_l = \frac{w_l}{e^{\alpha + \beta \varepsilon_l} + 1} \ .$$

Bose statistics. For a system composed of bosons, the particles are indistinguishable and the number of particles occupying a singlet quantum state is not limited. The average number of particles occupying energy level ε_l is

$$a_l = \frac{w_l}{e^{\alpha + \beta \varepsilon_l} - 1} \ .$$

(b) We see from (a) that when $e^\alpha \gg 1$, or $\exp(-\alpha) \ll 1$,

$$\frac{w_l}{e^{\alpha + \beta \varepsilon_l} \pm 1} \sim w_l e^{-\alpha - \beta \varepsilon_l} \ ,$$

and the distinction among the above three types of statistics vanishes.

From $e^{-\alpha} = n \left(\frac{h^2}{2\pi m k T} \right)^{3/2}$, ($n$ is the particle density), we see that the above condition is satisfied when $T \gg \frac{n^{2/3} h^2}{2\pi m k}$. So the distinction among the three types of statistics becomes unimportant in the limit of high temperatures.

It can also be understood from a physical point of view. When $e^\alpha \gg 1$, we have $a_l / w_l \ll 1$, which shows that the average number of particles in any quantum state is much less than 1. The reason is that the number of microstates available to the particles is very large, much larger than the total particle number. Hence the probability for two particles to occupy the same quantum state is very small and Pauli's exclusion principle is satisfied naturally. As a result, the distinction between Fermi and Bose statistics vanishes.

(c) The necessity of using quantum statistics arises from the following two points. One is the indistinguishability of particles and Pauli's exclusion principle, because of which $e^{-\alpha} = n\left(\dfrac{h^2}{2\pi mkT}\right)$ is not very much smaller than 1 (degenerate). The other is the quantization of energy levels, i.e., $\Delta E/kT$, where ΔE is the spacing between energy levels, is not very much smaller than 1 (discrete).

For a two-dimensional neutron system,

$$\frac{\Delta E}{kT} = \frac{h^2}{2mkTL^2} .$$

Taking $L \approx 1$ cm, we have $T \approx 10^{-13}$ K. So the energy levels are quasi-continuous at ordinary temperatures. Hence the necessity of using quantum statistics is essentially determined by the strong-degeneracy condition

$$e^{-\alpha} = n\left(\frac{h^2}{2\pi mkT}\right) \gtrsim 1 .$$

Substituting the quantities into the above expression, we see that quantum statistics must be used when $T \lesssim 10^{-2}$ K.

2015

(a) State the basic differences in the fundamental assumptions underlying Maxwell-Boltzman (MB) and Fermi-Dirac (FD) statistics.

(b) Make a rough plot of the energy distribution function at two different temperatures for a system of free particles governed by MB statistics and one governed by FD statistics. Indicate which curve corresponds to the higher temperature.

(c) Explain briefly the discrepancy between experimental values of the specific heat of a metal and the prediction of MB statistics. How did FD statistics overcome the difficulty?

(Wisconsin)

Solution:

(a) FD, as compared with MB, statistics has two additional assumptions:

1) The principle of indistinguishability: identical particles cannot be distinguished from one another.

2) Pauli's exclusion principle: Not more than one particle can occupy a quantum state.

In the limit of non-degeneracy, FD statistics gradually becomes MB statistics.

(b) $\rho(\varepsilon)$ gives the number of particles in unit interval of energy or at energy level ε. Figure 2.5 gives rough plots of the energy distributions ((a) MB, (b) FD).

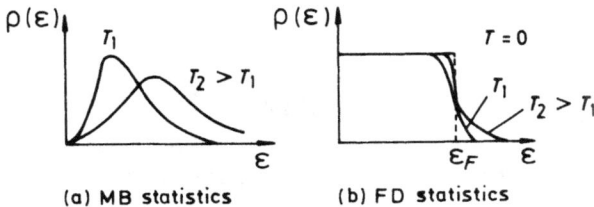

(a) MB statistics   (b) FD statistics

Fig. 2.5.

(c) According to MB statistics (or the principle of equipartition of energy), the contribution of an electron to the specific heat of a metal should be 1.5 K. This is not borne out by experiments, which shows that the contribution to specific heat of free electrons in metal can usually be neglected except for the case of very low temperatures. At low temperatures the contribution of electrons to the specific heat is proportional to the temperture T. FD statistics which incorporates Pauli's exclusion principle can explain this result.

2016

State which statistics (classical Maxwell-Boltzmann; Fermi-Dirac; or Bose-Einstein) would be appropriate in these problems and explain why (semi-quantitatively):

(a) Density of He^4 gas at room temperature and pressure.

(b) Density of electrons in copper at room temperature.

(c) Density of electrons and holes in semiconducting Ge at room temperature (Ge band-gap ≈ 1 volt).

(UC, Berkeley)

Solution:

(a) Classical Maxwell-Boltzmann statistics is appropriate because

$$n\lambda^3 = \frac{p}{kT} \cdot \left(\frac{h^2}{2\pi mkT}\right)^{3/2} \approx 3 \times 10^{-6} \ll 1 \ .$$

(b) Fermi-Dirac statistics is appropriate because electrons are Fermions and the Fermi energy of the electron gas in copper is about 1 eV which is equivalent to a high temperature of 10^4K. At room temperature (low temperature), the electron gas is highly degenerate.

(c) Classical Maxwell-Boltzmann statistics is appropriate because at room temperature the electrons and holes do not have sufficient average energy to jump over the 1 eV band-gap in appreciable numbers.

2017

Show that $\lambda = \exp(\mu/kT) = nV_Q$ for an ideal gas, valid where $\lambda \ll 1$; here μ is the chemical potential, n is the gas density and

$$V_Q = (h^2/2\pi mkT)^{3/2}$$

is the quantum volume. Even if you cannot prove this, this result will be useful in other problems.

(UC, Berkeley)

Solution:

In the approximation $\lambda \ll 1$, Fermi-Dirac and Bose-Einstein statistics both tend to Maxwell-Boltzmann statistics:

$$\frac{1}{\exp\dfrac{(\varepsilon - \mu)}{kT} \pm 1} \rightarrow e^{\mu/kT} \cdot e^{-\varepsilon/kT} \ .$$

The density of states of an ideal gas (spin states excluded) is

$$D(\varepsilon)d\varepsilon = \frac{2\pi}{h^3}(2m)^{3/2}\sqrt{\varepsilon}d\varepsilon \ .$$

Therefore,

$$n = \int_0^\infty D(\varepsilon)d\varepsilon \cdot e^{\mu/kT}e^{-\varepsilon/kT}$$

$$= \lambda \cdot \left(\frac{mkT}{2\pi\hbar}\right)^{3/2} = \frac{\lambda}{V_Q} \ .$$

That is, $\lambda = nV_Q$.

2018

A long, thin (i.e., needle-shaped) dust grain floats in a box filled with gas at a constant temperature T. On average, is the angular momentum vector nearly parallel to or perpendicular to the long axis of the grain? Explain.

(*MIT*)

Solution:

Let the long axis of the grain coincide with the z-axis. The shape of the grain indicates that the principal moments of inertia satisfy $I_z < I_x, I_y$. When thermal equilibrium is reached, we have

$$\frac{1}{2} I_z \omega_z^2 = \frac{1}{2} I_x \omega_x^2 = \frac{1}{2} I_y \omega_y^2 \ ,$$

so that $|\omega_z| = \left(\dfrac{I_x}{I_z}\right)^{1/2} |\omega_x| = \left(\dfrac{I_y}{I_z}\right)^{1/2} |\omega_y|$. Therefore

$$|I_z \omega_z| = \sqrt{I_z/I_x}|I_x \omega_x| < |I_x \omega_x| \ .$$

similarly

$$|I_z \omega_z| < |I_y \omega_y| \ .$$

So the angular momentum vector is nearly perpendicular to the long axis of the grain.

2019

A cubically shaped vessel 20 cm on a side constains diatomic H_2 gas at a temperature of 300 K. Each H_2 molecule consists of two hydrogen atoms with mass of 1.66×10^{-24} g each, separated by $\sim 10^{-8}$ cm. Assume that the gas behaves like an ideal gas. Ignore the vibrational degree of freedom.

(a) What is the average velocity of the molecules?

(b) What is the average velocity of rotation of the molecules around an axis which is the perpendicular bisector of the line joining the two atoms (consider each atom as a point mass)?

(c) Derive the values expected for the molar heat capacities C_p and C_v for such a gas.

(*Columbia*)

Solution:

(a) The number of the translational degrees of freedom is 3. Thus we have

$$\frac{3}{2}kT = \frac{1}{2}M\overline{v^2} \, ,$$

so $\bar{v} \approx \sqrt{\overline{v^2}} = \sqrt{\dfrac{3kT}{M}} \approx 2 \times 10^3$ m/s.

(b) The number of the rotational degrees of freedom is 2. Hence

$$\frac{1}{2}I\overline{\omega^2} = \frac{2}{2}kT \, ,$$

where $I = m \cdot \left(\dfrac{r}{2}\right)^2 \cdot 2 = \dfrac{1}{2}mr^2$ is the moment of inertia of the molecules H_2, m is the mass of the atom H and r is the distance between the two hydrogen atoms. Thus we get

$$\sqrt{\overline{\omega^2}} \approx 3.2 \times 10^{13}/s \, .$$

(c) The molar heat capacities are respectively

$$C_v = \frac{5}{2}R = 21 \text{ J/mol} \cdot \text{K} \, ,$$
$$C_p = \frac{7}{2}R = 29 \text{ J/mol} \cdot \text{K} \, .$$

2020

The circuit shown is in thermal equilibrium with its surroundings at a temperature T. Find the classical expression for the root mean square current through the inductor.

(*MIT*)

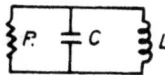

Fig. 2.6.

Solution:

Fluctuations in the motion of free electrons in the conductor give rise to fluctuation currents. If the current passing through the inductor is $I(t)$, then the average energy of the inductor is $\overline{W} = \dfrac{L}{2}\overline{I^2}$, where $\overline{I^2}$ is the mean-square current. According to the principle of equipartition of energy, we have $\overline{W} = \dfrac{1}{2}kT$. Hence

$$\sqrt{\overline{I^2}} = \sqrt{\frac{kT}{L}} \ .$$

2021

Energy probability.

Find and make careful sketch of the probability density, $\rho(E)$, for the energy E of a single atom in a classical non-interacting monatomic gas in thermal equilibrium.

(MIT)

Solution:

When the number of gas atoms is very large, we can represent the states of the system by a continuous distribution. When the system reaches thermal equilibrium, the probability of an atom having energy E is proportional to $\exp(-E/kT)$, where $E = p^2/2m$, p being the momentum of the atom. So the probability of an atom lying between $\mathbf{p}$ and $\mathbf{p} + d\mathbf{p}$ is

$$A \exp(-p^2/2mkT)d^3\mathbf{p} \ .$$

From

$$A \int \exp(-p^2/2mkT)d^3\mathbf{p} = 1 \ ,$$

we obtain

$$A = (2\pi mkT)^{-3/2} \ .$$

Therefore,

$$\int A e^{-p^2/2mkT} d^3\mathbf{p} = \frac{2\pi}{(\pi kT)^{3/2}} \int_0^\infty E^{1/2} e^{-E/kT} dE$$
$$\equiv \int_0^\infty \rho(E) dE \ ,$$

giving

$$\rho(E) = \frac{2}{\sqrt{\pi}(kT)^{3/2}} E^{1/2} e^{-E/kT} .$$

2022

Suppose that the energy of a particle can be represented by the expression $E(z) = az^2$ where z is a coordinate or momentum and can take on all values from $-\infty$ to $+\infty$.

(a) Show that the average energy per particle for a system of such particles subject to Boltzmann statistics will be $\overline{E} = kT/2$.

(b) State the principle of equipartition of energy and discuss briefly its relation to the above calculation.

(*Wisconsin*)

Solution:

(a) From Boltzmann statistics, whether z is position or momentum, its distribution function is

$$f(z) \propto \exp\left(-\frac{E(z)}{kT}\right) .$$

So the average energy of a single particle is

$$\overline{E} = \int_{-\infty}^{+\infty} f(z)E(z)dz = \frac{\displaystyle\int_{-\infty}^{+\infty} \exp\left(-\frac{E(z)}{kT}\right) E(z)dz}{\displaystyle\int_{-\infty}^{+\infty} \exp\left(-\frac{E(z)}{kT}\right) dz} .$$

Inserting $E(z) = az^2$ in the above, we obtain $\overline{E} = \dfrac{1}{2}kT$.

(b) Principle of equipartition of energy: For a classical system of particle in thermal equilibrium at temperature T, the average energy of each degree of freedom of a particle is equal to $\dfrac{1}{2}kT$.

There is only one degree of freedom in this problem, so the average energy is $\dfrac{1}{2}kT$.

2023

A system of two energy levels E_0 and E_1 is populated by N particles at temperature T. The particles populate the energy levels according to the classical distribution law.

(a) Derive an expression for the average energy per particle.

(b) Compute the average energy per particle vs the temperature as $T \to 0$ and $T \to \infty$.

(c) Derive an expression for the specific heat of the system of N particles.

(d) Compute the specific heat in the limits $T \to 0$ and $T \to \infty$.

(*Wisconsin*)

Solution:

(a) The average energy of a particle is

$$u = \frac{E_0 e^{-\beta E_0} + E_1 e^{-\beta E_1}}{e^{-\beta E_0} + e^{-\beta E_1}} .$$

Assuming $E_1 > E_0 > 0$ and letting $\Delta E = E_1 - E_0$, we have

$$u = \frac{E_0 + E_1 e^{-\beta \Delta E}}{1 + e^{-\beta \Delta E}} .$$

(b) When $T \to 0$, i.e., $\beta = 1/kT \to \infty$, one has

$$u \approx (E_0 + E_1 e^{-\beta \Delta E})(1 - e^{-\beta \Delta E}) = E_0 + \Delta E e^{-\beta \Delta E} .$$

When $T \to \infty$, or $\beta \to 0$, one has

$$u \approx \frac{1}{2}(E_0 + E_1 - \beta E_1 \Delta E)\left(1 + \frac{1}{2}\beta \Delta E\right) \approx \frac{1}{2}(E_0 + E_1) - \frac{\beta}{4}(\Delta E)^2 .$$

(c) The specific heat (per mole) is

$$C = N_A \frac{\partial u}{\partial T} = N_A \frac{\partial u}{\partial \beta} \cdot \frac{\partial \beta}{\partial T} = R \left(\frac{\Delta E}{kT}\right)^2 \frac{e^{-\Delta E/kT}}{(1 + e^{-\Delta E/kT})^2} .$$

(d) When $T \to 0$, one has

$$C \approx R \cdot \left(\frac{\Delta E}{kT}\right)^2 \cdot e^{-\Delta E/kT} .$$

When $T \to \infty$,

$$C \approx \frac{R}{4} \cdot \left(\frac{\Delta E}{kT} \right)^2 .$$

2024

Consider a glass in which some fraction of its constituent atoms may occupy either of two slightly different positions giving rise to two energy levels $\Delta_i > 0$ and $-\Delta_i$ for the ith atom.

(a) If each participating atom has the same levels Δ and $-\Delta$, calculate the contribution of these atoms to the heat capacity. (Ignore the usual Debye specific heat which will also be present in a real solid.)

(b) If the glass has a random composition of such atoms so that all values of Δ_i are equally likely up to some limiting value $\Delta_0 > 0$, find the behavior of the low temperature heat capacity, i.e., $kT \ll \Delta_0$. (Definite integrals need not be evaluated provided they do not depend on any of the parameters.)

(Princeton)

Solution:

(a) The mean energy per atom is $\bar{\varepsilon} = -\Delta \tanh \left(\dfrac{\Delta}{kT} \right)$. Its contribution to the specific heat is

$$c_v = \frac{d\bar{\varepsilon}}{dT} = 4k \left(\frac{\Delta}{kT} \right)^2 \frac{1}{(e^{\Delta/kT} + e^{-\Delta/kT})^2} .$$

Summing up the terms for all such atoms, we have

$$c_v = 4Nk \left(\frac{\Delta}{kT} \right)^2 \cdot \frac{1}{(e^{\Delta/kT} + e^{-\Delta/kT})^2} .$$

(b) The contribution to the specific heat of the ith atom is

$$c_i = 4k \left(\frac{\Delta_i}{kT} \right)^2 \frac{1}{(e^{\Delta_i/kT} + e^{-\Delta_i/kT})^2} .$$

When $kT \ll \Delta_i$, we have

$$c_i = 4k \left(\frac{\Delta_i}{kT} \right)^2 e^{-2\Delta_i/kT} .$$

Summing up the terms for all such atoms, we have

$$c = 4k \sum_i \left(\frac{\Delta_i}{kT}\right)^2 e^{-2\Delta_i/kT}$$

$$= 4k \int \left(\frac{\Delta}{kT}\right)^2 e^{-2\Delta/kT} \rho(\Delta) d\Delta \;,$$

where $\rho(\Delta)$ is the state density of distribution of Δ_i.

2025

The three lowest energy levels of a certain molecule are $E_1 = 0, E_2 = \varepsilon, E_3 = 10\varepsilon$. Show that at sufficiently low temperatures (how low?) only levels E_1, E_2 are populated. Find the average energy E of the molecule at temperature T. Find the contributions of these levels to the specific heat per mole, C_v, and sketch C_v as a function of T.

(*Wisconsin*)

Solution:

We need not consider energy levels higher than the three lowest energy levels for low temperatures. Assuming the system has N particles and according to the Boltzmann statistics, we have

$$N_1 + N_2 + N_3 = N \;,$$

$$\frac{N_2}{N_1} = e^{-\varepsilon/kT} \;,$$

$$\frac{N_3}{N_1} = e^{-10\varepsilon/kT} \;,$$

hence

$$N_3 = \frac{N}{1 + e^{9\varepsilon/kT} + e^{10\varepsilon/kT}} \;.$$

When $N_3 < 1$, there is no occupation at the energy level E_3. That is, when $T < T_c$, only the E_1 and E_2 levels are occupied, where T_c satisfies

$$\frac{N}{1 + e^{9\varepsilon/kT_c} + e^{10\varepsilon/kT_c}} = 1 \;.$$

If $N \gg 1$, we have

$$T_c \approx \frac{10\varepsilon}{k \ln N} \;.$$

The average energy of the molecule is

$$E = \frac{\varepsilon(e^{-\varepsilon/kT} + 10e^{-10\varepsilon/kT})}{1 + e^{-\varepsilon/kT} + e^{-10\varepsilon/kT}} \ .$$

The molar specific heat is

$$C_v = N_A \frac{\partial E}{\partial T} = \frac{R\varepsilon^2(e^{-\beta\varepsilon} + 100e^{-10\beta\varepsilon} + 81e^{-11\beta\varepsilon})}{(1 + e^{-\beta\varepsilon} + e^{-10\beta\varepsilon})^2}\beta^2 \ ,$$

where $\beta = 1/kT$ and N_A is Avogadro's number.
For high temperatures, $kT \gg \varepsilon$,

$$C_v \approx \frac{182}{9}R\left(\frac{\varepsilon}{kT}\right)^2 \propto \frac{1}{T^2} \ .$$

For low temperatures, $kT \ll \varepsilon$,

$$C_v \approx R\varepsilon^2 \frac{e^{-\varepsilon/kT}}{(kT)^2} \ .$$

The variation of C_v with T is shown in Fig. 2.7.

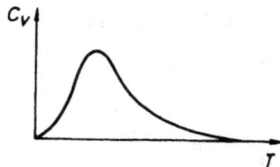

Fig. 2.7.

2026

Given a system of two distinct lattice sites, each occupied by an atom whose spin $(s = 1)$ is so oriented that its energy takes one of three values $\varepsilon = 1, 0, -1$ with equal probability. The atoms do not interact with each other. Calculate the ensemble average values $\overline{U}$ and $\overline{U}^2$ for the energy U of the system, assumed to be that of the spins only.

(UC, Berkeley)

Solution:

For a single atom, we have

$$\bar{\varepsilon} = -\frac{e^\beta - e^{-\beta}}{1 + e^\beta + e^{-\beta}} \ ,$$

$$\overline{\varepsilon^2} = \frac{e^\beta + e^{-\beta}}{1 + e^\beta + e^{-\beta}} \ .$$

For the system, we have

$$\overline{U} = \bar{\varepsilon}_1 + \bar{\varepsilon}_2 = -2\frac{e^{\beta\cdot} - e^{-\beta}}{1 + e^\beta + e^{-\beta}} \ ,$$

$$\overline{U^2} = \overline{(\varepsilon_1 + \varepsilon_2)^2} = \overline{\varepsilon_1^2} + \overline{\varepsilon_2^2} + \overline{2\varepsilon_1\varepsilon_2} \ .$$

Since $\overline{\varepsilon_1\varepsilon_2} = \bar{\varepsilon}_1 \cdot \bar{\varepsilon}_2$, it follows

$$\overline{U^2} = \frac{2[\exp(2\beta) + \exp(-2\beta)] + \exp(\beta) + \exp(-\beta)}{(1 + \exp(\beta) + \exp(-\beta))^2} \ .$$

2027

Obtain the temperature of each system:

(a) 6.0×10^{22} atoms of helium gas occupy 2.0 litres at atmospheric pressure. What is the temperature of the gas?

(b) A system of particles occupying single-particle levels and obeying Maxwell-Boltzmann statistics is in thermal contact with a heat reservoir at temperature T. If the population distribution in the non-degenerate energy levels is as shown, what is the temperature of the system?

Energy (eV)	population
30.1×10^{-3}	3.1%
21.5×10^{-3}	8.5%
12.9×10^{-3}	23%
4.3×10^{-3}	63%

(c) In a cryogenic experiment, heat is supplied to a sample at the constant rate of 0.01 watts. The entropy of the sample increases with time as shown in the table. What is the temperature of the sample at $t = 500$ sec?

Time:	100	200	300	400	500	600	700	(sec)
Entropy:	2.30	2.65	2.85	3.00	3.11	3.20	3.28	(J/K)

(UC, Berkeley)

Solution:

(a) Using the equation of state for an ideal gas, we get

$$T = pV/nk = 241 \text{ K} .$$

(b) The population distribution is given by

$$\frac{n_2}{n_1} = \exp((\varepsilon_1 - \varepsilon_2)/kT) .$$

Therefore

$$T = \frac{\varepsilon_1 - \varepsilon_2}{k} \frac{1}{\ln\left(\dfrac{n_2}{n_1}\right)} .$$

Using the given n_1 and n_2, we get T as follows:

$$99.2; \ 99.5; \ 99.0; \ 99.5; \ 100.2; \ 98.8 \text{ K} .$$

The mean value is $T = 99.4$ K.

(c) The rate of heat intake is $q = \dfrac{dQ}{dt} = T\dfrac{dS}{dt}$, giving

$$T = \frac{q}{\left(\dfrac{dS}{dt}\right)} .$$

We estimate $\dfrac{dS}{dt}$ by the middle differential at $t = 500$s, and get

$$\frac{dS}{dt} = \left(\frac{3.20 - 3.00}{600 - 400}\right) = 1.0 \times 10^{-3} \text{J/sec.K} .$$

Therefore $T = 10$K.

2028

Assume that the reaction $H \rightleftharpoons p + e$ occurs in thermal equilibrium at $T = 4000$ K in a very low density gas (no degeneracy) of each species with overall charge neutrality.

(a) Write the chemical potential of each gas in terms of its number density $[H]$, $[p]$, or $[e]$. For simplicity you may ignore the spectrum of excited bound states of H and consider only the ground state. Justify this assumption.

(b) Give the condition for thermal equilibrium and calculate the equilibrium value of [e] as a function of [H] and T.

(c) Estimate the nucleon density for which the gas is half-ionized at $T = 4000$ K. (Note that this is an approximate picture of the universe at a redshift $z = 10^3$.)

(*UC, Berkeley*)

Solution:

(a) From Boltzmann statistics, we have for an ideal gas without spin

$$n = e^{\mu/kT} \cdot (2\pi mkT/h^2)^{3/2} .$$

Both the proton and electron have spin $1/2$, therefore

$$[p] = 2(2\pi m_p kT/h^2)^{3/2} e^{\mu_p/kT}$$
$$[e] = 2(2\pi m_e kT/h^2)^{3/2} e^{\mu_e/kT} .$$

For the hydrogen atom, p and e can form four spin configurations with ionization energy E_d. Hence

$$[H] = 4(2\pi m_H kT/h^2)^{3/2} \exp(E_d/kT) \exp(\mu_H/kT) .$$

The chemical potentials μ_p, μ_e and μ_H are given by the above relations with the number densities.

(b) The equilibrium condition is $\mu_H = \mu_e + \mu_p$. Note that as $m_H \approx m_p$ and $[e] = [p]$ we have

$$[e] = \sqrt{[H]} \cdot (2\pi m_e kT/h^2)^{3/2} \cdot \exp(-E_d/2kT) .$$

(c) When the gas is half-ionized, $[e] = [p] = [H] = n$. Hence

$$n = (2\pi m_e kT/h^2)^{3/2} \cdot \exp(-E_d/kT) = 3.3 \times 10^{16} \text{ m}^{-3} .$$

2029

A piece of metal can be considered as a reservoir of electrons; the work function (energy to remove an electron from the metal) is 4 eV. Considering only the 1s orbital (which can be occupied by zero, one, or two electrons) and knowing that the hydrogen atom has an ionization energy of 13.6 eV and an electron affinity of 0.6 eV, determine for atomic hydrogen in chemical equilibrium at $T = 300$ K in the vicinity of a metal the probabilities of finding H^+, H^0 and H^-. Give only one significant figure.

What value of the work function would give equal probabilities to H^0 and H^-?

(*UC, Berkeley*)

Solution:

We have (see Fig. 2.8)

$$e + H^+ \leftrightarrows H ,$$
$$e + H \leftrightarrows H^- .$$

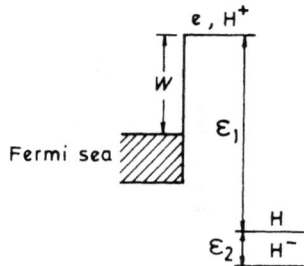

Fig. 2.8.

The chemical potential of the electron gas is $\mu_e = -W$. From classical statistics, we can easily obtain

$$[e] = 2e^{\mu_e/kT} \left(\frac{2\pi m_e kT}{h^2} \right)^{3/2} ,$$

$$[H^+] = 2e^{\mu_{H^+}/kT} \left(\frac{2\pi m_p kT}{h^2} \right)^{3/2} ,$$

where the factor 2 arises from the internal degrees of freedom of spin. For the hydrogen atom, electron and proton spins can have four possible spin states, hence

$$[H^0] = 4e^{\mu_{H^0}/kT} \left(\frac{2\pi m_p kT}{h^2} \right)^{3/2} e^{\epsilon_1/kT} .$$

For H^-, both electrons are in their ground state with total spin 0 (singlet), as the space wave function is symmetric when the particles are interexchanged. Therefore, the spin degrees of freedom of H^- correspond only to the two spin states of the nucleon; hence

$$[H^-] = 2e^{\mu_{H^-}/kT} \left(\frac{2\pi m_p kT}{h^2}\right)^{3/2} \exp\left(\frac{\varepsilon_1 + \varepsilon_2}{kT}\right) .$$

The conditions for chemical equilibrium are

$$\mu_{H^0} = \mu_e + \mu_{H^+} ,$$
$$\mu_{H^-} = \mu_e + \mu_{H^0} ,$$

so that

$$\frac{[H^0]}{[H^+]} = 2 \exp \frac{\mu_e + \varepsilon_1}{kT} ,$$
$$\frac{[H^-]}{[H^0]} = \frac{1}{2} \exp \frac{\mu_e + \varepsilon_2}{kT} .$$

Thus, the relative probabilities of finding H^+, H^0 and H^- are

$$P_{H^+} : P_{H^0} : P_{H^-} = [H^+] : [H^0] : [H^-] = 1 : 2 \exp \frac{\mu_e + \varepsilon_1}{kT} :$$
$$\exp \frac{2\mu_e + \varepsilon_1 + \varepsilon_2}{kT} = 1 : 2e^{371} : e^{240}$$

If $P_H = P_{H^-}$, or $[H^0] = [H^-]$, we have

$$W = -\mu_e = -\varepsilon_2 + kT \ln 2 \approx 0.6 \text{ eV} .$$

2030

The potential energy V between the two atoms ($m_H = 1.672 \times 10^{-24}$ g) in a hydrogen molecule is given by the empirical expression

$$V = D\{e^{-2a(r-r_0)} - 2e^{-a(r-r_0)}\} .$$

where r is the distance between the atoms.
$D = 7 \times 10^{-12}$ erg,

$$a = 2 \times 10^8 \text{ cm}^{-1}$$
$$r_0 = 8 \times 10^{-9} \text{ cm.}$$

Estimate the temperatures at which rotation (T_R) and vibration (T_V) begin to contribute to the specific heat of hydrogen gas. Give the approximate values of C_v and C_p (the molar specific heats at constant volume and at constant pressure) for the following temperatures:

$$T_1 = 25 \text{ K}, \; T_2 = 250 \text{ K}, \; T_3 = 2500 \text{ K}, \; T_4 = 10000 \text{ K}.$$

Neglect ionization and dissociation.

(*UC, Berkeley*)

Solution:

The average distance between the two atoms is approximately the equilibrium distance. From

$$\left(\frac{\partial V}{\partial r} \right)_{r=d} = 0 \,,$$

we obtain $d = r_0$. The frequency of the radial vibration of the two atoms is

$$\omega = \sqrt{\frac{k}{\mu}} \,,$$

where $\mu = m_H/2$ is the reduced mass and

$$k = \frac{\partial^2 V}{\partial r^2}\Big|_{r=d} = 2a^2 D \,.$$

So

$$\omega = \sqrt{\frac{4a^2 D}{m}} \,.$$

The characteristic energy of the rotational level is

$$k\theta_R = \frac{\hbar^2}{2\mu d^2} \,,$$

then

$$\theta_R = \frac{\hbar^2}{k m_H r_0^2} = 75 \text{ K} \,.$$

The characteristic energy of vibration is $k\theta_V = \hbar\omega$, then

$$\theta_V = \frac{\hbar\omega}{k} = \frac{2a\hbar}{k} \sqrt{\frac{D}{m_H}} = 6250 \text{ K} \,.$$

Thus, rotation begins to contribute to the specific heat at $T = 75$ K, and vibration does so at $T = 6250$ K.

When $T_1 = 25$ K, only the translational motion contributes to C, then

$$C_v = \frac{3}{2}R = 12.5 \text{ J/K}, \quad C_p = \frac{5}{2}R = 20.8 \text{ J/K} .$$

When $T_2 = 250$ K, only translation and rotation contribute to C, then

$$C_v = \frac{5}{2}R = 20.8 \text{ J/K}, \quad C_p = \frac{7}{2}R = 29.1 \text{ J/K} .$$

When $T_3 = 2500$ K, the result is the same as for $T_2 = 250$ K. When $T_4 = 10000$ K, vibration also contributes to C, then

$$C_v = \frac{7}{2}R = 29.1 \text{ J/K}, \quad C_p = \frac{9}{2}R = 37.4 \text{ J/K} .$$

2031

Derive an expression for the vibrational specific heat of a diatomic gas as a function of temperature. (Let $\hbar\omega_0/k = \theta$). For full credit start with an expression for the vibrational partition function, evaluate it, and use the result to calculate C_{vib}.

Describe the high and low T limits of C_{vib}.

(*Wisconsin*)

Solution:

The vibrational energy levels of a diatomic gas are

$$\varepsilon_v = \hbar\omega_0(v + 1/2), \quad v = 0, 1, 2, \ldots .$$

The partition function is

$$Z_{\text{vib}} = \sum_{v=0}^{\infty} \exp\left[-\beta\hbar\omega_0\left(v + \frac{1}{2}\right)\right] = \left(\frac{e^{-\frac{x}{2}}}{1 - e^{-x}}\right) .$$

where $x = \beta\hbar\omega_0$. The free energy of 1 mole of the gas is

$$F = -N_A kT \ln Z_{\text{vib}} = \frac{N_A}{2}\hbar\omega_0 + \frac{N_A}{\beta}\ln[1 - \exp(-\beta\hbar\omega_0)] .$$

and the internal energy is

$$U = F - T\frac{\partial F}{\partial T} = \frac{N_A}{2}\hbar\omega_0 + \frac{N_A \hbar\omega_0}{\exp(\beta\hbar\omega_0) - 1} .$$

The molar specific heat is

$$C_v = \frac{dU}{dT} = R\frac{x^2 e^x}{(e^x - 1)^2} , \qquad x = \frac{\hbar\omega_0}{kT} = \frac{\theta}{T} .$$

(a) In the limit of high temperatures, $T \gg \theta$, or $x \ll 1$, we have

$$C_v \approx R .$$

(b) In the limit of low temperatures, $T \ll \theta$, or $x \gg 1$, we have

$$C_v \approx R(\theta/T)^2 \exp(-\theta/T) .$$

2032

A one-dimensional quantum harmonic oscillator (whose ground state energy is $\hbar\omega/2$) is in thermal equilibrium with a heat bath at temperature T.

(a) What is the mean value of the oscillator's energy, $\langle E \rangle$, as a function of T?

(b) What is the value of ΔE, the root-mean-square fluctuation in energy about $\langle E \rangle$?

(c) How do $\langle E \rangle$ and ΔE behave in the limits $kT \ll \hbar\omega$ and $kT \gg \hbar\omega$?

(*MIT*)

Solution:

The partition function is

$$z = \sum_{n=0}^{\infty} \exp\left(\frac{-E_n}{kT}\right) = \sum_{n=0}^{\infty} \exp\left(-\left(n + \frac{1}{2}\right)\frac{\hbar\omega}{kT}\right) = \frac{2}{\sinh\left(\frac{\hbar\omega}{2kT}\right)} .$$

(a) The mean energy is

$$\langle E \rangle = kT^2 \frac{\partial}{\partial T}\ln z = \frac{\hbar\omega}{2}\coth\left(\frac{\hbar\omega}{2kT}\right) .$$

(b) The root-mean-square fluctuation is

$$\Delta E = T\sqrt{k\frac{\partial\langle E\rangle}{\partial T}} = \frac{\hbar\omega}{2\sinh\left(\frac{\hbar\omega}{2kT}\right)} \ .$$

(c) When $kT \ll \hbar\omega$,

$$\langle E\rangle \rightarrow \frac{\hbar\omega}{2} \ , \quad \Delta E \rightarrow \hbar\omega\exp\left(-\frac{\hbar\omega}{2kT}\right) \ .$$

When $kT \gg \hbar\omega$,

$$\langle E\rangle \rightarrow kT, \quad \Delta E \rightarrow kT \ .$$

2033

Consider a system of N_0 non-interacting quantum mechanical oscillators in equilibrium at temperature T. The energy levels of a single oscillator are

$$E_m = (m + 1/2)\gamma/V \quad \text{with } m = 0, 1, 2\ldots\text{etc.}$$

(γ is a constant, the oscillators and volume V are one dimensional.)

(a) Find U and C_v as functions of T.
(b) Sketch $U(T)$ and $C_v(T)$.
(c) Determine the equation of state for the system.
(d) What is the fraction of particles in the m-th level?

(SUNY, Buffalo)

Solution:

(a) The partition function is

$$z = \sum_{m=0}^{\infty} e^{-\beta(m+1/2)\gamma V^{-1}} = \frac{e^{(-\frac{\beta\gamma}{2V})}}{1 - e^{-\beta\gamma V^{-1}}}$$

$$= \frac{1}{2}\operatorname{csch}\frac{\gamma\beta}{2V} \ .$$

The internal energy is

$$U = -N_0\frac{\partial}{\partial\beta}\ln z = \frac{N_0\gamma}{2V}\coth\frac{\gamma\beta}{2V}$$

$$= \frac{N_0\gamma}{2V}\coth\frac{\gamma}{2VkT} \ .$$

The specific heat at constant volume is

$$c_v = \left(\frac{\partial U}{\partial T}\right)_v = N_0 k \left(\frac{\gamma}{2VkT}\right)^2 \mathrm{csch}^2 \left(\frac{\gamma}{2VkT}\right) \ .$$

(b) As shown in Fig. 2.9.

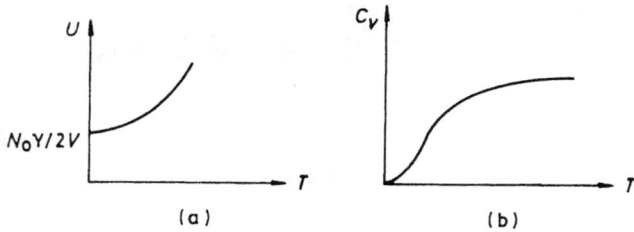

Fig. 2.9.

(c) The equation of state is

$$p = \frac{N_0}{\beta} - \frac{\partial}{\partial V} \ln z = \frac{N_0 \gamma}{2V^2} \coth \left(\frac{\gamma}{2VkT}\right) \ ,$$

where p is the pressure.

(d) The fraction of particles in the m-th level is

$$a_m = e^{-\alpha - \beta(m+1/2)\gamma V^{-1}} = \frac{N_0}{z} e^{-\beta(m+1/2)\gamma V^{-1}}$$

$$= 2 N_0 e^{-\beta(m+1/2)\gamma V^{-1}} \cdot \sinh \left(\frac{\gamma\beta}{2V}\right) \ .$$

2034

The molecules of a certain gas consist of two different atoms, each with zero nuclear spin, bound together. Measurements of the specific heat of this material, over a wide range of temperatures, give the graph shown below.

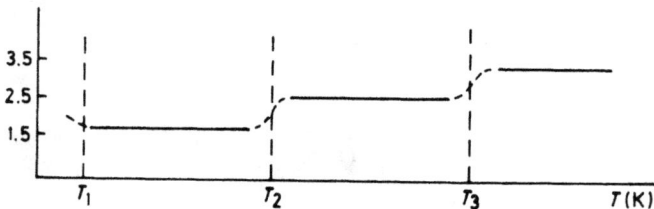

Fig. 2.10.

(The values marked on the vertical scale correspond to the height of the curve in each of the "plateau" regions.)

(a) Account for each of the different results found in the temperature regions: above T_3; between T_2 and T_3; between T_1 and T_2; below T_1,

(b) Given that the first excited state of the rotational spectrum of this molecule is at an energy kT_e above the ground rotational state, and $T_e = 64$ K, calculate from basic theory the rotational contribution to the specific heat capacity of this gas at 20K at 100K, at 300K.

(*UC, Berkeley*)

Solution:

(a) When $T > T_3$, the translational, rotational and vibrational motions are all excited, and $C_v = 7k/2$. When $T_2 < T < T_3$, the vibrational motion is not excited and $C_v = 5k/2$. When $T_1 < T < T_2$, only the translational motion contributes to the specific heat and $C_v = 3k/2$. When $T < T_1$, a phase transition occurs, and the gas phase no longer exists.

(b) When $T = 20$ K, neglect the higher rotational energy levels and consider only the ground state and the 1st excited state. We have

$$\overline{E} = \frac{3kT_e e^{-T_e/T}}{1 + 3e^{-T_e/T}} \ ,$$

$$C_v = \frac{d\overline{E}}{dT} = 3k \left(\frac{T_e}{T}\right)^2 \frac{e^{-T_e/T}}{(1 + 3e^{-T_e/T})^2} = 0.1k \ .$$

When $T = 100$ K, consider the first two excited states and we have

$$\overline{E} = kT_e \frac{3e^{-T_e/T} + 15e^{-3T_e/T}}{1 + 3e^{-T_e/T} + 5e^{-3T_e/T}} \ .$$

$$C_v = \frac{d\overline{E}}{dT} = 3k \left(\frac{T_e}{T}\right)^2 \cdot \frac{e^{-T_e/T} + 15e^{-3T_e/T} + 10e^{-4T_e/T}}{(1 + 3e^{-T_e/T} + 5e^{-3T_e/T})^2}$$

$$= 0.22k \ .$$

When $T = 300$ K, all the rotational energy levels are to be considered and

$$C_v = 1.0 \ k \ .$$

2035

The quantum energy levels of a rigid rotator are

$$\varepsilon_j = j(j+1)h^2/8\pi^2 ma^2 ,$$

where $j = 0, 1, 2, \ldots$. The degeneracy of each level is $g_j = 2j + 1$.

(a) Find the general expression for the partition function, and show that at high temperatures it can be approximated by an integral.

(b) Evaluate the high-temperature energy and heat capacity.

(c) Find the low-temperature approximations to z, U and C_v.

(SUNY, Buffalo)

Solution:

(a) The partition function is

$$z = \sum_{j=0}^{\infty} g_j e^{-\varepsilon_j/kT} = \sum_{j=0}^{\infty} (2j+1) e^{-j(j+1)h^2/8\pi^2 ma^2 kT} .$$

(b) At high temperatures $\Delta_x \equiv (h^2/8\pi^2 ma^2 kT)^{1/2} \ll 1$,

$$z = 2e^{h^2/32\pi^2 ma^2 kT} \sum_{j=0}^{\infty} \left(j + \frac{1}{2}\right) e^{-(j+1/2)^2 h^2/8\pi^2 ma^2 kT}$$

$$= 2e^{(\Delta_x)^2/4} \frac{1}{(\Delta_x)^2} \sum_{j=0}^{\infty} \varepsilon_j e^{-\varepsilon_j^2} \Delta\varepsilon_j ,$$

where

$$\varepsilon_j = \left(j + \frac{1}{2}\right)\Delta_x, \quad \Delta\varepsilon_j = \varepsilon_{j+1} - \varepsilon_j = \Delta_x .$$

Hence

$$z \approx \frac{2}{(\Delta_x)^2} e^{(\Delta_x)^2/4} \int_0^{\infty} \varepsilon e^{-\varepsilon^2} d\varepsilon = \frac{1}{(\Delta_x)^2} e^{(\Delta_x)^2/4}$$

$$\approx \frac{1}{(\Delta_x)^2} = 8\pi^2 ma^2 kT/h^2 .$$

The internal energy is

$$U = kT^2 \frac{\partial}{\partial T} \ln z = kT .$$

The heat capacity is

$$C_v = \frac{\partial U}{\partial T} = k \ .$$

(c) For low temperatures, we need only take the first two terms of z, i.e., $z \approx 1 + 3e^{-\theta/T}$, where $\theta = h^2/4\pi^2 ma^2 k$. So

$$U = \frac{3k\theta e^{-\theta/T}}{1 + 3e^{-\theta/T}}$$

$$C_v = \frac{3k(\theta/T)^2 e^{-\theta/T}}{(1 + 3e^{-\theta/T})^2} \ .$$

2036

The quantum energy levels of a rigid rotator are

$$\varepsilon_j = j(j+1)h^2/8\pi^2 ma^2 \ ,$$

where $j = 0, 1, 2, \ldots, m$ and a are positive constants. The degeneracy of each level is $g_j = 2j + 1$.

(a) Find the general expression for the partition function z_0.

(b) Show that at high temperatures it can be approximated by an integral.

(c) Evaluate the high-temperature energy U and heat capacity C_v.

(d) Also, find the low-temperature approximations to z_0, U and C_v.

(*SUNY, Buffalo*)

Solution:

(a) The partition function is

$$z_0 = \sum_{j=0}^{\infty} (2j+1) \exp\left(-\frac{j(j+1)h^2}{8\pi^2 ma^2 kT}\right)$$

$$= \sum_{j=0}^{\infty} (2j+1) \exp\left(-\frac{\theta j(j+1)}{T}\right)$$

where,

$$\theta = \frac{h^2}{8\pi^2 ma^2 k} \ .$$

(b) At high temperatures $\theta/T \ll 1$ and $\exp[-\theta j(j+1)/T)$ changes slowly as j changes, so that we can think of $(2j+1)\exp[-\theta j(j+1)/T]$ as a continuous function of j. Let $x = j(j+1)$, then $dx = 2j+1$, and we can write z_0 as an integral:

$$z_0 = \int_0^\infty e^{-\theta x/T} dx = \frac{T}{\theta} = \frac{8\pi^2 ma^2 kT}{h^2} \ .$$

(c) At high temperatures, the internal energy is

$$U = -\frac{\partial}{\partial \beta} \ln z_0 = kT \ .$$

The heat capacity is

$$C_v = k \ .$$

(d) At low temperatures, we have $T \ll \theta$, and $\exp[-\theta j(j+1)/T]$ is very small. We need only take the first two terms of z_0, so

$$z_0 \approx 1 + 3\exp(-2\theta/T) \ ,$$
$$U = \frac{6k\theta}{z_0} e^{-2\theta/T}$$
$$C_v = \frac{12k\theta^2}{z_0^2 T^2} e^{-2\theta/T} \ .$$

2037

The energy levels of a three-dimensional rigid rotor of moment of inertial I are given by

$$E_{J,M} = \hbar^2 J(J+1)/2I \ ,$$

where $J = 0, 1, 2, \ldots ; M = -J, -J+1, \ldots, J$. Consider a system of N rotors:

(a) Using Boltzmann statistics, find an expression for the thermodynamical internal energy of the system.

(b) Under what conditions can the sum in part (a) be approximated by an integral? In this case calculate the specific heat C_v of the system.

(*Wisconsin*)

Solution:

(a) The partition function of the system is

$$z = \sum_{J=0}^{\infty} (2J + 1) \exp[-\hbar^2 J(J+1)/2IkT] .$$

The internal energy is

$$U = NkT^2 \frac{d \ln z}{dT}$$

$$= N \frac{\sum_{J} \frac{\hbar^2}{2I} J(J+1)(2J+1) \exp\left[-\frac{\hbar^2}{2IkT} J(J+1)\right]}{\sum_{J} (2J+1) \exp\left[-\frac{\hbar^2}{2IkT} J(J+1)\right]} .$$

(b) In the limit of high temperatures, $kT \gg \hbar^2/2I$, and the above sum can be replaced by an integral. Letting $x = J(J+1)$, we have

$$z = \int_0^{\infty} \exp\left\{-\frac{\hbar^2}{2IkT} x\right\} dx = \frac{2IkT}{\hbar^2} ,$$

$$U = NkT .$$

Thus the molar specific heat is $C_v = N_A k = R$.

2038

Consider a heteronuclear diatomic molecule with moment of inertia I. In this problem, only the rotational motion of the molecule should be considered.

(a) Using classical statistical mechanics, calculate the specific heat $C(T)$ of this system at temperature T.

(b) In quantum mechanics, this system has energy levels

$$E_j = \frac{\hbar^2}{2I} j(j+1) , j = 0, 1, 2, \ldots .$$

Each j level is $(2j + 1)$-fold degenerate. Using quantum statistics, derive expressions for the partition function z and the average energy $\langle E \rangle$ of this system, as a function of temperature. Do not attempt to evaluate these expressions.

(c) By simplifying your expressions in (b), derive an expression for the specific heat $C(T)$ that is valid at very low temperatures. In what range of temperatures is your expression valid?

(d) By simplifying your answer to (b), derive a high temperature approximation to the specific heat $C(T)$. What is the range of validity of your approximation?

(Princeton)

Solution:

(a) For a classical rotator, one has

$$E = \frac{1}{2I}\left(p_\theta^2 + \frac{1}{\sin^2\theta}p_\varphi^2\right) ,$$

$$z = \int e^{-\beta E}\,dp_\theta dp_\varphi d\theta d\varphi = \frac{8\pi^2 I}{\beta} ,$$

$$\langle E\rangle = -\frac{\partial}{\partial\beta}\ln z = \frac{1}{\beta} = kT .$$

Thus $C(T) = k$.

(b) In quantum statistical mechanics,

$$z = \sum_{j=0}^{\infty}(2j+1)\exp\left[-\frac{\beta\hbar^2}{2I}j(j+1)\right]$$

$$\langle E\rangle = -\frac{1}{z}\frac{\partial z}{\partial\beta}$$

$$= \frac{\displaystyle\sum_{j=0}^{\infty}(2j+1)\frac{\hbar^2}{2I}j(j+1)\exp\left[-\frac{\beta\hbar^2}{2I}j(j+1)\right]}{\displaystyle\sum_{j=0}^{\infty}(2j+1)\exp\left[-\frac{\beta\hbar^2}{2I}j(j+1)\right]} .$$

(c) In the limit of low temperatures, $\dfrac{\beta\hbar}{2I} \gg 1$, or $\dfrac{\hbar^2}{2I} \gg kT$, so only the first two terms $j = 0$ and $j = 1$ are important. Thus

$$z = 1 + 3 \exp\left(-\frac{\beta\hbar^2}{I}\right) .$$

$$\langle E \rangle = \frac{3\hbar^2}{I} \cdot \frac{\exp\left(-\frac{\beta\hbar^2}{I}\right)}{1 + 3\exp\left(-\frac{\beta\hbar^2}{I}\right)} .$$

Hence

$$C(T) = 3k\left(\frac{\beta\hbar^2}{I}\right)^2 \frac{\exp\left(\frac{\beta\hbar^2}{I}\right)}{\left[3 + \exp\left(\frac{\beta\hbar^2}{I}\right)\right]^2}$$

$$= 3k\left(\frac{\hbar^2}{kTI}\right)^2 \frac{\exp\left(\frac{\hbar^2}{kTI}\right)}{\left[3 + \exp\left(\frac{\hbar^2}{kTI}\right)\right]^2} .$$

(d) In the limit of high temperatures, $\frac{\beta\hbar^2}{2I} \ll 1$ or $kT \gg \frac{\hbar^2}{2I}$, so the sum can be replaced by an integral, that is,

$$z = \int_0^\infty (2x + 1) \exp\left[-\frac{\beta\hbar^2}{2I} x(x + 1)\right] dx = \frac{2I}{\hbar^2} kT ,$$

$$\langle E \rangle = -\frac{\partial}{\partial\beta} \ln z = kT .$$

Thus $C(T) = k$.

2039

At the temperature of liquid hydrogen, 20.4K, one would expect molecular H_2 to be mostly (nearly 100%) in a rotational state with zero angular momentum. In fact, if H_2 is cooled to this temperature, it is found that more than half is in a rotational state with angular momentum $\hbar$. A catalyst must be used at 20.4K to convert it to a state with zero rotational angular momentum. Explain these facts.

(*Columbia*)

Solution:

The hydrogen molecule is a system of fermions. According to Pauli's exclusion principle, its ground state electron wave function is symmetric. So if the total nuclear spin I is zero, the rotational quantum number of angular momentum must be even and the molecule is called parahydrogen. If the total nuclear quantum spin I is one, the rotational quantum number of angular momentum must be odd and it is called orthohydrogen. Since the spin I has the $2I + 1$ orientation possibilities, the ratio of the number of orthohydrogen molecules to the number of parahydrogen molecules is 3:1 at sufficiently high temperatures. As it is difficult to change the total nuclear spin when hydrogen molecules come into collision with one another, the ortho- and parahydrogen behave like two independent components. In other words, the ratio of the number of orthohydrogen molecules to that of parahydrogen molecules is quite independent of temperature. So there are more orthohydrogen molecules than parahydrogen molecules even in the liquid state. A catalyst is needed to change this.

2040

A gas of molecular hydrogen H_2, is originally in equilibrium at a temperature of 1,000 K. It is cooled to 20K so quickly that the nuclear spin states of the molecules do not change, although the translational and rotational degrees of freedom do readjust through collisions. What is the approximate internal energy per molecule in terms of temperature units K?

Note that the rotational part of the energy for a diatomic molecule is $Al(l + 1)$ where l is the rotational quantum number and $A \sim 90K$ for H_2. Vibrational motion can be neglected.

(*MIT*)

Solution:

Originally the temperature is high and the para- and orthohydrogen molecules are in equilibrium in a ratio of about 1:3. When the system is quickly cooled, for a rather long period the nuclear spin states remain the same. The ratio of parahydrogen to orthohydrogen is still 1:3. Now the para- and orthohydrogen are no longer in equilibrium but, through collisions, each component is in equilibrium by itself. At the low temperature of 20 K, $\exp(-\beta A) \sim \exp(-90/20) \ll 1$, so that each is in its ground state.

Thus $\overline{E}_{r,p} = 0, \overline{E}_{r,o} = A(1+1) \cdot 1 = 2A = 180$ K, giving

$$\overline{E}_r = \frac{1}{4}\overline{E}_{r,p} + \frac{3}{4}\overline{E}_{r,o} = 135 \text{ K} .$$

From equipartition of energy, we have

$$\overline{E}_t = \frac{3}{2}kT = 30 \text{ K} .$$

The average energy of a molecule is

$$\overline{E} = \overline{E}_t + \overline{E}_r = 165 \text{ K} .$$

2041

The graph below shows the equilibrium ratio of the number of ortho-hydrogen molecules to the number of parahydrogen molecules, as a function of the absolute temperature. The spins of the protons are parallel in orthohydrogen and antiparallel in parahydrogen.

(a) Exhibit a theoretical expression for this ratio as a function of the temperature.

(b) Calculate the value of the ratio for 100 K, corresponding to the point P on the graph. The separation of the protons in the hydrogen molecule is 0.7415Å.

(*UC, Berkeley*)

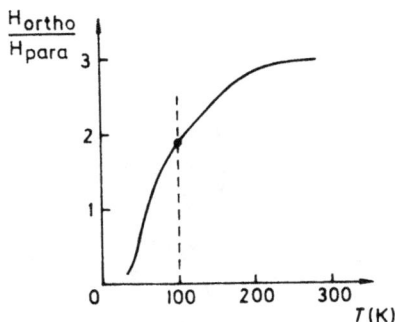

Fig. 2.11.

Solution:

(a) The moment of inertia of the hydrogen molecule is

$$I = \frac{m_H}{2} d^2 \, ,$$

and its rotational energy level is

$$E_l = \frac{L^2}{2I} = \frac{l(l+1)\hbar^2}{2I} \, ,$$

with degeneracy $(2l + 1), l = 0, 1, 2, \ldots$ .For ortho-H, $l = 1, 3, 5, \ldots$; for para-H, $l = 0, 2, 4, 6, \ldots$. Thus in hydrogen molecules, the ratio of the number of ortho-H to that of para-H is

$$f = \frac{3 \displaystyle\sum_{l=1,3,5\ldots} (2l + 1)e^{-l(l+1)\lambda}}{\displaystyle\sum_{l=0,2,4,\ldots} (2l + 1)e^{-l(l+1)\lambda}} \, ,$$

where the coefficient 3 results from spin degeneracy and

$$\lambda = \frac{\hbar^2}{md^2kT} \, .$$

(b) When $T = 100$ K, $\lambda = 0.88$, since as l increases the terms in the summations decrease rapidly, we need consider only the first two terms. Hence

$$f = 3\frac{3e^{-2\lambda} + 7e^{-12\lambda}}{1 + 5e^{-6\lambda}} = 1.52 \, .$$

2042

In hydrogen gas at low temperatures, the molecules can exist in two states: proton spins parallel (orthohydrogen) or anti-parallel (parahydrogen). The transition betwen these two molecular forms is slow. Experiments performed over a time scale of less than a few hours can be considered as if we are dealing with two separate gases, in proportions given by their statistical distributions at the last temperature at which the gas was allowed to come to equilibrium.

(a) Knowing that the separation between protons in a hydrogen molecule is 7.4×10^{-9} cm, estimate the energy difference between the ground state and the first excited rotational state of parahydrogen. Use degrees Kelvin as your unit of energy. Call this energy $k\theta_0$, so that rrors in (a) do not propagate into the other parts of the question.

(b) Express the energy difference between the ground and first excited rotational states of orthohydrogen, $k\theta_1$, in terms of $k\theta_0$. In an experiment to measure specific heats, the gas is allowed to come to equilibrium at elevated temperature, then cooled quickly to the temperature at which specific heat is measured. What will the constant-volume molar specific heat be at:

(c) temperatures well above θ_0 and θ_1, but not high enough to excite vibrational levels?

(d) temperatures much below θ_0 and θ_1 [include the leading temperature-dependent term]?

(e) $T = \theta_0/2$?

(UC, Berkeley)

Solution:

The hydrogen nucleus is a fermion. The total wave function including the motion of the nucleus is antisymmetric. The symmetry of the total wave function can be determined from the rotational and spin wave functions. For orthohydrogen, the spin wave function is symmetric when the nuclei are interchanged. Therefore, its rotational part is antisymmetric, i.e. l is odd. Similarly, for parahydrogen, l i.e. even. Then we have

$$\text{orthohydrogen: } E_l = \frac{l(l+1)\hbar^2}{2I} \;, \quad l = 1, 3, 5, \ldots$$

$$\text{parahydrogen: } E_l = \frac{l(l+1)\hbar^2}{2I} \;, \quad l = 0, 2, 4, \ldots$$

where I is the moment of inertia of the nucleus about the center of separation.

(a) $I = \dfrac{m}{2}d^2$,

$$k\theta_0 = \frac{2 \times (2+1)\,\hbar^2}{2}\frac{\hbar^2}{I} = 3\frac{\hbar^2}{I} ,$$

then $k\theta_0 = \dfrac{6\hbar}{md^2} = 7.3 \times 10^{-21}$ J, $\theta_0 = 530$ K.

(b) $k\theta_1 = \dfrac{3 \times (3+1)}{2I}\hbar^2 - \dfrac{1 \times (1+1)}{2I}\hbar^2 = \dfrac{5}{3}k\theta_0$.

As the hydrogen gas had reached thermal equilibrium at high temperature before the experiment, the ratio of the number of the para- to that of the orthohydrogen in the experiment is 1:3, which is the ratio of the degrees of freedom of the spins.

(c) When $T \gg \theta_0, \theta_1$, the rotational energy levels are completely excited. From equipartition of energy, $\overline{E} = nkT$, or $C_v = nk$, where n is the total number of the hydrogen molecules. (Note that here we only consider the specific heat associated with rotation.)

(d) When $T \ll \theta_0, \theta_1$, there are almost no hydrogen atoms in the highly excited states. Therefore, we consider only the 1st excited state for para- and orthohydrogen. Noting the degeneracy of the energy levels, we have for orthohydrogen

$$\overline{E}_o = n_o \frac{7k\theta_1 e^{-\theta_1/T}}{3 + 7e^{-\theta_1/T}} ,$$

$$C_v^{(o)} = \frac{d\overline{E}_o}{dT} = n_o k \frac{21\left(\dfrac{\theta_1}{T}\right)^2 e^{-\theta_1/T}}{(3 + 7e^{-\theta_1/T})^2} .$$

$$\approx n_o k \cdot \frac{7}{3}\left(\frac{\theta_1}{T}\right)^2 e^{-\theta_1/T} .$$

Similarly we have for parahydrogen

$$C_v^{(p)} \approx n_p k \cdot 5\left(\frac{\theta_0}{T}\right)^2 e^{-\theta_0/T} .$$

Note that

$$n_o = \frac{3}{4}n, \quad n_p = \frac{1}{4}n ,$$

then

$$C_v = nk\left[\frac{7}{4}\left(\frac{\theta_1}{T}\right)^2 e^{-\theta_1/T} + \frac{5}{4}\left(\frac{\theta_0}{T}\right)^2 e^{-\theta_0/T}\right] .$$

(e) When $T = \theta_0/2$, the partition functions for ortho- and parahydrogen are

$$z_o = \sum_{l=1,3,5,\ldots} (2l+1) \exp[-l(l+1)\lambda] \, ,$$

$$z_p = \sum_{l=0,2,4,\ldots} (2l+1) \exp[-l(l+1)\lambda]$$

where $\lambda = h^2/4\pi^2 md^2 kT$. It does not appear possible to solve these and calculate C_v accurately, but we can estimate them using the approximate results of (d).

2043

Molecular hydrogen is usually found in two forms, orthohydrogen ("parallel" nuclear spins) and parahydrogen ("anti-parallel" nuclear spins).

(a) After coming to equilibrium at "high" temperatures, what fraction of H_2 gas is parahydrogen (assuming that each variety of hydrogen is mostly in its lowest energy state)?

(b) At low temperatures orthohydrogen converts mostly to parahydrogen. Explain why the energy released by each converting molecule is much larger than the energy change due to the nuclear spin flip.

(*Wisconsin*)

Solution:

(a) For the two kinds of diatomic molecules of identical nuclei, the vibrational motion and the degeneracy of the lowest state of electron are the same for both while their rotational motions are different. The identical nuclei being fermions, antisymmetric nuclear spin states are associated with rotational states of even quantum number l, and symmetry nuclear spin states are associated with rotational states of odd quantum number l (the reverse of bosons). Thus

$$Z_{N-R} = s(2s+1)Z_{para} + (s+1)(2s+1)Z_{ortho} \, ,$$

where s is the half-integer spin of a nucleon (for the hydrogen nucleus, $s = 1/2$), $s(2s+1)$ is the number of antisymmetric spin states and $(s+1)(2s+1)$

is the number of symmetric spin states.

$$Z_{\text{para}} = \sum_{l=0,2,4}^{\infty} (2l + 1) \exp\left[-\frac{l(l + 1)\theta}{T}\right],$$

$$Z_{\text{ortho}} = \sum_{l=1,3,\ldots}^{\infty} (2l + 1) \exp\left[-\frac{l(l + 1)\theta}{T}\right],$$

where $\theta = h^2/8\pi^2 Ik$, I being the rotational moment of inertia. For high temperatures, we have $Z_{\text{para}} = Z_{\text{ortho}}$, and $n_{\text{para}}/n_{H_2} = 1/4$. According to the condition given in the problem (the temperature is not too high), only states $l = 0$ and $l = 1$ exist. The fraction of parahydrogen is then

$$\frac{n_{\text{para}}}{n_{H_2}} = \frac{Z_{\text{para}}}{Z_{N-R}} = \frac{1}{1 + 3e^{-2\theta/T}} .$$

(b) When $T \ll \theta$, orthohydrogen changes into parahydrogen. The energy corresponding to the change in nuclear spin direction is the coupling energy of the magnetic dipoles of the nuclei and the electrons $\Delta E_{SJ} \sim 10^8$ Hz. Since the rotational states are related to the nuclear spin states, the rotational states change too, the corresponding energy change being

$$\Delta E_R = \frac{h^2}{8\pi^2 I} \approx 10^{11} \text{ Hz} .$$

When orthohydrogen converts to parahydrogen, the total energy change is $\Delta E = \Delta E_R + \Delta_{SJ} \approx \Delta E_R$. Thus the released energy is much greater than ΔE_{SJ}.

2044

A $^7N_{14}$ nucleus has nuclear spin $I = 1$. Assume that the diatomic molecule N_2 can rotate but does not vibrate at ordinary temperatures and ignore electronic motion. Find the relative abundances of the ortho- and para-molecules in a sample of nitrogen gas. (Ortho = symmetric spin state; Para = antisymmetric spin state). What happens to the relative abundance as the temperature is lowered towards absolute zero? (Justify your answers!)

(SUNY, Buffalo)

Solution:

The wave function of N_2 is symmetric as $^7N_{14}$ is a boson. The spin wave functions of N_2 consist of six symmetric and three antisymmetric functions. We know that the rotating wave function is symmetric when the spin wave function is symmetric, and the rotating wavefunction is antisymmetric when the spin wave function is antisymmetric. Hence, the partition function of ortho-N_2 is

$$Z_{\text{ortho}} = \sum_{l=0,2,4,\ldots}^{\infty} 6(2l+1)e^{-\theta_r l(l+1)/T} ,$$

where $\theta_r = \dfrac{\hbar^2}{2kT}$, and I is the rotational moment of inertia of N_2. Similarly,

$$Z_{\text{para}} = \sum_{l=1,3,5,\ldots} 3(2l+1)e^{-\theta_r l(l+1)/T} .$$

As $\theta_r/T \ll 1$ at ordinary temperatures, the sums can be replaced by integrals:

$$Z_{\text{ortho}} = 3 \int_0^{\infty} e^{-\theta_r x/T} dx = \frac{3T}{\theta_r}$$

$$Z_{\text{para}} = 1.5 \int_0^{\infty} e^{-\theta_r x/T} dx = \frac{3T}{2\theta_r} .$$

Therefore, the relative abundance is given by

$$\frac{N_{\text{ortho}}}{N_{\text{para}}} = \frac{Z_{\text{ortho}} e^{\beta\mu_{\text{ortho}}}}{Z_{\text{para}} e^{\beta\mu_{\text{para}}}} .$$

At equilibrium, $\mu_{\text{ortho}} = \mu_{\text{para}}$, the above ratio is 2.

When the temperature is lowered towards the absolute zero, we have $\theta_r/T \gg 1, \exp[-\theta_r l(l+1)/T] \ll 1$. Hence

$$Z_{\text{ortho}} \approx 6 ,$$
$$Z_{\text{para}} \approx 9 \exp(-2\theta_r/T) .$$

The relative abundance is

$$\frac{N_{\text{ortho}}}{N_{\text{para}}} = \left(\frac{2}{3}\right) \exp(2\theta_r/T) .$$

When $T \rightarrow 0$, the relative abundance $\rightarrow \infty$. All the para-molecules become ortho-molecules.

2045

(a) Write down a simple expression for the internal part of the partition function for a single isolated hydrogen atom in very weak contact with a reservoir at temperature T. Does your expression diverge for $T = 0$, for $T \neq 0$?

(b) Does all or part of this divergence arise from your choice of the zero of energy?

(c) Show explicitly any effects of this divergence on calculations of the average thermal energy U.

(d) Is the divergence affected if the single atom is assumed to be confined to a box of finite volume L^3 in order to do a quantum calculation of the full partition function? Explain your answer.

(*UC, Berkeley*)

Solution:

(a) The internal energy levels of hydrogen are given by $-E_0/n^2$ with degeneracy $2n^2$, where $n = 1, 2, 3, \ldots$. Therefore

$$Z = \sum_{n=1}^{\infty} 2n^2 \exp \left(\frac{E_0}{n^2 kT} \right) .$$

When $T = 0$, the expression has no meaning; when $T \neq 0$, it diverges.

(b) The divergence has nothing to do with the choice of the zero of energy. If we had chosen

$$E = -\frac{E_0}{n^2} + E' ,$$

then

$$Z' = e^{-E'/kT} \left(\sum_{n=1}^{\infty} 2n^2 e^{E_0/n kT} \right)$$

which would still diverge.

(c) When $T \neq 0$,

$$\overline{E} = \frac{\sum\limits_{n=1}^{\infty} \left(-\frac{E_0}{n^2}\right) 2n^2 e^{E_0/n^2 kT}}{\sum\limits_{n=1}^{\infty} 2n^2 e^{E_0/n^2 kT}} = 0 \ .$$

That is to say, because of thermal excitation (no matter how low the temperature is, provided $T \neq 0$), the electrons cannot be bounded by the nuclei.

(d) The divergence has its origin in the large degeneracy of hydrogen's highly excited states. If we confine the hydrogen molecule in a box of volume L^3, these highly excited states no longer exist and there will be no divergence.

2046

The average kinetic energy of the hydrogen atoms in a certain stellar atmosphere (assumed to be in thermal equilibrium) is 1.0 eV.

(a) What is the temperature of the atmosphere in Kelvins?

(b) What is the ratio of the number of atoms in the second excited state $(n = 3)$ to the number in the ground state?

(c) Discuss qualitatively the number of ionized atoms. Is it likely to be much greater than or much less than the number in $n = 3$? Why?

(*Wisconsin*)

Solution:
(a) The temperature of the stellar atmosphere is

$$T = \frac{2\varepsilon}{3k} = \frac{2 \times 1.6 \times 10^{-19}}{3 \times 1.38 \times 10^{-23}} = 7.7 \times 10^3 \text{ K} \ .$$

(b) The energy levels for hydrogen atom are

$$E_n = \left(\frac{-13.6}{n^2}\right) \text{ eV} \ .$$

Using the Boltzmann distribution, we get

$$\frac{N_3}{N_1} = \exp\left[\frac{(E_1 - E_3)}{kT}\right] \ .$$

Inserting $E_1 = -13.6$ eV, $E_3 = (-13.6/9)$ eV, and $kT = (2/3)$ eV into the above, we have $N_3/N_1 \approx 1.33 \times 10^{-8}$.

(c) The number of ionized atoms is the difference between the total number of atoms and the total number of atoms in bound states, i.e., the number of atoms in the level $n = \infty$. Obviously, it is much smaller than the number in $n = 3$. Thus $\dfrac{N_{\text{ion}}}{N_3} = \exp\left(\dfrac{E_3}{kT}\right) \approx 0.1$, i.e., N_{ion} is about one-tenth of N_3.

2047

A monatomic gas consists of atoms with two internal energy levels: a ground state of degeneracy g_1 and a low-lying excited state of degeneracy g_2 at an energy E above the ground state. Find the specific heat of this gas.

(*CUSPEA*)

Solution:

According to the Boltzmann distribution, the average energy of the atoms is

$$\varepsilon = \frac{3}{2}kT + E_0 + \frac{g_2 E e^{-E/kT}}{g_1 + g_2 e^{-E/kT}},$$

where E_0 is the dissociation energy of the ground state (we choose the ground state as the zero point of energy). Thus

$$
\begin{aligned}
c_v = \frac{\partial \varepsilon}{\partial T} &= \frac{3}{2}k + \frac{\partial}{\partial T}\left(\frac{g_2 E e^{-E/kT}}{g_1 + g_2 e^{-E/kT}}\right) \\
&= \frac{3}{2}k + \frac{\partial}{\partial T}\left(\frac{g_2 E}{g_2 + g_1 e^{E/kT}}\right) \\
&= \frac{3}{2}k + \frac{g_1 g_2 E^2 e^{E/kT}}{kT^2\left(g_2 + g_1 e^{(E/kT)}\right)^2}.
\end{aligned}
$$

2048

Consider a system which has two orbital (single particle) states both of the same energy. When both orbitals are unoccupied, the energy of the system is zero; when one orbital or the other is occupied by one particle, the energy is ε. We suppose that the energy of the system is much higher, say infinitely high, when both orbitals are occupied. Show that the ensemble average number of particles in the level is

$$\langle N \rangle = \frac{2}{2 + e^{(\varepsilon - \mu)/\tau}} \ .$$

(*UC, Berkeley*)

Solution:

The probability that a microscopic state is occupied is proportional to its Gibbs factor $\exp[(\mu - \varepsilon)\tau]$. We thus have

$$\langle N \rangle = \frac{1 \cdot e^{(\mu - \varepsilon)/\tau} + 1 \cdot e^{(\mu - \varepsilon)/\tau}}{1 + e^{(\mu - \varepsilon)/\tau} + e^{(\mu - \varepsilon)/\tau}} = \frac{2}{e^{(\varepsilon - \mu)/\tau} + 2} \ .$$

2049

(a) State the Maxwell-Boltzmann energy distribution law. Define terms. Discuss briefly an application where the law fails.

(b) Assume the earth's atmosphere is pure nitrogen in thermodynamic equilibrium at a temperature of 300 K. Calculate the height above sea-level at which the density of the atmosphere is one half its sea-level value.

(*Wisconsin*)

Solution:

(a) The Maxwell-Boltzmann energy distribution law: For a system of gas in equilibrium, the number of particles whose coordinates are between $\mathbf{r}$ and $\mathbf{r} + d\mathbf{r}$ and whose velocities are between $\mathbf{v}$ and $\mathbf{v} + d\mathbf{v}$ is

$$dN = n_0 \left(\frac{m}{2\pi kT}\right)^{3/2} e^{-\varepsilon/kT} d\mathbf{v} d\mathbf{r} \ ,$$

where n_0 denotes the number of particles in a unit volume for which the potential energy ε_p is zero, $\varepsilon = \varepsilon_k + \varepsilon_p$ is the total energy, $d\mathbf{v} = dv_x dv_y dv_z$, $d\mathbf{r} = dxdydz$.

The MB distribution is a very general law. It is valid for localized systems, classical systems and non-degenerate quantum systems. It does not hold for degenerate non-localized quantum systems, e.g., a system of electrons of spin 1/2 at a low temperature and of high density.

(b) We choose the z-axis perpendicular to the sea level and $z = 0$ at the sea level. According to the MB distribution law, the number of molecules in volume element $dxdydz$ at height z is $dN' = n_0 e^{-mgz/kT} dxdydz$. Then the number of molecules per unit volume at height z is

$$n(z) = n_0 e^{-mgz/kT} .$$

Thus

$$z = \frac{kT}{mg} \ln \frac{n_0}{n} = \frac{RT}{\mu g} \ln \frac{n_0}{n} .$$

The molecular weight of the nitrogen gas is $\mu = 28\text{g/mol}$. With $g = 9.8\text{m/s}^2$, $R = 8.31\text{J/K·mole}$, $T = 300$ K, we find $z = 6297$ m for $n_0/n = 2$. That is, the density of the atmosphere at the height 6297m is one-half the sea level value.

2050

A circular cylinder of height L, cross-sectional area A, is filled with a gas of classical point particles whose mutual interactions can be ignored. The particles, all of mass m, are acted on by gravity (let g denote the gravitational acceleration, assumed constant). The system is maintained in thermal equilibrium at temperature T. Let c_v be the constant volume specific heat (per particle). Compute c_v as a function of T, the other parameters given, and universal parameters. Also, note especially the result for the limiting cases, $T \to 0, T \to \infty$.

(CUSPEA)

Solution:

Let z denote the height of a molecule of the gas. The average energy of the molecules is

$$e = 1.5 \, kT + mg\bar{z} ,$$

where $\bar{z}$ is the average height. According to the Boltzmann distribution, the probability density that the molecule is at height z is $p(z) \propto \exp(-mgz/kT)$. Hence

$$\bar{z} = \int_0^L z e^{-mgz/kT} dz \Big/ \int_0^L e^{-mgz/kT} dz$$

$$= \frac{kT}{mg} \left(1 - \frac{L}{e^{mgL/kT} - 1} \right) ,$$

and

$$e = \frac{5}{2}kT - \frac{mgL}{e^{mgL/kT} - 1} \,,$$

$$c_v = \frac{\partial e}{\partial T} = \frac{5}{2}k - \frac{k(mgL)^2}{(kT)^2}\frac{e^{mgL/kT}}{(e^{mgL/kT} - 1)^2}$$

$$= \begin{cases} \dfrac{5}{2}k, & \text{for } T \to 0 \,, \\ \dfrac{3}{2}k, & \text{for } T \to \infty \,. \end{cases}$$

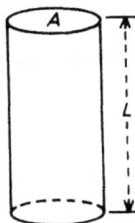

Fig. 2.12.

2051

Ideal monatomic gas is enclosed in cylinder of radius a and length L. The cylinder rotates with angular velocity ω about its symmetry axis and the ideal gas is in equilibrium at temperature T in the coordinate system rotating with the cylinder. Assume that the gas atoms have mass m, have no internal degrees of freedom, and obey classical statistics.

(a) What is the Hamiltonian in the rotating coordinates system?

(b) What is the partition function for the system?

(c) What is the average particle number density as a function of r?

(MIT)

Solution:

(a) The Hamiltonian for a single atom is

$$h' = \frac{p'^2}{2m} + \phi - \frac{1}{2}m\omega^2 r^2 \,,$$

$$\phi(r, \varphi, z) = \begin{cases} 0, & r \le a, |z| < \dfrac{L}{2} \,, \\ \infty, & \text{otherwise} \,. \end{cases}$$

The Hamiltonian for the system is

$$H' = \sum_i h'_i \ .$$

(b) The partition function is

$$z = \iint d^3p' d^3r e^{-\beta(p'^2/2m + \phi - m\omega^2 r^2/2)}$$

$$= \frac{L}{m^2\omega^2} (2\pi m kT)^{5/2} (e^{m\omega^2 a^2/2kT} - 1) \ .$$

(c) The average particle number density is

$$\Delta N/\Delta V = N \int d^3p' \exp[-\beta(p'^2/2m + \phi - m\omega^2 r^2/2)]/z$$

$$= \frac{N}{\pi L} \frac{\exp \dfrac{m\omega^2}{2kT} r^2}{\dfrac{2kT}{m\omega^2} \left(\exp \dfrac{m\omega^2 a^2}{2kT} - 1 \right)} \qquad (r < a) \ .$$

2052

Find the particle density as a function of radial position for a gas of N molecules, each of mass M, contained in a centrifuge of radius R and length L rotating with angular velocity ω about its axis. Neglect the effect of gravity and assume that the centrifuge has been rotating long enough for the gas particles to reach equilibrium.

(Chicago)

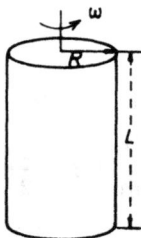

Fig. 2.13.

Solution:

In the rest system S, the energy E is independent of the radial distance r. But in the rotational system S', the energy of a particle is

$$E(r) = \frac{1}{2}I\omega^2 = \frac{1}{2}Mr^2\omega^2 .$$

The effect of rotation is the same as that of an additional external field acting on the system of

$$U(r) = -\frac{1}{2}Mr^2\omega^2 .$$

Using the Boltzmann distribution we get the particle number density

$$n(r) = A\exp\left(-\frac{U(r)}{kT}\right) = A\exp\left(\frac{M\omega^2 r^2}{2kT}\right)$$

where the normalization factor A can be determined by $\int n(rJ)dV = N$, giving

$$A = \frac{NQ}{\pi L[e^{QR^2} - 1]} , \quad \text{with} \quad Q = \frac{M\omega^2}{2kT} .$$

Thus we have

$$n(r) = \frac{NM\omega^2}{2\pi kTL} \frac{\exp\left(\dfrac{M\omega^2 r^2}{2kT}\right)}{\exp\left(\dfrac{M\omega^2 R^2}{2kT}\right) - 1} .$$

2053

Suppose that a quantity of neutral hydrogen gas is heated to a temperature T. T is sufficiently high that the hydrogen is completely ionized, but low enough that $kT/m_e c^2 \ll 1$ (m_e is the mass of the electron). In this gas, there will be a small density of positrons due to processes such as $e^- + e^- \leftrightarrow e^- + e^- + e^- + e^+$ or $e^- + p \leftrightarrow e^- + p + e^- + e^+$ in which electron-positron pairs are created and destroyed.

For this problem, you need not understand these reactions in detail. Just assume that they are reactions that change the number of electrons and positrons, but in such a way that charge is always conserved.

Suppose that the number density of protons is $10^{10}/\text{cm}^3$. Find the chemical potentials for the electrons and positrons. Find the temperature at which the positron density is $1/\text{cm}^3$. Find the temperature at which it is $10^{10}/\text{cm}^3$.

(Princeton)

Solution:

For $kT/m_e c^2 \ll 1$, nuclear reactions may be neglected. From charge conservation, we have $n_- = n_p + n_+$, where n_-, n_+ are the number densities of electrons and positrons respectively. For a non-relativistic non-degenerate case, we have

$$n_- = 2 \left(\frac{2\pi m_e kT}{h^2} \right)^{3/2} \exp\left(\frac{\mu_- - m_e c^2}{kT} \right),$$

$$n_+ = 2 \left(\frac{2\pi m_e kT}{h^2} \right)^{3/2} \exp\left(\frac{\mu_+ - m_e c^2}{kT} \right)$$

where μ_- and μ_+ are the chemical potentials of electrons and positrons respectively. From the chemical equilibrium condition, we obtain $\mu_- = -\mu_+ = \mu$. Hence

$$n_+/n_- = \exp(-2\mu/kT) \ .$$

For $n_+ = 1/\text{cm}^3$, $n_- \approx n_p = 10^{10} /\text{cm}^3$, we have $\exp(\mu/kT) = 10^5$ or $\mu/kT \approx 11.5$. Substituting these results into the expression of n_-, we have $T \approx 1.2 \times 10^8$ K, so $\mu \approx 1.6 \times 10^{-7}$ erg. For $n = 10^{10}/\text{cm}^3$, $\exp(\mu/kT) = \sqrt{2}$, $\mu/kT \approx 0.4$. Substituting these results into the expression of n_+, we get $T \approx 1.5 \times 10^8$ K, $\mu \approx 8.4 \times 10^{-9}$ erg.

2054

Consider a rigid lattice of distinguishable spin $1/2$ atoms in a magnetic field. The spins have two states, with energies $-\mu_0 H$ and $+\mu_0 H$ for spins up ($\uparrow$) and down ($\downarrow$), respectively, relative to **H**. The system is at temperature T.

(a) Determine the canonical partition function z for this system.

(b) Determine the total magnetic moment $M = \mu_0(N_+ - N_-)$ of the system.

(c) Determine the entropy of the system.

(Wisconsin)

Solution:

(a) The partition function is

$$z = \exp(x) + \exp(-x) ,$$

where $x = \mu_0 H/kT$.

(b) The total magnetic moment is

$$M = \mu_0(N_+ - N_-) = NkT\frac{\partial}{\partial H}\ln z$$
$$= N\mu_0 \tanh(x) .$$

(c) The entropy of the system is

$$S = Nk(\ln z - \beta\partial\ln z/\partial\beta)$$
$$= Nk(\ln 2 + \ln(\cosh x)) - x\tanh(x) .$$

2055

A paramagnetic system consists of N magnetic dipoles. Each dipole carries a magnetic moment μ which can be treated classically. If the system at a finite temperature T is in a uniform magnetic field H, find

(a) the induced magnetization in the system, and
(b) the heat capacity at constant H.

(UC, Berkeley)

Solution:

(a) The mean magnetic moment for a dipole is

$$\langle\mu\rangle = \frac{\int \mu\cos\theta\,\exp(x\cos\theta)d\Omega}{\int\exp(x\cos\theta)d\Omega}$$
$$= \frac{\mu\int_0^\pi \cos\theta\,\exp(x\cos\theta)\sin\theta\,d\theta}{\int_0^\pi \exp(x\cos\theta)\sin\theta\,d\theta}$$
$$= \mu\left[\coth x - \frac{1}{x}\right] ,$$

where $x = \mu H/kT$. Then the induced magnetization in the system is

$$\langle M\rangle = N\langle\mu\rangle = N\mu\left(\coth x - \frac{1}{x}\right) .$$

(b) $c = \dfrac{\partial \langle u \rangle}{\partial T} = -H \dfrac{\partial \langle M \rangle}{\partial T} = Nk(1 - x^2 \mathrm{csch}^2 x^2)$.

2056

Consider a gas of spin $1/2$ atoms with density n atoms per unit volume. Each atom has intrinsic magnetic moment μ and the interaction between atoms is negligible.

Assume that the system obeys classical statistics.

(a) What is the probability of finding an atom with μ parallel to the applied magnetic field **H** at absolute temperature T? With μ anti-parallel to **H**?

(b) Find the mean magnetization of the gas in both the high and low temperature limits?

(c) Determine the magnetic susceptibility χ in terms of μ.

(SUNY, Buffalo)

Solution:

(a) The interaction energy between an atom and the external magnetic field is $\varepsilon = -\boldsymbol{\mu} \cdot \mathbf{H}$. By classical Boltzmann distribution, the number of atoms per unit volume in the solid angle element $d\Omega$ in the direction (θ, φ), is

$$g \exp(-\beta\varepsilon)d\Omega = g \exp(\mu H \cos\theta/kT)d\Omega ,$$

where θ is the angle between $\boldsymbol{\mu}$ and $\mathbf{H}$ and g is the normalization factor given by

$$2\pi g \int_0^\pi e^{-\beta\varepsilon} \sin\theta d\theta = n ,$$

i.e.,

$$g = \frac{n\mu H}{4\pi kT \sinh \dfrac{\mu H}{kT}} .$$

Hence the probability density for the magnetic moment of an atom to be parallel to **H** is

$$\frac{g}{n} e^{\mu H/kT} = \frac{1}{4\pi} \left(\frac{\mu H}{kT}\right) e^{\mu H/kT} \Big/ \sinh\left(\frac{\mu H}{kT}\right) .$$

and that for the magnetic moment to be antiparallel to **H** is

$$\frac{g}{n} e^{-\mu H/kT} = \frac{1}{4\pi} \left(\frac{\mu H}{kT}\right) e^{-\mu H/kT} \Big/ \sinh\left(\frac{\mu H}{kT}\right) .$$

(b) The average magnetization of the gas at temperature T is

$$\overline{M} = 2\pi g \int_0^\pi e^{\mu H \cos\theta/kT} \mu \cos\theta \sin\theta \, d\theta$$

$$= n\mu \left[\coth\left(\frac{\mu H}{kT}\right) - \frac{kT}{\mu H} \right] .$$

At high temperatures, $\frac{\mu H}{kT} \ll 1$. Let $\frac{\mu H}{kT} = x$, and expand

$$\coth x - \frac{1}{x} = \frac{1}{x}\left(1 + \frac{x^2}{3} - \frac{x^4}{45} + \ldots\right) - \frac{1}{x} \approx \frac{1}{3}x .$$

so $\overline{M} \approx \frac{n\mu^2}{3kT} H$.

At low temperatures, $x \gg 1$, then

$$\coth x - \frac{1}{x} \approx 1 .$$

and $\overline{M} \approx n\mu$.

(c) The magnetic susceptibility of the system is

$$\chi(\mu) = \frac{\overline{M}}{H} \approx \begin{cases} n\mu^2/3kT, & \text{at high temperature} \\ \infty, & \text{at low temperature .} \end{cases}$$

There is spontaneous magnetization in the limit of low temperatures.

2057

A material consists of n independent particles and is in a weak external magnetic field H. Each particle can have a magnetic moment $m\mu$ along the magnetic field, where $m = J, J-1, \ldots, -J+1, -J$, J being an integer, and μ is a constant. The system is at temperature T.

(a) Find the partition function for this system.

(b) Calculate the average magnetization, $\overline{M}$, of the material.

(c) For large values of T find an asymptotic expression for $\overline{M}$.

(*MIT*)

Solution:

(a) The partition function is

$$z = \sum_{m=-J}^{J} e^{m\mu H/kT} = \sinh\left[\left(J + \frac{1}{2}\right)\mu H/kT\right] \bigg/ \sinh\left(\frac{1}{2}\mu H/kT\right) .$$

(b) The average magnetization is

$$\overline{M} = -\left(\frac{\partial F}{\partial H}\right)_T = NkT\left(\frac{\partial}{\partial H}\ln z\right)_T$$
$$= \frac{N\mu}{2}\left[(2J + 1)\coth\left[(2J + 1)\frac{\mu H}{2kT}\right] - \coth\frac{\mu H}{2kT}\right] .$$

(c) When $kT \gg \mu H$, using

$$\coth x \approx \frac{1}{x}\left(1 + \frac{x^2}{3}\right) , \quad \text{for} \quad x \ll 1$$

we get

$$\overline{M} \approx \frac{1}{3}NJ(J + 1)\frac{\mu^2 H}{kT} .$$

2058

Two dipoles, with dipole moments $\mathbf{M}_1$ and $\mathbf{M}_2$, are held apart at a separation R, only the orientations of the moments being free. They are in thermal equilibrium with the environment at temperature T. Compute the mean force $\mathbf{F}$ between the dipoles for the high temperature limit $\frac{M_1 M_2}{kT R^3} \ll 1$. The system is to be treated classically.

Remark: The potential energy between two dipoles is:

$$\phi = \frac{(3(\mathbf{M}_1 \cdot \mathbf{R})(\mathbf{M}_2 \cdot \mathbf{R}) - (\mathbf{M}_1 \cdot \mathbf{M}_2)R^2)}{R^5} .$$

(Princeton)

Solution:

Taking the z-axis along the line connecting M_1 and M_2, we have

$$\phi = \frac{M_1 M_2}{R^3}[2\cos\theta_1\cos\theta_2 - \sin\theta_1\sin\theta_2\cos(\varphi_1 - \varphi_2)] .$$

The classical partition function is

$$z = \int e^{-\beta\phi} d\Omega_1 d\Omega_2$$

$$= \int \exp\left[-\frac{\beta M_1 M_2}{R^3}\right.$$

$$\left.(2\cos\theta_1\cos\theta_2 - \sin\theta_1\sin\theta_2\cos(\varphi_1 - \varphi_2))\right] d\Omega_1 d\Omega_2 .$$

As $\lambda = \beta M_1 M_2/R^3 \ll 1$, expanding the integrand with respect to λ, retaining only the first non-zero terms, and noting that the integral of a linear term of $\cos\theta$ is zero, we have

$$z = \int \left[1 + \frac{\lambda^2}{2}(2\cos\theta_1\cos\theta_2 - \sin\theta_1\sin\theta_2\cos(\varphi_1 - \varphi_2))^2\right] d\Omega_1 d\Omega_2$$

$$= 16\pi^2 + \frac{32\pi^2}{9}\lambda^2 + \frac{4\pi^2}{9} = \frac{4\pi^2}{9}(37 + 8\lambda^2) ,$$

$$u = -\frac{1}{z}\frac{\partial z}{\partial \beta} = \frac{16\lambda}{37 + 8\lambda^2} \cdot \frac{M_1 M_2}{R^3} ,$$

$$F = -\frac{\partial u}{\partial R} = \frac{16kT}{R} \cdot \frac{74\lambda^2}{37 + 8\lambda^2} .$$

2059

The molecule of a perfect gas consists of two atoms, of mass m, rigidly separated by a distance d. The atoms of each molecule carry charges q and $-q$ respectively, and the gas is placed in an electric field ε. Find the mean polarization, and the specific heat per molecule, if quantum effects can be neglected.

State the condition for this last assumption to be true.

(*UC, Berkeley*)

Solution:

Assume that the angle between a molecular dipole and the external field is θ. The energy of a dipole in the field is

$$E = -E_0\cos\theta, \quad E_0 = dq\varepsilon .$$

Then

$$\bar{p} = \frac{\int dq \cos\theta e^{E_0 \cos\theta/kT} d\Omega}{\int e^{E_0 \cos\theta/kT} d\Omega}$$

$$= \frac{\int_0^\pi \cos\theta e^{E_0 \cos\theta/kT} \sin\theta d\theta}{\int_0^\pi e^{E_0 \cos\theta/kT} \sin\theta d\theta} dq$$

$$= dq \left[\coth\left(\frac{E_0}{kT}\right) \cdot - \frac{kT}{E_0}\right] ,$$

$$\bar{E} = -\bar{p}\varepsilon .$$

$$\chi = \frac{\partial \bar{p}}{\partial \varepsilon} = \frac{dq}{\varepsilon} \cdot \frac{kT}{E_0} \left[1 - \frac{\left(\dfrac{E_0}{kT}\right)^2}{\sinh^2\left(\dfrac{E_0}{kT}\right)} \right] ,$$

$$c = \frac{\partial \bar{E}}{\partial T} = k \left[1 - \frac{\left(\dfrac{E_0}{kT}\right)^2}{\sinh^2\left(\dfrac{E_0}{kT}\right)} \right] .$$

The condition for classical approximation to be valid is that the quantization of the rotational energy can be neglected, that is, $kT \gg \dfrac{\hbar^2}{md^2}$.

2060

The response of polar substances (e.g., HCl, H_2O, etc) to applied electric fields can be described in terms of a classical model which attributes to each molecule a permanent electric dipole moment of magnitude p.

(a) Write down a general expression for the average macroscopic polarization $\bar{p}$ (dipole moment per unit volume) for a dilute system of n molecules per unit volume at temperature T in a uniform electric field E.

(b) Calculate explicitly an approximate result for the average macroscopic polarization $\bar{p}$ at high temperatures $(KT \gg pE)$.

(*MIT*)

Solution:

(a) The energy of a dipole in electric field is

$$u_e = -\mathbf{p} \cdot E = -pE \cos\theta .$$

The partition function is then

$$z \approx \int_0^{\pi} e^{pE \cos\theta/kT} \sin\theta \, d\theta = \frac{2kT}{pE} \sinh\left(\frac{pE}{kT}\right) .$$

The polarization is

$$\bar{p} = -\left(\frac{\partial F}{\partial E}\right)_{V,T,N} = nkT \frac{\partial \ln z}{\partial E} = -\frac{nkT}{E} + np \coth \frac{pE}{kT} .$$

(b) Under the condition $x = \dfrac{pE}{kT} \ll 1, \coth x \approx \dfrac{1}{3}x + \dfrac{1}{x}$, and we have

$$\bar{p} \approx np^2 E/3kT .$$

2061

The entropy of an ideal paramagnet in a magnetic field is given approximately by

$$S = S_0 - CU^2 ,$$

where U is the energy of the spin system and C is a constant with fixed mechanical parameters of the system.

(a) Using the fundamental definition of the temperature, determine the energy U of the spin system as a function of T.

(b) Sketch a graph of U versus T for all values of T $(-\infty < T < \infty)$.

(c) Briefly tell what physical sense you can make of the negative temperature part of your result.

(*Wisconsin*)

Solution:

(a) From the definition of temperature,

$$T = \left(\frac{\partial U}{\partial S}\right)_V = -\frac{1}{2CU} ,$$

we have $U = -\dfrac{1}{2CT}.$

(b) We assume $C > 0$. The change of U with T is shown in the Fig. 2.14.

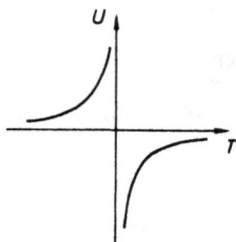

Fig. 2.14.

(c) Under normal conditions, the number of particles in higher energy states is smaller than that in lower energy states. The physical significance of a negative temperature is that under such condition the number of particles in an excited state is greater than that in the ground state. That is, there are more particles with magnetic moments anti-parallel to the magnetic field than those with magnetic moments parallel to the magnetic field.

2062

Consider a system of N non-interacting particles $(N \gg 1)$ in which the energy of each particle can assume two and only two distinct values, 0 and E $(E > 0)$. Denote by n_0 and n_1 the occupation numbers of the energy levels 0 and E, respectively. The fixed total energy of the system is U.

(a) Find the entropy of the system.

(b) Find the temperature as a function of U. For what range of values of n_0 is $T < 0$?

(c) In which direction does heat flow when a system of negative temperature is brought into thermal contact with a system of positive temperature? Why?

(*Princeton*)

Solution:

(a) The number of states is

$$\Omega = \frac{N!}{n_0! n_1!} \ .$$

Hence $S = k \ln \Omega = k \ln \dfrac{N!}{n_0! n_1!}$.

(b) $n_1/n_0 = \exp(-E/kT)$, where we have assumed the energy levels to be nondegenerate. Thus

$$T = \frac{E}{k} \cdot \frac{1}{\ln \dfrac{n_0}{n_1}} = \frac{E}{k} \cdot \frac{1}{\ln \left(\dfrac{NE - U}{U} \right)} .$$

When $n_0 < N/2$, we get $T < 0$.

(c) Heat will flow from a negative temperature system to a positive temperature system. This is because the negative temperature system has higher energy on account of population inversion, i.e., it has more particles in higher energy states than in lower energy states.

3. BOSE-EINSTEIN AND FERMI-DIRAC STATISTICS
(2063-2115)

2063

A system of N identical spinless bosons of mass m is in a box of volume $V = L^3$ at temperature $T > 0$.

(a) Write a general expression for the number of particles, $n(E)$, having an energy between ε and $\varepsilon + d\varepsilon$ in terms of their mass, the energy, the temperature, the chemical potential, the volume, and any other relevant quantities.

(b) Show that in the limit that the average distance, d, between the particles is very large compared to their de Broglie wavelength (i.e., $d \gg \lambda$) the distribution becomes equal to that calculated using the classical (Boltzmann) distribution function.

(c) Calculate the 1st order difference in average energy between a system of N non-identical spinless particles and a system of N identical spinless bosons when $d \gg \lambda$. For both systems the cubical box has volume $V = L^3$ and the particles have mass m.

(UC, Berkeley)

Solution:

(a) The number of particles is

$$n(\varepsilon) = \frac{2\pi V (2m)^{3/2}}{h^3} \cdot \frac{\sqrt{\varepsilon}}{e^{(\varepsilon-\mu)/kT} - 1} d\varepsilon \ .$$

(b) In the approximation of a dilute gas, we have $\exp(-\mu/kT) \gg 1$, and the Bose-Einstein distribution becomes the Boltzmann distribution. We will prove as follows that this limiting condition is just $d \gg \lambda$.

Since

$$\begin{aligned}
N &= \frac{2\pi V (2m)^{3/2}}{h^3} \cdot \int_0^\infty \sqrt{\varepsilon}\, e^{-\frac{\varepsilon}{kT}} e^{\frac{\mu}{kT}}\, d\varepsilon \\
&= V \left(\frac{2\pi m kT}{h^2} \right)^{3/2} e^{\frac{\mu}{kT}} \ ,
\end{aligned}$$

we have

$$e^{-\frac{\mu}{kT}} = \frac{V}{N} \cdot \frac{1}{\lambda^3} = \left(\frac{d}{\lambda} \right)^3 \ ,$$

where $\lambda = h/\sqrt{2\pi m kT}$ is the de Broglie wavelength of the particle's thermal motion, and $d = \sqrt[3]{V/N}$.

Thus the approximation $\exp(-\mu/kT) \gg 1$ is equivalent to $d \gg \lambda$.

(c) In the 1st order approximation

$$\frac{1}{e^{(\varepsilon-\mu)/kT} - 1} \approx e^{-(\varepsilon-\mu)/kT}\left(1 + e^{-(\varepsilon-\mu)/kT}\right) \ ,$$

the average energy is

$$\begin{aligned}
\overline{E} = \frac{2\pi V (2m)^{3/2}}{h^3} &\left[\int_0^\infty \varepsilon \sqrt{\varepsilon}\, e^{\mu/kT} e^{-\varepsilon/kT}\, d\varepsilon \right. \\
&\left. + \int_0^\infty \varepsilon \sqrt{\varepsilon}\, e^{2\mu/kT} e^{-2\varepsilon/kT}\, d\varepsilon \right] = \frac{3}{2} N kT \left(1 + \frac{1}{4\sqrt{2}} \frac{\lambda^3}{d^3} \right) \ .
\end{aligned}$$

2064

Consider a quantum-mechanical gas of non-interacting spin zero bosons, each of mass m which are free to move within volume V.

(a) Find the energy and heat capacity in the very low temperature region. Discuss why it is appropriate at low temperatures to put the chemical potential equal to zero.

(b) Show how the calculation is modified for a photon (mass $= 0$) gas. Prove that the energy is proportional to T^4.
Note: Put all integrals in dimensionless form, but do not evaluate.

(*UC, Berkeley*)

Solution:

(a) The Bose distribution

$$\frac{1}{e^{(\varepsilon-\mu)/kT} - 1}$$

requires that $\mu \leq 0$. Generally

$$n = \int \frac{1}{e^{(\varepsilon-\mu)/kT} - 1} \cdot \frac{2\pi}{h^3} (2m)^{3/2} \sqrt{\varepsilon} d\varepsilon .$$

When T decreases, the chemical potential μ increases until $\mu = 0$, for which

$$n = \int \frac{1}{e^{\varepsilon/kT} - 1} \frac{2\pi}{h^3} (2m)^{3/2} \sqrt{\varepsilon} d\varepsilon .$$

Bose condensation occurs when the temperature continues to decrease with $\mu = 0$. Therefore, in the limit of very low temperatures, the Bose system can be regarded as having $\mu = 0$. The number of particles at the non-condensed state is not conserved. The energy density u and specific heat c are thus obtained as follows:

$$u = \int \frac{\varepsilon}{e^{\varepsilon/kT} - 1} \cdot \frac{2\pi}{h^3} (2m)^{3/2} \sqrt{\varepsilon} d\varepsilon$$

$$= \frac{2\pi}{h^3} (2m)^{3/2} (kT)^{3/2} \int_0^\infty \frac{x^{3/2}}{e^x - 1} dx .$$

$$c = 5\pi k \left(\frac{2mkT}{h^2} \right)^{3/2} \int_0^\infty \frac{x^{3/2}}{e^x - 1} dx .$$

(b) For a photon gas, we have $\mu = 0$ at any temperature and $\varepsilon = \hbar\omega$. The density of states is $\dfrac{\omega^2 d\omega}{\pi^2 c^3}$, and the energy density is

$$u = \frac{1}{\pi^2 c^3} \int \frac{\hbar\omega^3}{e^{\hbar\omega/kT} - 1} d\omega = \frac{\hbar}{\pi^2 c^3} \left(\frac{kT}{\hbar} \right)^4 \int_0^\infty \frac{x^3 dx}{e^x - 1} .$$

2065

A gas of N spinless Bose particles of mass m is enclosed in a volume V at a temperature T.

(a) Find an expression for the density of single-particle states $D(\varepsilon)$ as a function of the single-particle energy ε. Sketch the result.

(b) Write down an expression for the mean occupation number of a single particle state, $\bar{n}_\varepsilon$ as a function of ε, T, and the chemical potential $\mu(T)$. Draw this function on your sketch in part (a) for a moderately high temperature, that is, a temperature above the Bose-Einstein transition. Indicate the place on the ε-axis where $\varepsilon = \mu$.

(c) Write down an integral expression which implicitly determines $\mu(T)$. Referring to your sketch in (a), determine in which direction $\mu(T)$ moves as T is lowered.

(d) Find an expression for the Bose-Einstein transition temperature, T_c, below which one must have a macroscopic occupation of some single-particle states. Leave your answer in terms of a dimensionless integral.

(e) What is $\mu(T)$ for $T < T_c$?

Describe $\bar{n}(\varepsilon, T)$ for $T < T_c$?

(f) Find an exact expression for the total energy, $U(T, V)$ of the gas for $T < T_c$. Leave your answer in terms of a dimensionless integral.

(*MIT*)

Solution:

(a) From $\varepsilon = p^2/2m$ and

$$D(\varepsilon)d\varepsilon = \frac{4\pi V}{h^3} p^2 dp$$

we find

$$D(\varepsilon) = \frac{2\pi V}{h^3} (2m)^{3/2} \varepsilon^{1/2} .$$

The result is shown in Fig. 2.15

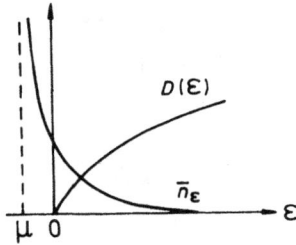

Fig. 2.15.

(b) $\bar{n}_\epsilon = \dfrac{1}{e^{(\epsilon-\mu)/kT} - 1}$ $\quad (\mu \leq 0)$.

(c) With $\epsilon = \dfrac{p^2}{2m}$ we have $N = \dfrac{4\pi V}{(2\pi\hbar)^3} \displaystyle\int_0^\infty p^2 \bar{n}_\epsilon \, dp$

$$= \frac{2\pi(2m)^{3/2}V}{(2\pi\hbar)^3} \int_0^\infty \epsilon^{1/2} \frac{d\epsilon}{e^{(\epsilon-\mu)/kT} - 1} \, ,$$

or $\qquad N/V = \dfrac{2\pi}{h^3}(2mkT)^{3/2} \displaystyle\int_0^\infty x^{1/2} \frac{dx}{e^{x-x_\mu} - 1} \, ,$

where $x_\mu = \mu/kT \leq 0$. As N/V remains unchanged when T decreases, $\mu(T)$ increases and approaches zero.

(d) Let n be the number density and T_c the critical temperature. Note that at temperature T_c the chemical potential μ is near to zero and the particle number of the ground state is still near to zero, so that we have

$$n = \frac{2\pi}{h^3}(2m)^{3/2} \int_0^\infty \frac{\epsilon^{1/2} d\epsilon}{e^{\epsilon/kT} - 1}$$

$$= \frac{2\pi}{h^3}(2mkT_c)^{3/2} \int_0^\infty \frac{x^{1/2}dx}{e^x - 1} \, ,$$

where the integral

$$A = \int_0^\infty \frac{x^{1/2}dx}{e^x - 1} = 1.306\sqrt{\pi} \, .$$

Hence

$$T_c = \frac{h^2}{2mk}\left(\frac{n}{2\pi A}\right)^{2/3} \, .$$

(e) For bosons, $\mu < 0$. When $T \leq T_c$, $\mu \approx 0$ and we have

$$\bar{n}_{\epsilon>0} = \frac{1}{\exp\left(\frac{\epsilon}{kT}\right) - 1} \, ,$$

and

$$\bar{n}_{s=0} = n\left[1 - \frac{2\pi}{h^3}(2mkT)^{3/2}\int_0^\infty x^{1/2}\frac{dx}{e^x - 1}\right]$$

$$= n\left[1 - \left(\frac{T}{T_c}\right)^{3/2}\right] .$$

(f) When $T < T_c$, we have

$$U = \frac{2\pi V}{h^3}(2m)^{3/2}(kT)^{5/2}\int_0^\infty \frac{x^{3/2}dx}{e^x - 1}$$

$$= 0.770NkT\left(\frac{T}{T_c}\right)^{3/2} .$$

2066

(a) In quantum statistical mechanics, define the one-particle density matrix in the r-representation where **r** is the position of the particle.

(b) For a system of N identical free bosons, let

$$\rho_1(\mathbf{r}) = \frac{1}{V}\sum_{\mathbf{k}}\langle N_k\rangle e^{i\mathbf{k}\cdot\mathbf{r}} ,$$

where $\langle N_k\rangle$ is the thermal averaged number of particles in the momentum state **k**. Discuss the limiting behavior of $\rho_1(\mathbf{r})$ as $r \to \infty$, when the temperature T passes from $T > T_c$ to $T < T_c$, where T_c is the Bose-Einstein condensation temperature. In the case $\lim_{r\to\infty}\rho_1(\mathbf{r})$ approaches zero, can you describe how it approaches zero as r becomes larger and larger?

(*SUNY, Buffalo*)

Solution:

(a) The one-particle Hamiltonian is $H = p^2/2m$, and the energy eigenstates are $|E\rangle$. The density matrix in the energy representation is then $\rho(E) = \exp(-E/k_BT)$, which can be transformed to the coordinate representation

$$\langle\mathbf{r}|\rho|\mathbf{r'}\rangle = \sum_{E,E'}\langle\mathbf{r}|E\rangle\langle E|e^{-H/k_BT}|E'\rangle\langle E'|\mathbf{r'}\rangle$$

$$= \sum_{E,E'}\varphi_E(\mathbf{r})e^{-E/k_BT}\delta_{EE'}\varphi_{E'}^*(\mathbf{r})$$

$$= \sum_E \varphi_E(\mathbf{r})e^{-E/k_BT}\varphi_E^*(\mathbf{r'}) .$$

where k_B is Boltzmann's constant. The stationary one-particle wavefunction is

$$\varphi_E(\mathbf{r}) = \frac{1}{\sqrt{V}} e^{i\mathbf{k}\cdot\mathbf{r} - iEt} ,$$

where $E = \hbar^2 k^2 / 2m$. Thus we obtain

$$\langle \mathbf{r}|\rho|\mathbf{r}'\rangle = \frac{1}{V} \sum_{\mathbf{k}} e^{i\mathbf{k}\cdot(\mathbf{r}-\mathbf{r}') - h^2 k^2 / 8\pi^2 m k_B T}$$

$$= \frac{1}{(2\pi)^3} \int d^3 k\, e^{i\mathbf{k}\cdot(\mathbf{r}-\mathbf{r}') - h^2 k^2 / 8\pi^2 m k_B T}$$

$$= \left(\frac{mk_B T}{2\pi\hbar^2}\right)^{3/2} e^{-mk_B T (\mathbf{r}-\mathbf{r}')^2 2\pi^2 / h^2} ,$$

(b) For free bosons, we have

$$\langle N_k \rangle = [\exp(\hbar^2 k^2 / 2m - \mu)/k_B T) - 1]^{-1} .$$

So

$$\rho_1(\mathbf{r}) = \frac{1}{V} \sum_{\mathbf{k}} e^{i\mathbf{k}\cdot\mathbf{r}} [e^{(h^2 k^2 / 8\pi^2 m - \mu)/k_B T} - 1]^{-1}$$

$$= \frac{1}{(2\pi)^3} \int d^3 k\, e^{i\mathbf{k}\cdot\mathbf{r}} [e^{(h^2 k^2 / 8\pi^2 m - \mu)/k_B T} - 1]^{-1} ,$$

$$= \frac{2}{(2\pi)^2} \frac{1}{r} \int_0^\infty dk \cdot k \sin kr [e^{(h^2 k^2 / 8\pi^2 m - \mu) k_B T} - 1]^{-1} .$$

$\mu = 0$ when the temperature T passes from $T > T_c$ to $T < T_c$, hence

$$\rho_1(\mathbf{r}) = \frac{2}{(2\pi)^2} \frac{1}{r} \int_0^\infty dk k \sin kr [e^{h^2 k^2 / 8\pi^2 m k_B T_c} - 1]^{-1} .$$

When $r \to \infty$, we have approximately

$$\rho(r) \approx \frac{2}{(2\pi)^2} \frac{1}{r^3} \int_0^{\sqrt{2mk_B T_c} r/\hbar} dx \cdot x \sin x \left[e^{h^2 x^2 / 8\pi^2 m k_B T_c r^2} - 1\right]^{-1}$$

$$\approx \frac{2}{(2\pi)^2} \frac{1}{r} \left(\int^{\sqrt{2mk_B T_c} r/\hbar} dx \frac{\sin x}{x}\right) \frac{2mk_B T_c}{\hbar^2}$$

$$\approx \frac{4mk_B T_c}{(2\pi)^2 \hbar^2 r} \int_0^\infty dx \frac{\sin x}{x}$$

$$\approx \frac{mk_B T_c}{2\pi\hbar^2} \frac{1}{r} .$$

2067

Consider a gas of non-interacting, non-relativistic, identical bosons. Explain whether and why the Bose-Einstein condensation effect that applies to a three-dimensional gas applies also to a two-dimensional gas and to a one-dimensional gas.

(Princeton)

Solution:

Briefly speaking, the Bose-Einstein condensation occurs when $\mu = 0$. For a two-dimensional gas, we have

$$
\begin{aligned}
N &= \frac{2\pi m A}{h^2} \int_0^\infty \frac{d\varepsilon}{e^{(\varepsilon-\mu)/kT} - 1} \\
&= \frac{2\pi m A}{h^2} \int_0^\infty \left(\sum_{l=1}^\infty e^{-l(\varepsilon-\mu)/kT} \right) d\varepsilon \\
&= \frac{2\pi m A}{h^2} kT \sum_{l=1}^\infty \frac{1}{l} e^{l\mu/kT} .
\end{aligned}
$$

If $\mu = 0$, the above expression diverges. Hence $\mu \neq 0$ and Bose-Einstein condensation does not occur.

For a one-dimensional gas, we have

$$
N = \frac{\sqrt{2mL}}{2h} \int_0^\infty \frac{d\varepsilon}{\sqrt{\varepsilon}(e^{(\varepsilon-\mu)/kT} - 1)} .
$$

If $\mu = 0$, the integral diverges. Again, Bose-Einstein condensation does not occur.

2068

Consider a photon gas enclosed in a volume V and in equilibrium at temperature T. The photon is a massless particle, so that $\varepsilon = pc$.

(a) What is the chemical potential of the gas? Explain.

(b) Determine how the number of photons in the volume depends upon the temperature.

(c) One may write the energy density in the form

$$
\frac{E}{V} = \int_0^\infty \rho(\omega) \, d\omega .
$$

Determine the form of $\rho(\omega)$, the spectral density of the energy.

(d) What is the temperature dependence of the energy $\overline{E}$?

(UC, Berkeley)

Solution:

(a) The chemical potential of the photon gas is zero. Since the number of photons is not conserved at a given temperature and volume, the average photon number is determined by the expression $\left(\dfrac{\partial F}{\partial \overline{N}}\right)_{T,V} = 0$, then

$$\mu = \left(\frac{\partial F}{\partial \overline{N}}\right)_{T,V} = 0 \ .$$

(b) The density of states is $8\pi V p^2 dp/h^3$, or $V\omega^2 d\omega/\pi^2 c^3$. Then the number of photons is

$$\overline{N} = \int \frac{V}{\pi^2 c^3} \omega^2 \frac{1}{e^{\hbar\omega/kT} - 1} d\omega$$

$$= \frac{V}{\pi^2 c^3} \left(\frac{kT}{\hbar}\right)^3 \int_0^\infty \frac{\alpha^2 d\alpha}{e^\alpha - 1} \propto T^3 \ .$$

(c), (d) $\dfrac{\overline{E}}{V} = \displaystyle\int \frac{\omega^2}{\pi^2 c^3} \frac{\hbar\omega}{e^{\hbar\omega/kT} - 1} d\omega$

$$= \frac{(kT)^4}{\pi^2 c^3 \hbar^3} \int \frac{\xi^3 d\xi}{e^\xi - 1} \ .$$

Hence

$$\rho(\omega) = \frac{\hbar}{\pi^2 c^3} \frac{\omega^3}{e^{\hbar\omega/kT} - 1} \ ,$$

and $\overline{E} \propto T^4$.

2069

(a) Show that for a photon gas $p = U/3V$.

(b) Using thermodynamic arguments (First and Second Laws), and the above relationship between pressure and energy density, obtain the dependence of the energy density on the temperature in a photon gas.

(UC, Berkeley)

Solution:

(a) The density of states is

$$D(\varepsilon)d\varepsilon = \alpha V \varepsilon^2 d\varepsilon ,$$

where α is a constant.

With

$$\ln \Xi = -\int D(\varepsilon) \ln(1 - e^{-\beta\varepsilon})d\varepsilon ,$$

$$p = \frac{1}{\beta}\frac{\partial}{\partial V} \ln \Xi$$

$$= -\frac{\alpha}{\beta}\int \varepsilon^2 \ln(1 - e^{-\beta\varepsilon})d\varepsilon ,$$

we have

$$p = \frac{1}{3V}\int_0^\infty V\alpha\varepsilon^2 \frac{\varepsilon}{e^{\beta\varepsilon} - 1}d\varepsilon = \frac{U}{3V} .$$

(b) For thermal radiation, we have

$$U(T,V) = u(T)V .$$

Using the following formula of thermodynamics

$$\left(\frac{\partial U}{\partial V}\right)_T = T\left(\frac{\partial p}{\partial T}\right)_V - p$$

we get $u = \dfrac{T}{3}\dfrac{du}{dT} - \dfrac{u}{3}$, i.e. $u = \gamma T^4$, where γ is a constant.

2070

Consider a cubical box of side L with no matter in its interior. The walls are fixed at absolute temperature T, and they are in thermal equilibrium with the electromagnetic radiation field in the interior.

(a) Find the mean electromagnetic energy per unit volume in the frequency range from ω to $\omega + d\omega$ as a function of ω and T. (If you wish to start with a known distribution function – e.g., Maxwell-Boltzmann, Planck, etc. – you need not derive that function.)

(b) Find the temperature dependence of the total electromagnetic energy per unit volume. (Hint: you do not have to actually carry out the integration of the result of part (a) to answer this question.)

(SUNY, Buffalo)

Solution:

(a) The mean electromagnetic energy in the momentum interval $p \rightarrow p + dp$ is given by

$$dE_p = 2 \cdot \frac{V}{(2\pi\hbar)^3} \cdot \frac{4\pi p^2 dp\hbar\omega}{e^{\hbar\omega/2\pi kT} - 1} \, ,$$

where the factor 2 corresponds to the two polarizations of electromagnetic waves and $V = L^3$.

Making use of $p = \hbar\omega/c$, we obtain the mean electromagnetic energy in the frequency interval $\omega \rightarrow \omega + d\omega$:

$$dE_\omega = \frac{V\hbar}{\pi^2 c^3} \frac{\omega^3 d\omega}{e^{\hbar\omega/2\pi kT} - 1} \, .$$

The corresponding energy density is

$$du_\omega = \frac{\hbar}{\pi^2 c^3} \frac{\omega^3 d\omega}{e^{\hbar\omega/2\pi kT} - 1} \, .$$

(b) The total electromagnetic energy per unit volume is

$$u = \int_0^\infty du_\omega = \frac{(kT)^4}{\pi^2(\hbar c)^3} \int_0^\infty \frac{x^3 dx}{e^x - 1} \, .$$

Thus $u \propto T^4$.

2071

A historic failure of classical physics is its description of the electromagnetic radiation from a black body. Consider a simple model for an ideal black body consisting of a cubic cavity of side L with a small hole in one side.

(a) Assuming the classical equipartition of energy, derive an expression for the average energy per unit volume and unit frequency range (Rayleigh-Jeans' Law). In what way does this result deviate from actual observation?

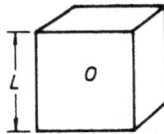

Fig. 2.16.

(b) Repeat the calculation, now using quantum ideas, to obtain an expression that properly accounts for the observed spectral distribution (Planck's Law).

(c) Find the temperature dependence of the total power emitted from the hole.

(*CUSPEA*)

Solution:

(a) For a set of three positive integers (n_1, n_2, n_3), the electromagnetic field at thermal equilibrium in the cavity has two modes of oscillation with the frequency $\nu(n_1, n_2, n_3) = \dfrac{c}{2L}(n_1^2 + n_2^2 + n_3^2)^{1/2}$. Therefore, the number of modes within the frequency interval $\Delta\nu$ is

$$\left(\frac{4\pi}{8}\nu^2\Delta\nu\right)\left(\frac{2L}{c}\right)^2 \cdot 2 \; .$$

Equipartition of energy then gives an energy density

$$u_\nu = \frac{1}{L^3}\frac{dE}{d\nu} = \frac{1}{L^3} \cdot \frac{kT \cdot \dfrac{4\pi}{8}\nu^2\Delta\nu \cdot \left(\dfrac{2L}{c}\right)^2 \cdot 2}{\Delta\nu}$$
$$= 8\pi\nu^2 kT/c^3 \; .$$

When ν is very large, this expression does not agree with experimental observations since it implies $u_\nu \propto \nu^2$.

(b) For oscillations of freqeuncy ν, the average energy is

$$\frac{\displaystyle\sum_{n=0}^{\infty} nh\nu e^{-nh\nu/kT}}{\displaystyle\sum_{n=0}^{\infty} e^{-nh\nu/kT}} = -\frac{\partial}{\partial\beta}\ln\sum_{n=0}^{\infty} e^{-\beta nh\nu}\Big|_{\beta=\frac{1}{kT}}$$

$$= h\nu e^{-h\nu\beta}/(1 - e^{-h\nu\beta}) = \frac{h\nu}{e^{h\nu\beta} - 1} \; ,$$

which is to replace the classical quantity kT to give

$$u_\nu = \frac{8\pi\nu^2}{c^3} \cdot \frac{h\nu}{e^{h\nu\beta} - 1} \; .$$

(c) The energy radiated from the hole per unit time is

$$u \propto \int_0^\infty u_\nu d\nu \propto T^4 .$$

2072

Electromagnetic radiation following the Planck distribution fills a cavity of volume V. Initially ω_i is the frequency of the maximum of the curve of $u_i(\omega)$, the energy density per unit angular frequency versus ω. If the volume is expanded quasistatically to $2V$, what is the final peak frequency ω_f of the $u_f(\omega)$ distribution curve? The expansion is adiabatic.

(*UC, Berkeley*)

Solution:

As the Planck distribution is given by $1/[\exp(\hbar\omega/kT) - 1]$ and the density of states of a photon gas is

$$D(\omega)d\omega = a\omega^2 d\omega \quad (a = \text{const}) ,$$

the angular frequency ω which makes $u(\omega)$ extremum is $\omega = \gamma T$, where γ is a constant. On the other hand, from $dU = TdS - pdV$ and $U = 3pV$, we obtain $V^4 p^3 = \text{const}$ when $dS = 0$. Since $p \propto T^4$, we have

$$VT^3 = \text{const.},$$

$$T_f = \left(\frac{V_i}{V_f}\right)^{1/3} T_i = \frac{T_i}{\sqrt[3]{2}} ,$$

$$\omega_f = \frac{\omega_i}{\sqrt[3]{2}} .$$

2073

A He-Ne laser generates a quasi-monochromatic beam at 6328Å. The beam has an output power of 1mw $(10^{-3}$ watts), a divergence angle of 10^{-4} radians, and a spectral linewidth of 0.01Å. If a black body with an area of 1 cm^2 were used to generate such a beam after proper filtering, what should its temperature be approximately?

(*UC, Berkeley*)

Solution:

Considering black body radiation in a cavity we get the number density of photons in the interval $d\varepsilon d\Omega$:

$$dn = \frac{2}{h^3} \frac{1}{e^{\varepsilon/kT} - 1} \cdot \frac{\varepsilon^2}{c^3} d\varepsilon d\Omega .$$

The number of photons in the laser beam flowing through an area A per unit time is $dn' = cAdn$, and the output power is $W = \varepsilon dn'$.

Introducing $\varepsilon = hc/\lambda$ and $d\Omega = \pi(d\theta)^2$ into the expression, we obtain

$$W = W_0 \frac{1}{e^{hc/\lambda kT} - 1} ,$$

where

$$W_0 = \frac{2\pi Ahc^2 d\lambda (d\theta)^2}{\lambda^5} .$$

Therefore

$$T = \frac{hc}{\lambda k} \cdot \frac{1}{\ln\left(\dfrac{W_0}{W} + 1\right)} .$$

Using the known quantities, we get

$$W_0 = 3.60 \times 10^{-9} \text{ W} , \quad T = 6 \times 10^9 \text{ K} .$$

2074

(a) Show that the number of photons in equilibrium at temperature T in a cavity of volume V is $N = V(kT/\hbar c)^3$ times a numerical constant.

(b) Use this result to obtain a qualitative expression for the heat capacity of a photon gas at constant volume.

(UC, Berkeley)

Solution:

(a) The density of states of the photon gas is given by

$$dg = \frac{V}{\pi^2 \hbar^3 c^3} \varepsilon^2 d\varepsilon .$$

Thus

$$N = \int \frac{V}{\pi^2 \hbar^3 c^3} \varepsilon^2 \frac{1}{e^{\varepsilon \beta} - 1} d\varepsilon$$

$$= V \left(\frac{kT}{\hbar c} \right)^2 \cdot \alpha \, ,$$

where

$$\beta = \frac{1}{kT}, \quad \alpha = \frac{1}{\pi^2} \int_0^\infty \frac{\lambda^2}{e^\lambda - 1} d\lambda \, .$$

(b) The energy density is

$$u = \int \frac{V}{\pi^2 \hbar^3 c^3} \varepsilon^2 \frac{\varepsilon}{c^{\varepsilon \beta} - 1} d\varepsilon$$

$$= kTV \left(\frac{kT}{\hbar c} \right)^3 \cdot \frac{1}{\pi^2} \int_0^\infty \frac{\lambda^3 d\lambda}{e^\lambda - 1} \, ,$$

therefore $C_v \propto T^3$.

2075

As you know, the universe is pervaded by 3K black body radiation. In a simple view, this radiation arose from the adiabatic expansion of a much hotter photon cloud which was produced during the big bang.

(a) Why is the recent expansion adiabatic rather than, for example, isothermal?

(b) If in the next 10^{10} years the volume of the universe increases by a factor of two, what then will be the temperature of the black body radiation? (Show your work.)

(c) Write down an integral which determines how much energy per cubic meter is contained in this cloud of radiation. Estimate the result within an order of magnitude in joules per (meter)^3.

(*Chicago*)

Solution:
(a) The photon cloud is an isolated system, so its expansion is adiabatic.

(b) Th energy density of black body radiation is $u = aT^4$, so that the total energy $E \propto VT^4$. From the formula $TdS = dE + pdV$, we have

$$T\left(\frac{\partial S}{\partial T}\right)_V = \left(\frac{\partial E}{\partial T}\right)_V \propto VT^3 \ .$$

Hence $S = VT^3 \cdot$ const.

For a reversible adiabatic expansion, the entropy S remains unchanged. Thus when V doubles T will decrease by a factor $(2)^{-1/3}$. So after another 10^{10} years, the temperature of black body radiation will become $T = 3\text{K}/2^{1/3}$.

(c) The black body radiation obeys the Bose-Einstein Statistics:

$$\frac{E}{V} = 2 \int \frac{d^3p}{h^3} pc \frac{1}{e^{\beta pc} - 1} = \frac{8\pi c}{h^3} \frac{1}{(\beta c)^4} \int_0^\infty \frac{x^2 dx}{e^x - 1} \ ,$$

where the factor 2 is the number of polarizations per state. Hence

$$\frac{E}{V} = \frac{8\pi^5}{15} \frac{(kT)^4}{(hc)^3} = 10^{-14} \ \text{J/m}^3 \ .$$

2076

Our universe is filled with black body radiation (photons) at a temperature $T = 3$ K. This is thought to be a relic, of early developments following the "big bang".

(a) Express the photon number density n analytically in terms of T and universal constants. Your answer should explicitly show the dependence on T and on the universal constants. However, a certain numerical cofactor may be left in the form of a dimensionless integral which need not be evaluated at this stage.

(b) Now estimate the integral roughly, use your knowledge of the universal constants, and determine n roughly, to within about two orders of magnitude, for $T = 3$ K.

(*CUSPEA*)

Solution:

(a) The Bose distribution is given by

$$n(k) = 1/[\exp(\beta\varepsilon(k)) - 1] \ .$$

The total number of photons is then

$$N = 2 \cdot V \int \frac{d^3k}{(2\pi)^3} \frac{1}{e^{\beta\hbar ck/2\pi} - 1} \; ,$$

where $\varepsilon(k) = \hbar kc$ for photons and $\beta = \frac{1}{k_B T}$. The factor 2 is due to the two directions of polarization. Thus

$$n = \frac{N}{V} = \frac{1}{\pi^2} \left(\frac{k_B T}{\hbar c} \right)^3 \cdot I \; ,$$

where

$$
\begin{aligned}
I &= \int_0^\infty dx \frac{x^2}{e^x - 1} \\
&= \sum_{n=1}^\infty \int_0^\infty dx \cdot x^2 e^{-nx} = 2 \sum_{n=1}^\infty \frac{1}{n^3} \approx 2.4 \; .
\end{aligned}
$$

(b) When $T = 3$ K, $n \approx 1000/\text{cm}^3$.

2077

We are surrounded by black body photon radiation at 3K. Consider the question of whether a similar bath of thermal neutrinos might exist.

(a) What kinds of laboratory experiments put the best limits on how hot a neutrino gas might be? How good are these limits?

(b) The photon gas makes up 10^{-6} of the energy density needed to close the universe. Assuming the universe is no more than just closed, what order of magnitude limit does this consideration place on the neutrino's temperature?

(c) In a standard big-bang picture, what do you expect the neutrino temperature to be (roughly)?

(*Princeton*)

Solution:

(a) These are experiments to study the neutral weak current reaction between neutrinos and electrons, $\nu_\mu + e^- \rightarrow \nu_\mu + e^-$, using neutrinos created by accelerator at CERN. No such reactions were detected above the background and the confidence limit of measurements was

90%. This gives an upper limit to the weak interaction cross section of $\sigma < 2.6 \times 10^{-42} E_\nu$ cm^2/electron. With $E_\nu \sim kT$ we obtain $T < 10^6$ K.

(b) The energy density of the neutrino gas is $\rho_\nu \approx aT_\nu^4$, and that of the photon gas is $\rho_\gamma = aT^4$. As $\rho_\nu \leq 10^{-6} \rho_\gamma$ we have $T_\nu \leq T/10^{1.5}$. For $T \simeq 3$ K, we get $T_\nu \leq 0.1$ K.

(c) At the early age of the universe (when $kT \gtrsim m_\mu c^2$) neutrinos and other substances such as photons are in thermal equilibrium with $T_\nu = T_\gamma, \rho_\nu \approx \rho_\gamma$ and both have energy distributions similar to that of black body radiation. Afterwards, the neutrino gas expands freely with the universe and its energy density has functional dependence $\rho_\nu(\nu/T)$, where the frequency $\nu \propto \dfrac{1}{R}$, the temperature $T \propto \dfrac{1}{R}$, R being the "radius" of the universe. Hence the neutrino energies always follw the black body spectrum, just like the photons. However, because of the formation of photons by the annihilation of electron-position pairs, $\rho_\gamma > \rho_\nu$, and the temperature of the photon gas is slightly higher than that of the neutrino gas. As the photon temperature at present is 3 K, we expect $T_\nu < 3$ K.

2078

Imagine the universe to be a spherical cavity, with a radius of 10^{28} cm and impenetrable walls.

(a) If the temperature inside the cavity is 3K, estimate the total number of photons in the universe, and the energy content in these photons.

(b) If the temperature were 0 K, and the universe contained 10^{80} electrons in a Fermi distribution, calculate the Fermi momentum of the electrons.

(Columbia)

Solution:

(a) The number of photons in the angular frequency range from ω to $\omega + d\omega$ is

$$dN = \frac{V}{\pi^2 c^3} \frac{\omega^2 \, d\omega}{e^{\beta\hbar\omega} - 1}, \qquad \beta = \frac{1}{kT}.$$

The total number of photons is

$$
\begin{aligned}
N &= \frac{V}{\pi^2 c^3} \int_0^\infty \frac{\omega^2\, d\omega}{e^{\beta\omega h/2\pi} - 1} = \frac{V}{\pi^2 c^3} \frac{1}{(\beta\hbar)^3} \int_0^\infty \frac{x^2\, dx}{e^x - 1} \\
&= \frac{V}{\pi^2}\left(\frac{kT}{\hbar c}\right)^3 \cdot 2 \sum_{n=1}^\infty \frac{1}{n^3} \approx \frac{2 \times 1.2}{\pi^2} \cdot V \left(\frac{kT}{\hbar c}\right)^3 \\
&= \frac{2.4}{\pi^2} \frac{4}{3}\pi \cdot (10^{28})^3 \cdot \left(\frac{1.38 \times 10^{-16} \times 3}{1.05 \times 10^{-27} \times 3 \times 10^{10}}\right)^3 \\
&\approx 2.5 \times 10^{87} \ .
\end{aligned}
$$

The total energy is

$$
\begin{aligned}
E &= \frac{V\hbar}{\pi^2 c^3} \int_0^\infty \frac{\omega^3\, d\omega}{e^{\beta\omega h/2\pi} - 1} = \frac{\pi^2 k^4}{15(hc/2\pi)^3} V T^4 \\
&\approx 2.6 \times 10^{72} \text{ ergs} \ .
\end{aligned}
$$

(b) The Fermi momentum of the electrons is

$$
p_F = \hbar \left(3\pi^2 \frac{N}{V}\right)^{1/3} = 2 \times 10^{-26} \text{ g} \cdot \text{cm/s} \ .
$$

2079

An n-dimensional universe.

In our three-dimensional universe, the following are well-known results from statistical mechanics and thermodynamics:

(a) The energy density of black body radiation depends on the temperature as T^α, where $\alpha = 4$.

(b) In the Debye model of a solid, the specific heat at low temperatures depends on the temperature as T^β, where $\beta = 3$.

(c) The ratio of the specific heat at constant pressure to the specific heat at constant volume for a monatomic ideal gas is $\gamma = 5/3$.

Derive the analogous results (i.e., what are γ, α and β) in the universe with n dimensions.

(MIT)

Solution:

(a) The energy of black body radiation is

$$E = 2 \iint \frac{d^n p \, d^n q}{(2\pi\hbar)^n} \frac{\hbar\omega}{e^{\hbar\omega/2\pi kT} - 1}$$

$$= \frac{2V}{(2\pi\hbar)^n} \int d^n p \frac{\hbar\omega}{e^{\hbar\omega/2\pi kT} - 1} .$$

For the radiation we have $p = \hbar\omega/c$, so

$$\frac{E}{V} = 2 \left(\frac{k}{2\pi\hbar c} \right)^n k \int d^n x \frac{x}{e^x - 1} \cdot T^{n+1} ,$$

where $x = \hbar\omega/kT$. Hence $\alpha = n + 1$.

(b) The Debye Model regards solid as an isotropic continuous medium with partition function

$$Z(T,V) = \exp\left[-\hbar \sum_{i=1}^{nN} \omega_i / 2kT \right] \prod_{j=1}^{nN} [1 - \exp(-\hbar\omega_j/kT)]^{-1} .$$

The Holmholtz free energy is

$$F = -kT \ln Z = \frac{\hbar}{2} \sum_{i=1}^{nN} \omega_i + kT \sum_{i=1}^{nN} \ln[1 - \exp(-\hbar\omega_i/kT)] .$$

When N is very large,

$$\sum_{i=1}^{nN} \rightarrow \frac{n^2 N}{\omega_D^n} \int_0^{\omega_D} \omega^{n-1} d\omega ,$$

where ω_D is the Debye frequency. So we have

$$F = \frac{n^2 N}{2(n+1)} \hbar\omega_D + (kT)^{n+1} \frac{n^2 N}{(\hbar\omega_D)^n} \int_0^{x_D} x^{n-1} \ln[1 - \exp(-x)] dx ,$$

where $x_D = \hbar\omega_D/kT$. Hence

$$c_v = -T \left(\frac{\partial^2 F}{\partial T^2} \right)_V \propto T^n ,$$

i.e., $\beta = n$.

(c) The theorem of equipartition of energy gives the constant volume specific heat of a molecule as $c_v = \frac{l}{2}k$ where l is the number of degrees of freedom of the molecule. For a monatomic molecule in a space of n dimensions, $l = n$. With $c_p = c_v + k$, we get

$$\gamma = \frac{c_p}{c_v} = \frac{(n+2)}{n} \ .$$

2080

(a) Suppose one carries out a measurement of the specific heat at constant volume, C_v, for some solid as a function of temperature, T, and obtains the results:

T	C_v(arbitrary units)
1000K	20
500 K	20
40 K	8
20 K	1

Is the solid a conductor or an insulator? Explain.

(b) If the displacement of an atom about its equilibrium position in a harmonic solid is denoted by U, then the average displacement squared is given by

$$\langle U^2 \rangle = \frac{\hbar^2}{2M} \int_0^\infty \frac{d\varepsilon}{\varepsilon} g(\varepsilon)[1 + 2n(\varepsilon)] \ ,$$

where M is the mass of the atom, $g(\varepsilon)$ is a suitably normalized density of energy states and $n(\varepsilon)$ is the Bose-Einstein occupation factor for phonons of energy ε. Assuming a Debye model for the density of states:

$$g(\varepsilon) = 9\varepsilon^2/(\hbar\omega_D)^3 \quad \text{for } \varepsilon < \hbar\omega_D \ ,$$
$$g(\varepsilon) = 0 \quad \text{for } \varepsilon > \hbar\omega_D \ ,$$

where ω_D is the Debye frequency, determine the temperature dependence of $\langle U^2 \rangle$ for very high and very low temperatures. Do your results make sense?

(*Chicago*)

Solution:

(a) For the solid we have $C_v \propto T^3$ at low temperatures and $C_v =$ constant at high temperatures. So it is an insulator.

(b) The phonon is a boson. The Bose-Einstein occupation factor for phonons of energy ε is

$$n(\varepsilon) = \frac{1}{e^{\varepsilon/kT} - 1} \ .$$

So

$$
\begin{aligned}
\langle U^2 \rangle &= \frac{\hbar^3}{2M} \int_0^{\hbar\omega_D/2\pi} \frac{d\varepsilon}{\varepsilon} \frac{9\varepsilon^2}{(\hbar\omega_D)^3} \left[1 + \frac{2}{e^{\varepsilon/kT} - 1} \right] \\
&= \frac{9\hbar^2}{4M} \cdot \frac{1}{\hbar\omega_D} + \frac{9\hbar^2}{M} \cdot \frac{1}{(\hbar\omega_D)^3} \int_0^{\hbar\omega_D/2\pi} \frac{\varepsilon}{e^{\varepsilon/kT} - 1} d\varepsilon \ .
\end{aligned}
$$

If the temperature is high, i.e., $kT \gg \varepsilon$,

$$
\begin{aligned}
\langle U^2 \rangle &\approx \frac{9\hbar^2}{4M} \cdot \frac{1}{\hbar\omega_D} + \frac{9\hbar^2}{M} \cdot \frac{1}{(\hbar\omega_D)^3} \int_0^{\hbar\omega_D/2\pi} \varepsilon \cdot \frac{kT}{\varepsilon} d\varepsilon \ . \\
&\approx \frac{9\hbar^2}{M} \cdot \frac{kT}{(\hbar\omega_D)^2} \propto T \ .
\end{aligned}
$$

If the temperature is low, i.e., $kT \ll \varepsilon$.

$$\langle U^2 \rangle \approx \frac{9\hbar^2}{4M} \cdot \frac{1}{\hbar\omega_D} \ .$$

These results show that the atoms are in motion at $T = 0$, and the higher the temperature the more intense is the motion.

2081

Graphite has a layered crystal structure in which the coupling between the carbon atoms in different layers is much weaker than that between the atoms in the same layer. Experimentally it is found that the specific heat is proportional to T at low temperatures. How can the Debye theory be adapted to provide an explanation?

(SUNY, Buffalo)

Solution:

Graphite is an insulator and its specific heat is contributed entirely by the crystal lattice. When the temperature T increases from zero, the vibrational modes corresponding to the motion between layers is first excited since the coupling between the carbon atoms in different layers is much weaker. By the Debye model, we have

$$\omega = ck .$$

The number of longitudinal waves in the interval k to $k+dk$ is $(L/2\pi)^2 2\pi k \, dk$, where L is the length of the graphite crystal. From this, we obtain the number of the longitudinal waves in the interval ω to $\omega + d\omega$, $L^2 \omega \, d\omega / 2\pi c_{\parallel}^2$, where $c_{\parallel}$ is the velocity of longitudinal waves. Similarly, the number of transversal waves in the interval ω to $\omega + d\omega$ is $\dfrac{L^2 \omega \, d\omega}{\pi c_{\perp}^2}$.

Therefore, the Debye frequency spectrum is given by

$$g(\omega) \, d\omega = \frac{L^2}{2\pi} \left(\frac{1}{c_{\parallel}^2} + \frac{2}{c_{\perp}^2} \right) \omega \, d\omega ,$$

$$\omega < \omega_D \text{ (Debye frequency) .}$$

So,

$$C_v = k_B \int_0^{\omega_D} \frac{x^2 e^x}{(e^x - 1)^2} g(\omega) \, d\omega$$

$$= \frac{k_B^3 L^2 \left(\dfrac{1}{c_{\parallel}^2} + \dfrac{2}{c_{\perp}^2} \right)}{2\pi \hbar^2} T^2 \int_0^{x_D} \frac{x^3 e^x}{(e^x - 1)^2} \, dx .$$

where

$$x = \frac{\hbar\omega}{k_B T}, \quad x_D = \frac{\hbar\omega_D}{k_B T} , \quad k_B \text{ being Boltzmann's constant.}$$

At low temperatures, $\hbar\omega_D \gg k_B T$, i.e., $x_D \gg 1$, then,

$$\int_0^{x_D} \frac{x^3 e^x}{(e^x - 1)^2} \, dx \approx \int_0^{\infty} \frac{x^3 e^x}{(e^x - 1)^2} \, dx$$

$$= \int_0^{\infty} \frac{x^3}{4 \sinh^2 \left(\dfrac{1}{2} x \right)} \, dx = 6\varsigma(3)$$

where

$$\varsigma(3) = \sum_{n=1}^{\infty} n^{-3} \approx 1.2 \ .$$

So that the specific heat is proportional to T^2 at low temperatures, or more precisely,

$$C_v = \frac{3k_B^3 L^2 (c_\parallel^{-2} + 2c_\perp^{-2})}{\pi \hbar^2} \varsigma(3) T^2 \ .$$

2082

One Dimensional Debye Solid.

Consider a one dimensional lattice of N identical point particles of mass m, interacting via nearest-neighbor spring-like forces with spring constant $m\omega^2$. Denote the lattice spacing by a. As is easily shown, the normal mode eigenfrequencies are given by

$$\omega_k = \omega\sqrt{2(1 - \cos ka)}$$

with $k = 2\pi n/aN$, where the integer n ranges from $-N/2$ to $+N/2$ ($N \gg 1$). Derive an expression for the quantum mechanical specific heat of this system in the Debye approximation. In particular, evaluate the leading non-zero terms as functions of temperature T for the two limits $T \to \infty$, $T \to 0$.

(*Princeton*)

Solution:

Please refer to Problem 2083.

2083

A one dimensional lattice consists of a linear array of N particles ($N \gg 1$) interacting via spring-like nearest neighbor forces. The normal mode frequencies (radians/sec) are given by

$$\omega_n = \overline{\omega}\sqrt{2(1 - \cos(2\pi n/N))} \ ,$$

where $\overline{\omega}$ is a constant and n an integer ranging from $-N/2$ to $+N/2$. The system is in thermal equilibrium at temperature T. Let c_v be the constant "volume" (length) specific heat.

(a) Compute c_v for the regime $T \to \infty$.

(b) For $T \to 0$

$$c_v \to A\omega^{-\alpha} T^\gamma \ ,$$

where A is a constant that you need not compute. Compute the exponents α and γ.

The problem is to be treated quantum mechanically.

(*Princeton*)

Solution:

(a) $U = \displaystyle\sum_{n=-\frac{N}{2}}^{\frac{N}{2}} \hbar\omega_n \frac{1}{e^{\hbar\omega_n/kT} - 1} \ .$

When $kT \gg \hbar\omega_n$

$$U \approx \sum \hbar\omega_n \cdot \frac{kT}{\hbar\omega_n} = NkT \ .$$

Hence $c_v = \dfrac{dU}{dT} = Nk.$

(b) When $kT \ll \hbar\omega$, we have

$$U = \sum_{n=-\frac{N}{2}}^{\frac{N}{2}} \hbar\omega_n \frac{1}{e^{\hbar\omega_n/kT} - 1} \approx 2 \sum_{n=0}^{N/2} \hbar\omega_n e^{-\hbar\omega_n/kT} \ .$$

So

$$c_v = \frac{dU}{dT} = 2 \sum_{n=0}^{N/2} \frac{(\hbar\omega_n)^2}{kT^2} e^{-\hbar\omega_n/kT} \ .$$

Notice that as $N \gg 1$ we have approximately

$$c_v = 2 \frac{\hbar^2\omega^2}{kT^2} \int_0^{N/2} 2 \left(1 - \cos\frac{2\pi x}{N} \right) e^{-(\hbar\omega/\pi kT)\sin(\pi x/N)} dx$$

$$= 8 \frac{\hbar^2\omega^2}{kT^2} \int_0^{N/2} \frac{\sin^2\frac{\pi x}{N} e^{-(\hbar\omega/\pi kT)\sin(\pi x/N)}}{\cos\frac{\pi x}{N}} \cdot \frac{N}{\pi} d \left(\sin\frac{\pi x}{N} \right)$$

$$= \frac{8\hbar^2\omega^2 N}{kT^2\pi} \int_0^1 \frac{t^2 e^{-(t\hbar\omega/\pi kT)}}{\sqrt{1-t^2}} dt \ .$$

Because $\exp(-th\omega/\pi kT)$ decreases rapidly as $t \to 1$, we have

$$c_v \approx \frac{8\hbar^2\omega^2}{kT^2} \cdot \frac{N}{\pi} \int_0^1 t^2 e^{-(th\omega/\pi kT)} dt$$

$$= \frac{8\hbar^2\omega^2}{kT^2} \cdot \frac{N}{\pi} \cdot \left(\frac{kT}{\hbar\omega}\right)^3 \cdot \int_0^\infty \xi^2 e^{-\xi} d\xi = A\omega^{-1}T ,$$

where $A = (16Nk^2/h) \int_0^\infty \xi^2 \exp(-\xi)d\xi$.
Hence $\alpha = \gamma = 1$.

2084

Given the energy spectrum

$$\varepsilon_p = [(pc)^2 + m_0^2 c^4]^{1/2} \to pc \quad \text{as} \quad p \to \infty .$$

(a) Prove that an ultrarelativistic ideal fermion gas satisfies the equation of state $pV = E/3$, where E is the total energy.

(b) Prove that the entropy of an ideal quantum gas is given by

$$S = -k \sum_i [n_i \ln(n_i) \mp (1 \pm n_i) \ln(1 \pm n_i)]$$

where the upper (lower) signs refer to bosons (fermions).

(*SUNY, Buffalo*)

Solution:

(a) The number of states in the momentum interval p to $p + dp$ is $\frac{8\pi V}{h^3}p^2 dp$ (taking $S = \frac{1}{2}$). From $\varepsilon = cp$, we obtain the number of states in the energy interval ε to $\varepsilon + d\varepsilon$:

$$N(\varepsilon)d\varepsilon = \frac{8\pi V}{c^3 h^3}\varepsilon^2 d\varepsilon .$$

So the total energy is

$$E = \frac{8\pi V}{c^3 h^3} \int_0^\infty \frac{\varepsilon^3 d\varepsilon}{e^{\beta(\varepsilon - \mu)} + 1} .$$

In terms of the thermodynamic potential $\ln \Xi$,

$$
\begin{aligned}
pV = KT \ln \Xi &= kT \int_0^\infty \frac{8\pi V}{c^3 h^3} \varepsilon^2 \ln[1 + e^{-\beta(\varepsilon-\mu)}] d\varepsilon \\
&= \frac{8\pi V}{3c^3 h^3} \cdot kT \int_0^\infty \ln[1 + e^{-\beta(\varepsilon-\mu)}] d\varepsilon^3 \\
&= \frac{8\pi V}{3c^3 h^3} \int_0^\infty \frac{\varepsilon^3 e^{-\beta(\varepsilon-\mu)}}{1 + e^{-\beta(\varepsilon-\mu)}} d\varepsilon \\
&= \frac{1}{3} E .
\end{aligned}
$$

Note that this equation also applies to an ultrarelativistic boson gas.

(b) The average number of particles in the quantum state i is given by $n_i = 1/[\exp(\alpha + \beta\varepsilon_i) \mp 1]$, from which we have

$$
\exp(\alpha + \beta\varepsilon_i) = (1 \pm n_i)/n_i ,
$$

or

$$
\alpha + \beta\varepsilon_i = \ln(1 \pm n_i) - \ln(n_i) ,
$$

and

$$
\begin{aligned}
\ln \Xi = \mp \sum_i \ln(1 \mp e^{-\alpha-\beta\varepsilon_i}) &= \mp \sum_i \ln \frac{1}{1 \pm n_i} \\
&= \pm \sum_i \ln(1 \pm n_i) .
\end{aligned}
$$

By

$$
S = k \left[\ln \Xi - \alpha \left(\frac{\partial}{\partial \alpha} \ln \Xi \right)_\beta - \beta \left(\frac{\partial}{\partial \beta} \ln \Xi \right)_\alpha \right]
$$

we have

$$
S = -k \sum_i [n_i \ln(n_i) \mp (1 \pm n_i) \ln(1 \pm n_i)] .
$$

2085

Consider an ideal quantum gas of Fermi particles at a temperature T.

(a) Write the probability $p(n)$ that there are n particles in a given single particle state as a function of the mean occupation number, $\langle n \rangle$.

(b) Find the root-mean-square fluctuation $\langle (n - \langle n \rangle)^2 \rangle^{1/2}$ in the occupation number of a single particle state as a function of the mean occupation number $\langle n \rangle$. Sketch the result.

$$(MIT)$$

Solution:

(a) Let ε be the energy of a single particle state, μ be the chemcial potential. The partition function is

$$z = \sum_n \exp[n(\mu - \varepsilon)/kT] = 1 + \exp[(\mu - \varepsilon)/kT] .$$

The mean occupation number is

$$\langle n \rangle = kT \frac{\partial}{\partial \mu} \ln z = \frac{1}{e^{(\varepsilon - \mu)/kT} + 1} .$$

The probability is

$$p(n) = \frac{1}{z} e^{n(\mu - \varepsilon)/kT}$$
$$= \frac{(1 - \langle n \rangle)^n}{\langle n \rangle^{n-1}} .$$

(b) $\langle (n - \langle n \rangle)^2 \rangle = kT \frac{\partial \langle n \rangle}{\partial \mu} = \langle n \rangle (1 - \langle n \rangle) .$

So we have $\langle (n - \langle n \rangle)^2 \rangle^{1/2} = \sqrt{\langle n \rangle (1 - \langle n \rangle)}$.
The result is shown in Fig. 2.17.

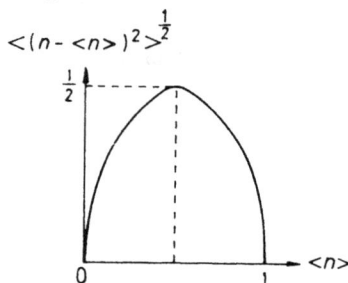

Fig. 2.17.

2086

In a perfect gas of electrons, the mean number of particles occupying a single-particle quantum state of energy E_i is:

$$N_i = \frac{1}{\exp[(E_i - \mu)/kT] + 1} .$$

(a) Obtain a formula which could be used to determine μ in terms of the particle density n and various constants.

(b) Show that the expression above reduces to the Maxwell-Boltzmann distribution in the limit $n\lambda^3 \ll 1$, where λ is the thermal de Broglie wavelength.

(c) Sketch N_i versus E_i for $T = 0$ K and for $T = \mu/5$ K. Label significant points along both axes.

(UC, Berkeley)

Solution:

(a) The particle number density is

$$n = \int \frac{4x}{h^3}(2m)^{3/2}\sqrt{\varepsilon}\frac{1}{e^{(\varepsilon-\mu)/kT} + 1}d\varepsilon$$

$$= \left(\frac{2\pi mkT}{h^2}\right)^{3/2} \frac{4}{\sqrt{\pi}} \int \frac{\sqrt{x}}{e^{-\mu/kT} \cdot e^x + 1}dx .$$

As

$$\frac{1}{\lambda} = \left(\frac{2\pi mkT}{h^2}\right)^{1/2} ,$$

$$n\lambda^3 = \frac{4}{\sqrt{\pi}} \int_0^\infty \frac{\sqrt{x}}{e^{-\mu/kT}e^x + 1}dx .$$

This formula can be used to determine μ.

(b) When $n\lambda^3 \ll 1$, we must have in the above integral

$$\exp\left(\frac{-\mu}{kT}\right) \gg 1 .$$

It follows that

$$N_i = \frac{1}{e^{(E_i-\mu)/kT} + 1} \approx e^{\mu/kT} \cdot e^{-E_i/kT} .$$

i.e., it reduces to the Boltzmann distribution.

(c) The variation of N_i versus E_i is as shown in Fig. 2.18.

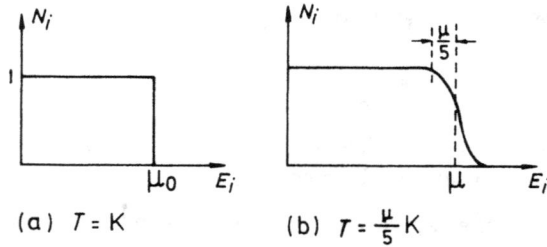

(a) $T = K$ (b) $T = \frac{\mu}{5} K$

Fig. 2.18.

2087

Suppose that in some sample the density of states of the electrons $D(\varepsilon)$ is a constant D_0 for energy $\varepsilon > 0$ $(D(\varepsilon) = 0$ for $\varepsilon < 0)$ and that the total number of electrons is equal to N.

(a) Calculate the Fermi potential μ_0 at 0 K.

(b) For non-zero temperatures, derive the condition that the system is non-degenerate.

(c) Show that the electronic specific heat is proportional to the temperature, T, when the system is highly degenerate.

(*UC, Berkeley*)

Solution:

(a) When $T = 0$ K, all the low lying energy levels are occupied, while those levels whose energies ε are greater than μ_0 are all vacant. Taking the 1/2 spin of electrons into consideration, every state can accomodate two electrons, and hence $2D_0\mu_0 V = N$, or

$$\mu_0 = \frac{N}{2V D_0} ,$$

where V is the volume of the sample.

(b) The non-degeneracy condition requires that $\exp\left(\frac{\mu}{kT}\right) \ll 1$, then

$$\frac{1}{e^{(\varepsilon-\mu)/kT} + 1} \approx e^{\mu/kT} \cdot e^{-\varepsilon/kT} .$$

In this approximation,

$$\frac{N}{V} = \int_0^\infty \frac{2D_0}{e^{(\varepsilon-\mu)/kT} + 1} d\varepsilon = 2D_0 \cdot kT \cdot e^{\mu/kT} .$$

That is, the non-degeneracy condition is $kT \gg \left(\frac{N}{V}\right) / 2D_0 = \mu_0$.

(c) When $T = 0$ K, the electrons are in the ground state without excitation. When $T \neq 0$ K, but $T \ll \mu_0/k$, only those electrons near the Fermi surface are excited, $N_{\text{eff}} \approx kT D_0$, and the specific heat contributed by each electron is $C_0 = \frac{3}{2}k$. Therefore, when the system is highly degnerate, the specific heat $C \propto T$.

2088

Consider a system of N "non-interacting" electrons/cm^3, each of which can occupy either a bound state with energy $\varepsilon = -E_d$ or a free-particle continuum with $\varepsilon = \frac{p^2}{2m}$. (This could be a semiconductor like Si with N shallow donors/cm^3.)

(a) Compute the density of states as a function of ε in the continuum.

(b) Find an expression for the chemical potential in the low temperature limit.

(c) Compute the number of free electrons (i.e., electrons in the continuum) as a function of T in the low temperature limit.

(*UC, Berkeley*)

Solution:

Suppose that each bound state can at most contain a pair of electrons with anti-parallel spins, and that the number of bound states is $\frac{N}{2}$. That is, when $T = 0$ K, all the bound states are filled up with no free electrons. When T is quite low, only a few electrons are in the free particle continuum so that we can use the approximation of weak-degeneracy.

(a) The density of states in the continuum is

$$D(\varepsilon) = \frac{4\pi}{h^3}(2m)^{3/2}\sqrt{\varepsilon} .$$

(b), (c) The number of electrons in the bound states are

$$N_b = \frac{N}{e^{-(E_d+\mu)/kT} + 1} ,$$

or

$$\frac{N - N_b}{N_b} = e^{-E_d/kT} e^{-\mu/kT} . \tag{1}$$

The number of electrons in the continuum (weak-degeneracy approximation) is

$$N_f = \int_0^\infty D(\varepsilon) e^{\mu/kT} e^{-\varepsilon/kT} d\varepsilon = N_c e^{\mu/kT} , \tag{2}$$

where

$$N_c = 2 \left(\frac{2\pi mkT}{h^2} \right)^{3/2} .$$

From (1) and (2), we get

$$\frac{N_f(N - N_b)}{N_b} = N_c e^{-E_d/kT} .$$

Since $N_b + N_f = N, N_b \approx N$, then

$$N_f^2 = N N_c e^{-E_d/kT} , \quad \text{or} \quad N_f = \sqrt{N N_c} e^{-E_d/2kT} . \tag{3}$$

Substitute (3) in (2), we get

$$\mu = kT \ln \frac{N_f}{N_c} = \frac{kT}{2} \ln \frac{N}{N_c} - \frac{E_d}{2} .$$

2089

(a) For a system of electrons, assumed non-interacting, show that the probability of finding an electron in a state with energy Δ above the chemical potential μ is the same as the probability of finding an electron absent from a state with energy Δ below μ at any given temperature T.

(b) Suppose that the density of states $D(\varepsilon)$ is given by

$$D(\varepsilon) = \begin{cases} a(\varepsilon - \varepsilon_g)^{1/2} , & \varepsilon > \varepsilon_g , \\ 0 , & 0 < \varepsilon < \varepsilon_g , \\ b(-\varepsilon)^{1/2} , & \varepsilon < 0 , \end{cases}$$

as shown in Fig. 2.19,

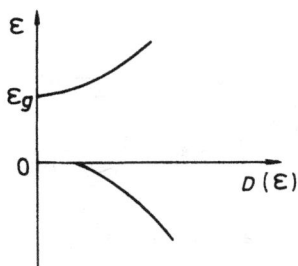

Fig. 2.19.

and that at $T = 0$ all states with $\varepsilon < 0$ are occupied while the other states are empty. Now for $T > 0$, some states with $\varepsilon > 0$ will be occupied while some states with $\varepsilon < 0$ will be empty. If $a = b$, where is the position of μ? For $a \neq b$, write down the mathematical equation for the determination of μ and discuss qualitatively where μ will be if $a > b$? $a < b$?

(c) If there is an excess of n_d electrons per unit volume than can be accommodated by the states with $\varepsilon < 0$, what is the equation for μ for $T = 0$? How will μ shift as T increases?

(SUNY, Buffalo)

Solution:

(a) By the Fermi distribution, the probability for a level ε to be occupied is

$$F(\varepsilon) = \frac{1}{e^{\beta(\varepsilon - \mu)} + 1} \, ,$$

so the probability for finding an electron at $\varepsilon = \mu + \Delta$ is

$$F(\mu + \Delta) = \frac{1}{e^{\beta \Delta} + 1} \, ,$$

and the probability for not finding electrons at $\varepsilon = \mu - \Delta$ is given by

$$1 - F(\mu - \Delta) = \frac{1}{e^{\beta \Delta} + 1} \, .$$

The two probabilities have the same value as required.

(b) When $T > 0$ K, some electrons with $\varepsilon < 0$ will be excited to states of $\varepsilon > \varepsilon_g$. That is to say, vacancies are produced in the some states of $\varepsilon < 0$

while some electrons occupy states of $\varepsilon > \varepsilon_g$. The number of electrons with $\varepsilon > \varepsilon_g$ is given by

$$
\begin{aligned}
n_e &= \int_{\varepsilon_g}^{\infty} D(\varepsilon) \frac{1}{e^{\beta(\varepsilon-\mu)} + 1} d\varepsilon \\
&= \int_{\varepsilon_g}^{\infty} a(\varepsilon - \varepsilon_g)^{1/2} \frac{1}{e^{\beta(\varepsilon-\mu)} + 1} d\varepsilon .
\end{aligned}
$$

The number of vacancies for $\varepsilon < 0$ is given by

$$
\begin{aligned}
n_p &= \int_{-\infty}^{0} D(\varepsilon)[1 - F(\varepsilon)] d\varepsilon \\
&= \int_{-\infty}^{0} b(-\varepsilon)^{1/2} \frac{1}{e^{-\beta(\varepsilon-\mu)} + 1} d\varepsilon .
\end{aligned}
$$

By $n_e = n_p$, we have $\mu = \varepsilon_g/2$ when $a = b$. We also obtain the equation to determine μ when $a \neq b$,

$$
\frac{a}{b} = \frac{e^{\beta(\varepsilon + \varepsilon_g - \mu)} + 1}{e^{\beta(\varepsilon + \mu)} + 1} .
$$

For $a > b$, we have

$$
\frac{e^{\beta(\varepsilon + \varepsilon_g - \mu)} + 1}{e^{\beta(\varepsilon + \mu)} + 1} > 1 ,
$$

so that $\varepsilon + \varepsilon_g - \mu > \varepsilon + \mu$, i.e., $\mu < \varepsilon_g/2$. Hence μ shifts to lower energies. For $a < b, \mu > \varepsilon_g/2$, μ shifts to higher energies.

(c) When $T = 0$, by

$$
\int_{\varepsilon_g}^{\mu} a(\varepsilon - \varepsilon_g)^{1/2} d\varepsilon = n_d ,
$$

we obtain

$$
\mu = \varepsilon_g + \left(\frac{3n_d}{2a} \right)^{2/3} .
$$

μ shifts to lower energies as T increases.

2090

(a) Calculate the magnitude of the Fermi wavevector for 4.2×10^{21} electrons confined in a box of volume 1 cm^3.

(b) Compute the Fermi energy (in eV) for this system.

(c) If the electrons are replaced by neutrons, compute the magnitude of the Fermi wavevector and the Fermi energy.

(UC, Berkeley)

Solution:

(a) The total number of particles is

$$N = \frac{2V}{h^3} \frac{4\pi}{3} p_F^3 .$$

The Fermi wavelength is

$$\lambda_F = \frac{h}{p_F} = \left(\frac{8\pi V}{3N}\right)^{1/3} = 1.25 \times 10^{-9} \text{m} = 12.5\text{Å} .$$

(b) The Fermi energy is

$$\varepsilon_F = \frac{p_F^2}{2m} = \frac{1}{2m}\left(\frac{h}{\lambda_F}\right)^2 = 1.54 \times 10^{-19}\,\text{J} = 0.96 \text{ eV} .$$

(c) If the electrons are replaced by neutrons, we find that

$$\lambda_F' = \lambda_F = 12.5\text{Å} ,$$

and $\varepsilon_F' = \dfrac{m}{m'}\varepsilon_F = 5.2 \times 10^{-4}\text{eV}.$

2091

Calculate the average energy per particle, ε, for a Fermi gas at $T = 0$, given that ε_F is the Fermi energy.

(UC, Berkeley)

Solution:

We consider two cases separately, non-relativistic and relativistic.

(a) For a non-relativistic particle, $p \ll mc$ (p is the momentum and m is the rest mass), it follows that

$$\varepsilon = \frac{p^2}{2m} .$$

We have $D(\varepsilon) = \sqrt{\varepsilon} \cdot$ const.
Then

$$\bar{\varepsilon} = \frac{\int_0^{\varepsilon_F} \varepsilon \sqrt{\varepsilon}\, d\varepsilon}{\int_0^{\varepsilon_F} \sqrt{\varepsilon}\, d\varepsilon} = \frac{3}{5}\varepsilon_F .$$

(b) For $p \gg mc$, we have $\varepsilon = pc$, and $D(\varepsilon) = \varepsilon^2 \cdot$ const. Therefore,

$$\bar{\varepsilon} = \frac{\int_0^{\varepsilon_F} \varepsilon^2 \cdot \varepsilon\, d\varepsilon}{\int_0^{\varepsilon_F} \varepsilon^2 d\varepsilon} = \frac{3}{4}\varepsilon_F .$$

2092

Derive the density of states $D(\varepsilon)$ as a function of energy ε for a free electron gas in one-dimension. (Assume periodic boundary conditions or confine the linear chain to some length L.) Then calculate the Fermi energy ε_F at zero temperature for an N electron system.

(Wisconsin)

Solution:

The energy of a particle is $\varepsilon = p^2/2m$. Thus,

$$dp = \left(\frac{m}{2\varepsilon}\right)^{1/2} d\varepsilon .$$

Taking account of the two states of spin, we have

$$D(\varepsilon)d\varepsilon = \frac{2L \cdot dp}{h} = \frac{L(2m)^{1/2}}{h\varepsilon^{1/2}}\, d\varepsilon ,$$

or

$$D(\varepsilon) = L\left(\frac{2m}{\varepsilon}\right)^{1/2} \Big/ h .$$

At temperature 0 K, the electrons will occupy all the states whose energy is from 0 to the Fermi energy ε_F. Hence

$$N = \int_0^{\varepsilon_F} D(\varepsilon)d\varepsilon ,$$

giving

$$\varepsilon_F = \frac{h^2}{2m}\left(\frac{N}{2L}\right)^2.$$

2093

Consider a Fermi gas at low temperatures $kT \ll \mu(0)$, where $\mu(0)$ is the chemical potential at $T = 0$. Give qualitative arguments for the leading value of the exponent of the temperature-dependent term in each of the following quantities: (a) energy; (b) heat capacity; (c) entropy; (d) Helmholtz free energy; (e) chemical potential. The zero of the energy scale is at the lowest orbital.

(*UC, Berkeley*)

Solution:

At low temperatures, only those particles whose energies fall within a thickness $\sim kT$ near the Fermi surface are thermally excited. The energy of each such particle is of the order of magnitude kT.

(a) $E = E(0) + \alpha kT \cdot kT$, where α is a proportionality constant. Hence $E - E(0) \propto T^2$.

(b) $C_v = \left(\dfrac{\partial E}{\partial T}\right)_V \propto T$.

(c) From $dS = \dfrac{C_v}{T}dT$, we have

$$S = \int_0^T \frac{C_v}{T}dT \propto T.$$

(d) From $F = E - TS$, we have $F - F(0) \propto T^2$.

(e) From $\mu = (F + pV)/N$ and $p = 2E/3V$, where N is the total number of particles, we have $\mu - \mu(0) \propto T^2$.

2094

Derive an expression for the chemical potential of a free electron gas with a density of N electrons per unit volume at zero temperature ($T = 0$ K). Find the chemical potential of the conduction electrons (which can be considered as free electrons) in a metal with $N = 10^{22}$ electrons/cm^3 at $T = 0$ K.

(UC, Berkeley)

Solution:

From the density of states

$$D(\varepsilon)d\varepsilon = 4\pi(2m)^{3/2}\sqrt{\varepsilon}d\varepsilon/h^3 \ ,$$

we get

$$N = \int_0^{\mu_0} \frac{4\pi}{h^3}(2m)^{3/2}\sqrt{\varepsilon}d\varepsilon = \frac{8\pi}{3}\left(\frac{2m\mu_0}{h^2}\right)^{3/2} \ .$$

Therefore, $\mu_0 = \dfrac{h^2}{2m}\left(\dfrac{3N}{8\pi}\right)^{2/3}.$

For $N = 10^{22}$ electrons/cm^3 = 10^{28} electrons/m^3, it follows that

$$\mu_0 = 2.7 \times 10^{-19} \text{ J} = 1.7 \text{ eV} \ .$$

2095

$D(E)$ is the density of states in a metal, and E_F is the Fermi energy. At the Fermi energy $D(E_F) \neq 0$.

(a) Give an expression for the total number of electrons in the system at temperature $T = 0$ in terms of E_F and $D(E_F)$.

(b) Give an expression of the total number of electrons in the system at $T \neq 0$ in terms of the chemical potential μ and $D(E)$.

(c) Calculate the temperature dependence of the chemical potential at low temperatures, i.e., $\mu \gg kT$.

(Remember: $\displaystyle\int_{-\infty}^{+\infty} \frac{xe^x}{(e^x+1)^2}dx = \frac{\pi^2}{3}$.)

(Chicago)

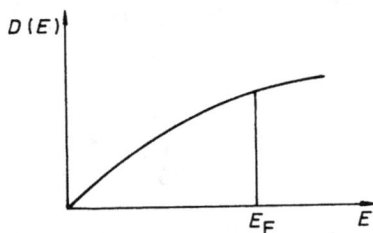

Fig. 2.20.

Solution:

The density of states is

$$D(E) = \frac{4\pi V (2m)^{3/2}}{h^3} E^{1/2} \ .$$

(a) If $T = 0$, the total number of electrons is

$$N = \int_0^{E_F} D(E)dE = \frac{2}{3}D(E_F)E_F \ .$$

(b) If $T \neq 0$,

$$N = \int_0^\infty \frac{D(E)}{e^{(E-\mu)/kT} + 1}dE \ .$$

(c) At low temperatures $\mu \gg kT$,

$$N = \int_0^\infty \frac{D(E)dE}{e^{(E-\mu)/kT} + 1}$$

$$= \int_0^\mu D(E)dE + \frac{\pi^2}{6}(kT)^2 D'(\mu) + \frac{7\pi^4}{360}(kT)^4 D'''(\mu) + \cdots$$

$$\approx \frac{8\pi V (2m)^{3/2}}{3h^3}\mu^{3/2}\left[1 + \frac{\pi^2}{8}\left(\frac{kT}{\mu}\right)^2\right] \ .$$

Thus, we get

$$\mu = E_F\left[1 - \frac{\pi^2}{12}\left(\frac{kT}{E_F}\right)^2\right] \ ,$$

where

$$E_F = \mu_0 = \frac{1}{2m} \left(\frac{3h^3 N}{8\pi V} \right)^{2/3} .$$

2096

For Na metal there are approximately 2.6×10^{22} conduction electrons/cm^3, which behave approximately as a free electron gas. From these facts,

(a) give an approximate value (in eV) of the Fermi energy in Na,

(b) give an approximate value for the electronic specific heat of Na at room temperature.

(*UC, Berkeley*)

Solution:

(a) The Fermi energy is

$$E_F = \frac{\hbar^2}{2m} \left(3\pi^2 \frac{N}{V} \right)^{2/3} .$$

We substitute $\hbar = 6.58 \times 10^{-16}$ eV·s, $m = 0.511$ MeV/c^2 and $\frac{N}{V} = 2.6 \times 10^{22}/cm^3$ into it and obtain $E_F \approx 3.2$ eV.

(b) The specific heat is

$$C \approx \frac{1}{M} \frac{N}{E_F} k^2 T = \frac{k}{m_e} \cdot \frac{kT}{E_F} ,$$

where $m_e = 9.11 \times 10^{-31}$ kg is the mass of the electron, $k = 1.38 \times 10^{-23}$ J/K is Boltzmann's constant, and $kT \approx \frac{1}{40}$ eV at room temperature. We substitute E_F and the other quantities in the above expression and obtain $C \approx 11.8$ J/K·g.

2097

The electrons in a metallic solid may be considered to be a three-dimensional free electron gas. For this case:

(a) Obtain the allowed values of k, and sketch the appropriate Fermi sphere in k-space. (Use periodic boundary conditions with length L).

(b) Obtain the maximum value of k for a system of N electrons, and hence an expression for the Fermi energy at $T = 0K$.

(c) Using a simple argument show that the contribution the electrons make to the specific heat is proportional to T.

(*Wisconsin*)

Solution:

(a) The periodic condition requires that the length of the container L is an integral multiple of the de Broglie wavelength for the possible states of motion of the particle, that is,

$$L = |n_x|\lambda , \quad |n_x| = 0, 1, 2, \ldots .$$

Utilizing the relation between the wavelength and the wave vector, $k = 2\pi/\lambda$, and taking into account the two propagating directions for each dimension, we obtain the allowed values of k_x

$$k_x = \frac{2\pi}{L} n_x , \quad n_x = 0, \pm 1, \pm 2, \ldots .$$

Similarly we have

$$k_y = \frac{2\pi}{L} n_y , \quad n_y = 0, \pm 1, \pm 2, \ldots$$

$$k_z = \frac{2\pi}{L} n_z , \quad n_z = 0, \pm 1, \pm 2, \ldots$$

Thus the energies

$$\varepsilon = \frac{p^2}{2m} = \frac{\hbar^2 k^2}{2m}$$

are discrete. The Fermi sphere shell is shown in Fig. 2.21.

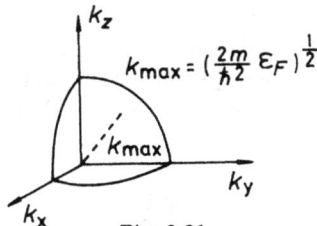

Fig. 2.21.

(b) $dn_x = \dfrac{L}{2\pi} dk_x$,

$\qquad dn_y = \dfrac{L}{2\pi} dk_y$,

$\qquad dn_z = \dfrac{L}{2\pi} dk_z$.

Thus, in the volume $V = L^3$, the number of quantum states of free electrons in the region $k_x \to k_x + dk_x, k_y \to k_y + dk_y, k_z \to k_z + dk_z$ is (considering the two directions of spin)

$$dn = dn_x dn_y dn_z = 2 \left(\frac{L}{2\pi}\right)^3 dk_x dk_y dk_z = \frac{V}{4\pi^3} dk_x dk_y dk_z \ .$$

At $T = 0$ K, the electrons occupy the lowest states. According to the Pauli exclusion principle, there is at most one electron in a quantum state. Hence

$$N = \frac{V}{4\pi^3} \iiint dk_x dk_y dk_z = \frac{V}{4\pi^3} \int_0^{k_{\max}} 4\pi k^2 dk \ ,$$

so that

$$k_{\max} = \left(3\pi^2 \frac{N}{V}\right)^{1/3} \ .$$

The Fermi energy is

$$\varepsilon_F = \frac{\hbar^2}{2m} k_{\max}^2 = \frac{\hbar^2}{2m} \left(3\pi^2 \frac{N}{V}\right)^{2/3} \ .$$

(c) At $T = 0$ K, the electrons occupy all the quantum states of energies from 0 to ε_F. When the temperature is increased, some of the electrons can be excited into states of higher energies that are not occupied, but they must absorb much energy to do so, so that the probability is very small. Thus the occupancy situation of most of the states do not change, except those with kT near the Fermi energy ε_F. Therefore, only the electrons in such states contribute to the specific heat. Let N_{eff} denote the number of such electrons, we have $N_{\mathrm{eff}} = kTN/\varepsilon_F$. Thus the molar specific heat contributed by the electrons is

$$C_v = \frac{3}{2} R \left(\frac{kT}{\varepsilon_F}\right) \propto T \ .$$

2098

Sketch the specific heat curve at constant volume, C_v, as a function of the absolute temperature, T, for a metallic solid. Give an argument showing why the contribution to C_v from the free electrons is proportional to T.

(Wisconsin)

Solution:

As shown in Fig. 2.22, the specific heat of a metal is

$$C_v = \gamma T + AT^3$$

where the first term on the right hand side is the contribution of the free electrons and the second term is the contribution of lattice oscillation.

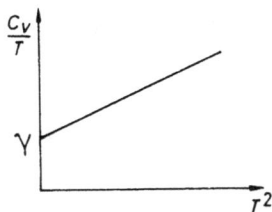

Fig. 2.22.

For a quantitative discussion of the contribution to C_v of the free electrons see answer to Problem 2097(a).

2099

(a) Derive a formula for the maximum kinetic energy of an electron in a non-interacting Fermi gas consisting of N electrons in a volume V at zero absolute temperature.

(b) Calculate the energy gap between the ground state and first excited state for such a Fermi gas consisting of the valence electrons in a 100Å cube of copper.

(c) Compare the energy gap with kT at 1 K.

The mass density and atomic weight of copper are 8.93 g/cm^3 and 63.6 respectively.

(UC, Berkeley)

Solution:

(a) When $T = 0$ K, the Fermi distribution is

$$f = \begin{cases} 1, & \varepsilon < \varepsilon_F , \\ 0, & \varepsilon > \varepsilon_F . \end{cases}$$

The density of quantum states is

$$\frac{4\pi}{h^3}(2m)^{3/2}\sqrt{\varepsilon}d\varepsilon .$$

Therefore, $\frac{N}{V} = \int_0^{\varepsilon_F} \frac{4\pi}{h^3}(2m)^{3/2}\sqrt{\varepsilon}d\varepsilon$, giving

$$\varepsilon_F = \frac{h^2}{2m}\left(\frac{3N}{8\pi V}\right)^{2/3} ,$$

i.e.,

$$\varepsilon_{\max} = \varepsilon_F = 1.1 \times 10^{-18} \text{ J} .$$

(b) As $n\lambda/2 = a$ and $p = h/\lambda$, the quantum levels of the valence electrons in the cube of copper are given by

$$\varepsilon = \frac{h^2}{8ma^2}(n_1^2 + n_2^2 + n_3^2) ,$$

where $n_1, n_2, n_3 = 0, 1, 2, \ldots$ (not simultaneously 0). The 1st excited state of the Fermi gas is such that an electron is excited from the Fermi surface to the nearest higher energy state. That is

$$(n, 0, 0) \rightarrow (n, 1, 0) .$$

Hence

$$\Delta\varepsilon = \frac{h^2}{8ma^2} = 6.0 \times 10^{-30} \text{ J} .$$

(c) $\dfrac{\Delta\varepsilon}{k} = 4.4 \times 10^{-7}\text{K} \ll 1\text{K}.$

2100

(a) For a degenerate, spin $\frac{1}{2}$, non-interacting Fermi gas at zero temperature, find an expression for the energy of a system of N such particles confined to a volume V. Assume the particles are non-relativistic.

(b) Given such an expression for the internal energy of a general system (not necessarily a free gas) at zero temperature, how does one determine the pressure?

(c) Hence calculate the pressure of this gas and show that it agrees with the result given by the kinetic theory.

(d) Cite, and explain briefly, two phenomena which are at least qualitatively explained by the Fermi gas model of metals, but are not in accord with classical statistical mechanics. Cite one phenomenon for which this simple model is inadequate for even a qualitative explanation.

(*UC, Berkeley*)

Solution:

(a) The density of states is given by

$$D(\varepsilon)d\varepsilon = \frac{4\pi V}{h^3}(2m)^{3/2}\sqrt{\varepsilon}d\varepsilon .$$

Hence

$$N = \frac{4\pi V}{h^3}(2m)^{3/2}\int_0^{\varepsilon_F}\sqrt{\varepsilon}d\varepsilon = \frac{8\pi V}{3}\left(\frac{2m\varepsilon_F}{h^2}\right)^{3/2}$$

and

$$E = \left(\frac{4\pi V}{h^3}\right)\int_0^{\varepsilon_F}(2m\varepsilon)^{3/2}d\varepsilon$$
$$= 3N\varepsilon_F/5 .$$

(b) From the thermodynamic relation

$$\left(\frac{\partial E}{\partial V}\right)_T = T\left(\frac{\partial p}{\partial T}\right)_V - p ,$$

and $T = 0$ K, we have

$$p = -\left(\frac{\partial E}{\partial V}\right)_T = \frac{2E}{3V} .$$

(c) Assume that the velocity distribution is $D(\mathbf{v})d\mathbf{v}$, then the number of the molecules which collide with a unit area of the walls of the container in a unit time, with velocities between $\mathbf{v}$ and $\mathbf{v} + d\mathbf{v}$ is $nv_x D(\mathbf{v})d\mathbf{v}$. The force that the unit areas suffers due to the collisions is

$$dp = 2mv_x^2 nD(\mathbf{v})d\mathbf{v} .$$

Hence the pressure is

$$
\begin{aligned}
p &= \int_{v_x>0} nD(\mathbf{v}) \cdot 2mv_x^2 d\mathbf{v} = \int_{-\infty<v_x<+\infty} nD(\mathbf{v}) \cdot mv_x^2 d\mathbf{v} \\
&= \frac{2}{3} \int nD(\mathbf{v}) \frac{1}{2} mv^2 dv = \frac{2}{3} \frac{E}{V} .
\end{aligned}
$$

For an electron gas

$$p = \frac{2E}{3V} = \frac{N}{5V} \cdot \frac{h^2}{m} \left(\frac{3N}{8\pi V} \right)^{2/3} .$$

(d) The specific heat and the paramagnetic magnetization of metals can be qualitatively explained by the Fermi gas model.

Superconductivity cannot be explained by the Fermi gas model.

2101

The free-electron model of the conduction electrons in metals seems naive but is often successful. Among other things, it gives a reasonably good account of the compressibility for certain metals. This prompts the following question. You are given the number density n and the Fermi energy ε of a non-interacting Fermi gas at zero absolute temperature, $T = 0$ K. Find the isothermal compressibility

$$\kappa = -\frac{1}{V} \left(\frac{\partial V}{\partial p} \right)_T ,$$

where V is volume, p is pressure.

Hint: Recall that $pV = \frac{2}{3}E$, where E is the total energy.

(*CUSPEA*)

Solution:

$$p = -\left(\frac{\partial F}{\partial V}\right)_T,$$ where F is the free energy, $F = E - TS$. When $T = 0$ K, $F = E$, and

$$p = -\left(\frac{\partial E}{\partial V}\right)_T.$$

Using $pV = \frac{2}{3}E$, we have

$$p = -\left(\frac{\partial E}{\partial V}\right)_T = -\left[\frac{\partial}{\partial V}\left(\frac{3}{2}pV\right)\right]_T = -\frac{3}{2}\left[V\left(\frac{\partial p}{\partial V}\right)_T + p\right],$$

or

$$V\left(\frac{\partial p}{\partial V}\right)_T = -\frac{5}{3}p.$$

Hence $\kappa = -\frac{1}{V}\left(\frac{\partial V}{\partial p}\right)_T = \frac{3}{5p}$ $(T = 0$ K$)$.

At $T = 0$ K,

$$p = \frac{2E}{3V} = \frac{2}{3V} \cdot 2 \cdot \frac{V}{(2\pi)^3} \int_{k<k_F} d^3k \frac{\hbar^2 k^2}{2m},$$

$$= \frac{\hbar^2 k_F^5}{15m\pi^2}.$$

With

$$nV = N = 2 \cdot \frac{V}{(2\pi)^3} \int_{k<k_F} d^3k,$$

we obtain

$$n = \frac{k_F^3}{3\pi^2}.$$

For an ideal gas, the energy of a particle is

$$\varepsilon(k) = \frac{\hbar^2 k^2}{2m}.$$

Thus

$$\varepsilon_F = \frac{\hbar^2 k_F^2}{2m}.$$

Therefore,

$$p = \frac{2}{5}n \cdot \varepsilon_F, \quad (T = 0 \text{ K}),$$

and

$$\kappa = \frac{3}{2n\varepsilon_F}. \quad (T = 0 \text{ K}).$$

2102

Fermi gas. Consider an ideal Fermi gas whose atoms have mass $m = 5 \times 10^{-24}$ grams, nuclear spin $I = \frac{1}{2}$, and nuclear magnetic moment $\mu = 1 \times 10^{-23}$ erg/gauss. At $T = 0$ K, what is the largest density for which the gas can be completely polarized by an external magnetic field of 10^5 gauss? (Assume no electronic magnetic moment).

(*MIT*)

Solution:

After the gas is completely polarized by an external magnetic field, the Fermi energy is $\varepsilon_F = \dfrac{\hbar^2}{2m}(6\pi^2 n)^{2/3}$, where n is the particle density.

With $\varepsilon_F \leq 2\mu H$, we have

$$n \leq \frac{1}{6\pi^2}\left(\frac{4m\mu H}{\hbar^2}\right)^{3/2}.$$

Hence, $n_{max} = \dfrac{1}{6\pi^2}\left(\dfrac{4m\mu H}{\hbar^2}\right)^{3/2} \approx 2 \times 10^{17} \text{atoms/cm}^3$.

2103

State and give a brief justification for the leading exponent n in the temperature dependence of the following quantities in a highly degenerate three-dimensional electron gas:

(a) the specific heat at constant volume;

(b) the spin contribution to the magnetic moment M in a fixed magnetic field H.

(*MIT*)

Solution:

Let us first consider the integral I:

$$
\begin{aligned}
I &= \int_0^\infty \frac{f(\varepsilon)d\varepsilon}{e^{(\varepsilon-\mu)/kT}+1} \\
&= \int_{-\mu/kT}^\infty \frac{f(\mu+kTz)}{e^z+1}kT\,dz \\
&= kT\int_0^{\mu/kT}\frac{f(\mu-kTz)}{e^{-z}+1}dz + kT\int_0^\infty \frac{f(\mu+kTz)}{e^z+1}dz \\
&= \int_0^\mu f(x)dx - kT\int_0^{\mu/kT}\frac{f(\mu-kTz)}{e^z+1}dz + kT\int_0^\infty \frac{f(\mu+kTz)dz}{e^z+1}
\end{aligned}
$$

where $kTz = -\mu + \varepsilon$. As $\mu/kT \gg 1$, we can substitute ∞ for the upper limit of the second integral in above expression so that

$$
I = \int_0^\mu f(x)dx + kT \int_0^\infty \frac{f(\mu + kTz) - f(\mu - kTz)}{e^z + 1} dz
$$

$$
= \int_0^\mu f(x)dx + 2(kT)^2 f'(\mu) \int_0^\infty \frac{zdz}{e^z + 1} + \cdots , \quad \text{where } f'(\mu) = \frac{df(\mu)}{d\mu} .
$$

(a) Let $f(\varepsilon) = \varepsilon^{3/2}$, then the internal energy $E \sim I, C_v = \left(\dfrac{\partial E}{\partial T}\right)_V \sim$ T, i.e., $n = 1$. In fact, when $T = 0$ K, because the heat energy is so small, only those electrons which lie in the transition band of width about kT on the Fermi surface can be excited into energy levels of energies $\approx kT$. The part of the internal energy directly related to T is then

$$
\overline{N}T \sim T^2, \quad \text{i.e.,} \quad C_v \sim T .
$$

(b) Let $f(\varepsilon) = \varepsilon^{1/2}$, then $M \sim I$, hence $M \approx M_0(1 - \alpha T^2)$, i.e., $n = 0$. When $T = 0$ K, the Fermi surface ε_F with spin direction parallel to $\mathbf{H}$ is $\varepsilon_{F\uparrow} = \mu + \mu_B H$ (μ_B is the Bohr magneton) while the Fermi surface ε_F with spin direction opposite to $\mathbf{H}$ is $\varepsilon_{F\downarrow} = \mu - \mu_B H$. Therefore, there exists a net spin magnetic moment parallel to $\mathbf{H}$. Hence $n = 0$.

2104

Take a system of $N = 2 \times 10^{22}$ electrons in a "box" of volume $V = 1$ cm^3. The walls of the box are infinitely high potential barriers. Calculate the following within a factor of five and show the dependence on the relevant physical parameters:

(a) the specific heat, C,

(b) the magnetic susceptibility, χ,

(c) the pressure on the walls of the box, p,

(d) the average kinetic energy, $\langle E_k \rangle$.

(*Chicago*)

Solution:
The density of states in $\mathbf{k}$ space is given by

$$
D(k)dk = 2V \cdot \frac{4\pi k^2}{8\pi^3} dk ,
$$

and the kinetic energy of an electron is $\varepsilon = \dfrac{\hbar^2}{2m}k^2$. Combining, we get

$$D(\varepsilon) = \frac{V}{2\pi^2}\left(\frac{2m}{\hbar^2}\right)^{3/2}\varepsilon^{1/2}.$$

At $T = 0$ K, the N electrons fill up the energy levels from zero to $E_F = \dfrac{\hbar^2}{2m}k_F^2$, i.e.,

$$N = \int_0^{E_F} D(\varepsilon)d\varepsilon = \frac{2}{3}D(E_F)E_F$$

whence $E_F = \dfrac{\hbar^2}{2m}\left(3\pi^2\dfrac{N}{V}\right)^{\frac{2}{3}}$.

(a) The specific heat is

$$C \approx k_B^2 TD(E_F) \approx \frac{N}{E_F}k_B^2 T ,$$

where k_B is Boltzmann's constant.

(b) The magnetic susceptibility is

$$\chi = \mu_B^2 D(E_F) \approx \frac{N}{E_F}\mu_B^2 ,$$

where μ_B is the Bohr magneton.

(c), (d) The average kinetic energy is

$$\langle E_k \rangle = \int_0^{E_F} \varepsilon D(\varepsilon)d\varepsilon = \frac{2}{5}D(E_F)E_F^2 = \frac{3}{5}NE_F ,$$

and the pressure on the walls of the box is

$$p = \frac{d\langle E_k \rangle}{dV} = \frac{3}{5}\frac{NE_F}{V} .$$

2105

An ideal gas of N spin $\frac{1}{2}$ fermions is confined to a volume V. Calculate the zero temperature limit of (a) the chemical potential, (b) the average energy per particle, (c) the pressure, (d) the Pauli spin susceptibility. Show that in Gaussian units the susceptibility can be written as $3\mu_B^2 N/2\mu(0)V$, where $\mu(0)$ is the chemical potential at zero temperature. Assume each fermion has interaction with an external magnetic field of the form $2\mu_0 H S_z$, where μ_B is the Bohr magneton and S_z is the z-component of the spin.

(*Wisconsin*)

Solution:

As the spin of a fermion is $\frac{1}{2}$, its z component has two possible directions with respect to the magnetic field: up ($\uparrow$) and down ($\downarrow$). These correspond to energies $\pm \mu_B H$, respectively. Thus the energy of a particle is

$$\varepsilon = \frac{p^2}{2m} \pm \mu_B H .$$

At $T = 0$ K, the particles considered occupy all the energy levels below the Fermi energy $\mu(0)$. Therefore, the kinetic energies of the particles of negative spins distribute between 0 and $\mu(0) - \mu_B H$, those of positive spins distribute between 0 and $\mu(0) + \mu_B H$, their numbers being

$$N_- = \frac{4\pi V}{3h^3} p_-^3 \quad \left(\frac{1}{2m} p_-^2 = \mu(0) + \mu_B H \right) ,$$

$$N_+ = \frac{4\pi V}{3h^3} p_+^3 \quad \left(\frac{1}{2m} p_+^2 = \mu(0) - \mu_B H \right) .$$

(a) The total number of particles is

$$N = N_+ + N_- = \frac{4\pi V (2m)^{3/2}}{3h^3} \{ [\mu(0) - \mu_B H]^{3/2} + [\mu(0) + \mu_B H]^{3/2} \} .$$

With $H = 0$, we obtain the chemical potential

$$\mu(0) = \frac{\hbar^2}{2m} \left(3\pi^2 \frac{N}{V} \right)^{2/3} .$$

(b) For particles with z-components of spin, $\frac{1}{2}$ and $-\frac{1}{2}$, the Fermi momenta are respectively

$$p_+ = \{ 2m[\mu(0) - \mu_B H] \}^{1/2}$$
$$p_- = \{ 2m[\mu(0) + \mu_B H] \}^{1/2} .$$

The corresponding total energies are

$$E_+ = \frac{4\pi V}{h^3} \int_0^{p_+} \left(\frac{p^2}{2m} + \mu_B H \right) p^2 dp$$

$$= \frac{4\pi V}{h^3} \left[\frac{p_+^5}{10m} + \frac{\mu_B H}{3} p_+^3 \right] ,$$

$$E_- = \frac{4\pi V}{h^3} \left[\frac{p_-^5}{10m} - \frac{\mu_B H}{3} p_-^3 \right] .$$

Hence the average energy per particle is

$$\frac{E}{N} = \frac{E_+ + E_-}{N} = \frac{4\pi V}{h^3 N} \left[\frac{1}{10m}(p_+^5 + p_-^5) + \frac{\mu_B H}{3}(p_+^3 - p_-^3) \right] .$$

For $\mu(0) \gg \mu_B H$,

$$\frac{E}{N} \approx \frac{3}{5}\mu(0) \left[1 - \frac{5}{2} \left(\frac{\mu_B H}{\mu_0} \right)^2 \right] .$$

(c) The pressure is

$$p = -\left(\frac{\partial E}{\partial V} \right)_T = -\frac{\partial E}{\partial \mu(0)} \cdot \frac{\partial \mu(0)}{\partial V} = \frac{2N}{5V}\mu(0) = \frac{2}{5}n\mu(0) .$$

(d) For $\mu(0) \gg \mu_B H$, the magnetization is given by

$$M = \mu_B(N_- - N_+)/V = \frac{3\mu_B^2 N}{2\mu(0)V} H = \chi H .$$

Hence $\chi = \dfrac{3N\mu_B^2}{2\mu(0)V}$.

2106

Consider a Fermi gas model of nuclei.

Except for the Pauli principle, the nucleons in a heavy nucleus are assumed to move independently in a sphere corresponding to the nuclear volume V. They are considered as a completely degenerate Fermi gas. Let $A = N$ (the number of neutrons) $+Z$ (the number of protons), assume $N = Z$, and compute the kinetic energy per nucleon, E_{kin}/A, with this model.

The volume of the nucleus is given by $V = \dfrac{4\pi}{3} R_0^3 A$, $R_0 \approx 1.4 \times 10^{-13}$ cm.
Please give the result in MeV.

(*Chicago*)

Solution:

In the momentum space,

$$dn = \frac{4V}{h^3} 4\pi p^2 \, dp \; ,$$

where n is the number density of neutrons.

The total number of neutrons is

$$A = \int dn = 16\pi V \int_0^{p_F} \frac{p^2}{h^3} \, dp$$
$$= \frac{16\pi V}{3h^3} p_F^3 \; ,$$

where p_F is the Fermi momentum.

The total kinetic energy of the neutrons is

$$E_{\text{kin}} = \int \frac{p^2}{2m} dn = \frac{16\pi V}{5h^3} \frac{p_F^5}{2m} \; .$$

Hence,

$$\frac{E_{\text{kin}}}{A} = \frac{3}{5} \frac{p_F^2}{2m} \; .$$

The volume V can be expressed in two ways:

$$V = \frac{4\pi}{3} R_0^3 A = \frac{3(2\pi)^3}{16\pi} p_F^{-3} A, \quad \left(\text{putting } \hbar \equiv \frac{h}{2\pi} = 1 \right)$$

giving $p_F = R_0^{-1} \left(\dfrac{9\pi}{8} \right)^{1/3}$, and

$$\frac{E_{\text{kin}}}{A} = \frac{3}{10} \left(\frac{9\pi}{8} \right)^{2/3} \frac{1}{mR_0^2} \approx 16 \text{ MeV} \; .$$

2107

At low temperatures, a mixture of ^{3}He and ^{4}He atoms form a liquid which separates into two phases: a concentrated phase (nearly pure ^{3}He), and a dilute phase (roughly 6.5% ^{3}He for $T \leq 0.1$ K). The lighter ^{3}He floats on top of the dilute phase, and ^{3}He atoms can cross the phase boundary (see Fig. 2.23).

The superfluid ^{4}He has negligible excitation, and the thermodynamics of the dilute phase can be represented as an ideal degenerate Fermi gas of particles with density n_d and effective mass m^* (m^* is larger than m_3, the mass of the bare ^{3}He atom, due to the presence of the liquid ^{4}He, actually $m^* = 2.4m_3$). We can crudely represent the concentrated phase by an ideal degenerate Fermi gas of density n_c and particle mass m_3.

(a) Calculate the Fermi energies for the two fluids.

(b) Using simple physical arguments, make an estimate of the very low temperature specific heat of the concentrated phase $c_c(T, T_{Fc})$ which explicitly shows its functional dependence on T and T_{Fc} (where T_{Fc} is the Fermi temperature of the concentrated phase, and any constants independent of T and T_{Fc} need not be determined). Compare the specific heats of the dilute and concentrated phases.

(c) How much heat is required to warm each phase from $T = 0$ K to T?

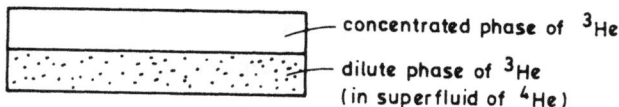

concentrated phase of ^{3}He

dilute phase of ^{3}He (in superfluid of ^{4}He)

Fig. 2.23.

(d) Suppose the container in the figure is now connected to external plumbing so that ^{3}He atoms can be transferred from the concentrated phase to the dilute phase at a rate of $\dot{N}_s$ atoms per second (as in a dilution refrigerator). For fixed temperature T, how much power can this system absorb?

(*Princeton*)

Solution:

(a) As $n = \dfrac{8\pi}{3}\left(\dfrac{2mE_F}{h^2}\right)^{3/2}$, we have $E_{Fc} = \dfrac{h^2}{2m_3}\left(\dfrac{3n_c}{8\pi}\right)^{2/3}$, and

$$E_{\mathrm{Fd}} = \frac{h^2}{2m^*} \left(\frac{3n_{\mathrm{d}}}{8\pi}\right)^{2/3}.$$

(b) For an ideal degenerate Fermi gas at low temperatures, only those particles whose energies are within $(E_F - kT)$ and $(E_F + kT)$ contribute to the specific heat. The effective particle number is $n_{\mathrm{eff}} = n\frac{kT}{E_F}$, so

$$c_v \propto n_{\mathrm{eff}} \propto \frac{T}{E_F} = \alpha_c \frac{T}{T_{\mathrm{Fc}}},$$

where α_c is a constant.

(c) $Q_c = \int_0^T c_v \, dT = \frac{\alpha_c T^2}{2T_{\mathrm{Fc}}},$

$$Q_{\mathrm{d}} = \int_0^T c_v \, dT = \frac{\alpha_{\mathrm{d}} T^2}{2T_{\mathrm{Fd}}}.$$

(d) The entropy per particle at low temperature is

$$S(T) = \int_0^T \frac{c_v}{T} \, dT = \lambda \frac{T}{T_F}, \quad \text{where } \lambda \text{ is a constant.}$$

The power absorbed is converted to latent heat, being

$$W = \dot{N}_s(S_{\mathrm{d}}(T) - S_c(T))T = \dot{N}_s T^2 \left(\frac{\lambda_{\mathrm{d}}}{T_{\mathrm{Fd}}} - \frac{\lambda_c}{T_{\mathrm{Fc}}}\right).$$

2108

A white-dwarf star is thought to constitute a degenerate electron gas system at a uniform temperature much below the Fermi temperature. This system is stable against gravitational collapse so long as the electrons are non-relativistic.

(a) Calculate the electron density for which the Fermi momentum is one-tenth of the electron rest mass $\times c$.

(b) Calculate the pressure of the degenerate electron gas under these conditions.

(*UC, Berkeley*)

Solution:

(a) $N = \dfrac{2V}{h^3} \displaystyle\iiint\limits_{p \leq p_F} d\mathbf{p}$,

giving $n = \dfrac{N}{V} = \dfrac{8\pi}{3} \left(\dfrac{p_F}{h}\right)^3$.

With

$$p_F = \frac{m_e c}{10}$$

we have

$$n = \frac{8\pi}{3}\left(\frac{m_e c}{10h}\right)^3 = 5.8 \times 10^{32} \ /\text{m}^3 \ .$$

(b) For a strong degenerate Fermi gas (under the approximation of zero valence), we get

$$\overline{E} = \frac{3}{5}N\mu_0 \ ,$$

and

$$p = \frac{2}{3}\frac{\overline{E}}{V} = \frac{2}{5}n\mu_0 = \frac{2}{5}n \cdot \frac{p_F^2}{2m} = 9.5 \times 10^{16} \ \text{N/m}^2 \ .$$

2109

A white dwarf is a star supported by the pressure of degenerate electrons. As a simplified model for such an object, consider a sphere of an ideal gas consisting of electrons and completely ionized Si^{28}, and of constant density throughout the star. (Note that the assumption of a constant density is inconsistent with hydrostatic equilibrium, since the pressure is then also constant. The assumption that the gas is ideal is also not really tenable. These shortcomings of the model are, however, not crucial for the issues which we wish to consider.) Let n_i denote the density of the silicon ions, and let $n_e = 14n_i$ denote the electron density. (The atomic number of silicon is 14).

(a) Find the relation between the mean kinetic energy $\overline{E}_e$ of the electrons and the density n_e, assuming that the densities are such that the electrons are "extremely relativistic," i.e., such that the rest energy is negligible compared with the total energy.

(b) Compute $\overline{E}_e$ (in MeV) in the case that the (rest mass) density of the gas equals $\rho = 10^9$ g/cm^3. Also compute the mean kinetic energy $\overline{E}_i$ of the silicon ions in the central region of the dwarf, assuming that the temperature is 10^8 K and assuming that the "ion gas" can be regarded as a Maxwell-Boltzmann gas, and hence convince yourself that $\overline{E}_e \gg \overline{E}_i$.

(c) If M is the mass of the star, and if R is its radius, then the gravitational potential energy is given by

$$U_G = \frac{3GM^2}{5R} .$$

In the case in which the internal energy is dominated by extremely relativistic electrons (as in part (b) above), the virial theorem implies that the total internal energy is approximately equal to the gravitational potential energy. Assuming equality, and assuming that the electrons do not contribute significantly to the mass of the star, show that the stellar mass can be expressed in terms of fundamental physical constants alone. Evaluate your answer numerically and compare it with the mass of the sun, 2×10^{30} kg. (It can be shown that this is approximately the maximum possible mass of a white dwarf.)

(UC, Berkeley)

Solution:

(a) Use the approximation of strong degenerate electron gas and $\varepsilon = pc$. From the quantum state density of electrons, it follows

$$\frac{2}{h^3} dp = \frac{8\pi}{h^3 c^3} \varepsilon^2 d\varepsilon ,$$

then

$$n_e = \int_0^{\varepsilon_F} \frac{8\pi}{h^3 c^3} \varepsilon^2 d\varepsilon$$

$$= \frac{8\pi}{3h^3 c^3} \varepsilon_F^3 .$$

Therefore

$$\overline{E}_e = \frac{\int_0^{\varepsilon_F} \varepsilon \cdot \varepsilon^2 d\varepsilon}{\int_0^{\varepsilon_F} \varepsilon^2 d\varepsilon} = \frac{3}{4}\varepsilon_F = \frac{3}{4} hc \left(\frac{3n_e}{8\pi}\right)^{1/3} .$$

(b) When $\rho = 10^9$ g/cm^3,

$$n_e = 14n_i = 3 \times 10^{32}\text{cm}^{-3} = 3 \times 10^{38} \text{ m}^{-3} ,$$

$$\overline{E}_e = 5 \times 10^{-13} \text{ J} = 3 \text{ MeV} ,$$

$$\overline{E}_i = \frac{3}{2}kT = 2 \times 10^{-15} \text{ J} = 1.3 \times 10^{-2} \text{ MeV} .$$

Obviously, $\overline{E}_i \ll \overline{E}_e$.

(c) From the virial theorem, we have

$$\left(\frac{4\pi}{3}R^3 n_e\right) \cdot \frac{3}{4}hc\left(\frac{3n_e}{8\pi}\right)^{1/3} = \frac{3}{5}\frac{GM^2}{R} .$$

Noting that

$$M = \frac{4\pi}{3}R^3 \frac{n_e}{14}m_i = \frac{8\pi}{3}R^3 n_e m_p ,$$

we obtain

$$M = \frac{15}{128\pi} \cdot \frac{hc}{Gm_p^2}\sqrt{\frac{5hc}{2G}} = 8.5 \times 10^{30} \text{ kg} = 4.1M_\odot ,$$

where $M_\odot$ is the mass of the sun.

2110

(a) Given that the mass of the sun is 2×10^{33} g, estimate the number of electrons in the sun. Assume the sun is largely composed of atomic hydrogen.

(b) In a white dwarf star of one solar mass the atoms are all ionized and contained in a sphere of radius 2×10^9 cm. Find the Fermi energy of the electrons in eV.

(c) If the temperature of the white dwarf is 10^7 K, discuss whether the electrons and/or nucleons in the star are degenerate.

(d) If the above number of electrons were contained in a pulsar of one solar mass and of radius 10 km, find the order of magnitude of their Fermi energy.

(Columbia)

Solution:

(a) The number of electrons is

$$N = \frac{2 \times 10^{33}}{1.67 \times 10^{-24}} \approx 1.2 \times 10^{57} .$$

(b) The Fermi energy of the electrons is

$$E_{Fe} = \frac{h^2}{2m_e}\left(\frac{3}{8\pi}\frac{N}{V}\right)^{2/3} = \frac{h^2}{2m_e}\left(\frac{9}{32\pi^2}\frac{N}{R^3}\right)^{2/3} \approx 4 \times 10^4 \text{ eV} .$$

The Fermi energy of the nucleons is

$$E_{Fn} = E_{Fe}\frac{m_e}{m_n} = \frac{1}{1840}E_{Fe} \, .$$

(c) $E_{Fe}/k = 4 \times 10^8$ K $> 10^7$ K.

$E_{Fn}/k \ll 10^7$ K.
Therefore, in a white dwarf, the electrons are strongly degenerate while the nucleons are weakly degenerate.

(d) The Fermi energy of the electrons if contained in a pulsar is

$$E'_{Fe} = \left(\frac{R}{R'}\right)^2 E_{Fe} = 4 \times 10^6 E_{Fe} = 1.6 \times 10^5 \text{ MeV} \, .$$

2111

At what particle density does a gas of free electrons (considered at $T = 0$ K) have enough one-particle kinetic energy (Fermi energy) to permit the reaction

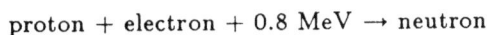

$$\text{proton} + \text{electron} + 0.8 \text{ MeV} \rightarrow \text{neutron}$$

to proceed from left to right? Using the result above estimate the minimum density of a neutron star.

(*UC, Berkeley*)

Solution:
When $T = 0$ K, the Fermi energy and the number density of the electron gas are related as follows:

$$n = \frac{8\pi}{3}\left(\frac{2m\varepsilon_F}{h^2}\right)^{3/2} \, .$$

The condition for the reaction to proceed is $\varepsilon_F \geq 0.8$ MeV, then

$$n_{min} = 3.24 \times 10^{36} \text{ m}^{-3} \, .$$

Hence the minimum mass density of a neutron star is

$$\rho_{min} = m_n n_{min} = 5.4 \times 10^9 \text{ kg/m}^3 \, .$$

2112

Assume that a neutron star is a highly degenerate non-relativistic gas of neutrons in a spherically symmetric equilibrium configuration. It is held together by the gravitational pull of a heavy object with mass M and radius r_0 at the center of the star. Neglect all interactions among the neutrons. Calculate the neutron density as a function of the distance from the center, r, for $r > r_0$.

(Chicago)

Solution:

For a non-relativistic degenerate gas, the density $\rho \propto \mu^{3/2}$, the pressure $p \propto \mu^{5/2}$, where μ is the chemical potential. Therefore, $p = a\rho^{5/3}$, where a is a constant. Applying it to the equation

$$\frac{dp}{\rho} = MGd\left(\frac{1}{r}\right) ,$$

we find $a \cdot \frac{5}{2} d\rho^{2/3} = MGd\left(\frac{1}{r}\right)$ and hence

$$\rho(r) = \left[\frac{2MG}{5a} \cdot \frac{1}{r} + \text{const}\right]^{3/2} .$$

As $r \to \infty, \rho(r) \to 0$, we find const. $= 0$. Finally, with $r > r_0$, we have

$$\rho(r) = \left[\frac{2MG}{5a} \cdot \frac{1}{r}\right]^{3/2} .$$

2113

Consider a degenerate (i.e., $T = 0$ K) gas of N non-interacting electrons in a volume V.

(a) Find an equation relating pressure, energy and volume of this gas for the extreme relativistic case (ignore the electron mass).

(b) For a gas of real electrons (i.e., of mass m), find the condition on N and V for the result of part (a) to be approximately valid.

(MIT)

Solution:

The energy of a non-interacting degenerate electron gas is:

$$E = 8\pi V \int_0^{p_F} \frac{\varepsilon p^2}{h^3} dp$$

where ε is the energy of a single electron, p_F is the Fermi momentum,

$$p_F = (3N/8\pi V)^{1/3} h .$$

(a) For the extreme relativistic case, $\varepsilon = cp$, so we have energy

$$E = \frac{2\pi c V}{h^3} p_F^4 ,$$

and pressure $p = -\left(\dfrac{\partial E}{\partial V}\right)_{T=0} = \dfrac{1}{3}\dfrac{E}{V}$, which gives the equation of state

$$pV = \frac{1}{3} E .$$

(b) For a real electron,

$$\varepsilon = \sqrt{(mc^2)^2 + (pc)^2} \approx pc\left[1 + \frac{1}{2}\left(\frac{mc}{p}\right)^2\right] ,$$

where p is its momentum, giving

$$E \approx 2\pi c V [p_F^4 + (mcp_F)^2]/h^3 .$$

The condition for the result of part (a) to be approximately valid is $p_F \gg mc$, or

$$\frac{N}{V} \gg \frac{8\pi}{3}\left(\frac{mc}{h}\right)^3 .$$

Either $N \to \infty$ or $V \to 0$ will satisfy this condition.

2114

Consider a box of volume V containing electron-positron pairs and photons in equilibrium at a temperature $T = 1/k\beta$. Assume that the equilibrium is established by the reaction

$$\gamma \leftrightarrow e^+ + e^- .$$

The reaction does not occur in free space, but one may think of it as catalyzed by the walls of the box. Ignoring the walls except insofar as they allow the reaction to occur, find

(a) The chemical potentials for the fermions.

(b) The average number of electron-positron pairs, in the two limits $kT \gg m_e c^2$ and $kT \ll m_e c^2$. (You may leave your answers in terms of dimensionless definite integrals.)

(c) The neglect of the walls is not strictly permissible if they contain a matter-antimatter imbalance. Supposing that this imbalance creates a net chemical potential $\mu \neq 0$ for the electrons, what is then the chemical potential of the positrons?

(d) Calculate the net charge of the system in the presence of this imbalance in the limit $kT \gg \mu \gg m_e c^2$. (Again, your answer may be left in terms of a dimensionless definite integral.)

(*Chicago*)

Solution:

(a) For a chemical reaction $A \leftrightarrow B + C$ at equilibrium, $\mu_A = \mu_B + \mu_C$. As the chemical potential of the photon gas $\mu_\gamma = 0$, we obtain

$$\mu_{e^+} + \mu_{e^-} = 0 .$$

Considering the symmetry between particle and antiparticle, we have

$$\mu_{e^+} = \mu_{e^-} .$$

Hence $\mu_{e^+} = \mu_{e^-} = 0$.

(b) At the limit $kT \gg m_e c^2$, neglecting the electron mass and letting $E = cp$, we obtain

$$
\begin{aligned}
N_{e^-} &= \frac{2V}{(2\pi\hbar)^3} \int d^3p \, \frac{1}{e^{\beta cp} + 1} \\
&= \frac{V}{\pi^2} \frac{(kT)^3}{(\hbar c)^3} \int_0^\infty \frac{x^2 dx}{e^x + 1} = N_{e^+} .
\end{aligned}
$$

At the limit $kT \ll m_e c^2$, the "1" in denominator of the Fermi factor

$$\frac{1}{[\exp(\beta\sqrt{(cp)^2 + (m_e c^2)^2}) + 1]}$$

can be neglected and we also have

$$\varepsilon_p \equiv \sqrt{(cp)^2 + (m_e c^2)^2} \approx m_e c^2 + p^2/2m \ .$$

Thus

$$N_{e\pm} = \frac{2V}{(2\pi\hbar)^3} \int_0^\infty e^{-\beta m_e c^2} e^{-\beta p^2/2m_e} 4\pi p^2 \, dp$$

$$= 2V \left(\frac{2\pi m_e kT}{h^2} \right)^{3/2} e^{-m_e c^2/kT} \ .$$

(c) As $\mu_{e+} + \mu_{e-} = 0$, $\mu_{e+} = -\mu_{e-} = -\mu$.

(d) The net charge of the system is $q = (-e)(n_{e-} - n_{e+})$, where

$$n_{e\pm} = \frac{8\pi V}{h^3} \int_0^\infty \frac{p^2 \, dp}{e^{\beta(\varepsilon_p \mp \mu)} + 1} \ .$$

As $\beta\mu \ll 1$, $e^{\pm\beta\mu} \approx 1 \pm \beta\mu$, and

$$q = \frac{-e \cdot 8\pi V}{h^3} \int_0^\infty p^2 \, dp \left[\frac{1}{e^{-\beta\mu} e^{\beta\varepsilon_p} + 1} - \frac{1}{e^{\mu\beta} e^{\beta\varepsilon_p} + 1} \right]$$

$$= -\frac{8\pi e V}{h^3 c^3} (kT)^2 \mu \int_0^\infty \frac{x^2 e^x}{(e^x + 1)^2} \, dx \ .$$

2115

In the very early stages of the universe, it is usually a good approximation to neglect particle masses and chemical potential compared with kT.

(a) Write down the average number and energy densities of a gas of non-interacting fermions in thermal equilibrium under these conditions. (You need not evaluate dimensionless integrals of order 1.)

(b) If the gas expands adiabatically while remaining in equilibrium, how do the average number and energy densities depend on the dimensions of the system?

Assume that the fermions are predominantly electrons and positrons when $T \simeq 10^{11}$ K in parts (c) and (d) below.

(c) Is the assumption made in (a) that the particles are non-interacting reasonable? Why? [Hint: What is the average coulomb interaction energy?

Positron charge $= 1.6 \times 10^{-19}$ coulomb; Boltzmann's constant $k = 1.38 \times 10^{-16}$ erg/K].

(d) If the interaction cross sections in the electron-positron gas are typically of order of magnitude of the Thompson cross section $\sigma_T = 8\pi r_0^2/3$ (classical electron radius $r_0 = 2.8 \times 10^{-13}$ cm), estimate the mean free time between collisions of the particles. If the expansion rate in part (b) $\approx 10^4$ sec^{-1}, is the assumption that the gas remains in equilibrium reasonable? Why?

(SUNY, Buffalo)

Solution:

(a) In the stated approximation, we have

$$\varepsilon = pc, \qquad \frac{\mu}{kT} \approx 0 \ .$$

So

$$N = \frac{V}{(2\pi\hbar)^3} 4\pi \int_0^\infty \frac{p^2}{e^{pc/kT} + 1} dp \ .$$

The average number density is

$$n = \frac{1}{2\pi^2} \left(\frac{kT}{\hbar c} \right)^3 \int_0^\infty \frac{x^2 dx}{e^x + 1} \ .$$

The average energy density is

$$\rho = \frac{1}{2\pi^2} \left(\frac{kT}{\hbar c} \right)^3 kT \int_0^\infty \frac{x^3 dx}{e^x + 1} \ .$$

(b) The quasi-static adiabatic expansion process satisfies the equation $d(\rho V) = -pdV$. Neglecting the particle mass, we have $p = \rho/3$ (analogous to a photon gas), then

$$\frac{d\rho}{\rho} = -\frac{4}{3} \frac{dV}{V} \ ,$$

giving

$$\rho \propto V^{-4/3} \ ,$$

from which we obtain $T \propto V^{-1/3}$. Hence the particle number density $n \propto V^{-1}$.

(c) The average distance between particles $r \propto n^{-1/3}$. The ratio of the Coulomb interaction energy per particle to the particle kinetic energy is

$$\frac{e^2/r}{kT} \sim \frac{e^2 n^{1/3}}{kT} \sim \frac{e^2}{\hbar c} \approx \frac{1}{137} \ .$$

This implies that the interaction energy is much less than the kinetic energy, which makes the approximation in (a) reasonable.

(d) The mean free time is $t \sim 1/n\sigma_T v$, where the average speed

$$v \sim \left(\frac{kT}{m_e}\right)^{1/2} ,$$

Hence $\quad t \sim \left(\frac{kT}{\hbar c}\right)^{-3} \sigma_T^{-1} \left(\frac{kT}{m_e}\right)^{-1/2} \sim 10^{-23}$ s.

The assumption that the gas remains in equilibrium is reasonable for the mean free time is much shorter than the expansion time which is of the order of 10^{-4}s.

4. ENSEMBLES (2116 - 2148)

2116

Heat Capacity.

The constant volume heat capacity of a system with average energy $\langle E \rangle$ is given by $C_v = \left(\dfrac{\partial \langle E \rangle}{\partial T}\right)_{N,V}$.

Use the canonical ensemble to prove that: C is related to the mean-square fluctuation in the energy as follows:

$$C_v = \frac{1}{kT^2}\langle (E - \langle E \rangle)^2 \rangle .$$

(*MIT*)

Solution:

The partition function is

$$Z = \sum \exp(-E_n/kT) .$$

Therefore, $\langle E \rangle = \dfrac{1}{Z}\sum E_n e^{-E_n/kT}$. Then

$$C_v = \left.\frac{\partial \langle E \rangle}{\partial T}\right|_{N,V} = -\frac{\partial \ln Z}{\partial T}\langle E \rangle + \frac{1}{kT^2}\langle E^2 \rangle$$

$$= \frac{1}{kT^2}[\langle E^2 \rangle - \langle E \rangle^2] = \frac{1}{kT^2}\langle (E - \langle E \rangle)^2 \rangle .$$

2117

(a) Give the thermodynamic definition of the Helmholtz free energy F, the classical statistical mechanical definition of the partition function Z, and the relationship between these quantities. Define all symbols.

(b) Using these expressions and thermodynamic arguments show that the heat capacity at consant volume c_v is given by

$$c_v = kT \left[\frac{\partial^2}{\partial T^2} (T \ln Z) \right]_V .$$

(c) Consider a classical system that has two discrete total energy states E_0 and E_1. Find Z and c_v.

(SUNY, Buffalo)

Solution:

(a) $F = U - TS$, $Z = \int \exp(-\beta E(p,q)) d\omega$, where U is the internal energy, T the absolute temperature, S the entropy, $\beta = 1/kT$, $E(p,q)$ the energy of the system and $d\omega = dp\,dq$ an infinitesimal volume element in the phase space, p and q being the generalized momentum and coordinate respectively, and k Boltzmann's constant.

The relation between F and Z is

$$F = -kT \ln Z .$$

(b) From $dF = -SdT - pdV$, we have

$$S = - \left(\frac{\partial F}{\partial T} \right)_V .$$

Hence

$$c_v = T \left(\frac{\partial S}{\partial T} \right)_V = -T \left(\frac{\partial^2 F}{\partial T^2} \right)_V$$

$$= kT \left[\frac{\partial^2}{\partial T^2} (T \ln Z) \right]_V .$$

(c) $Z = e^{-\beta E_0} + e^{-\beta E_1}$,

$$c_v = kT \left\{ \frac{\partial^2}{\partial T^2} [T \ln(e^{-\beta E_0} + e^{-\beta E_1})] \right\}_V$$

$$= \frac{(E_1 - E_0)^2}{4kT^2 \cosh^2 \left(\dfrac{E_1 - E_0}{2kT} \right)} .$$

2118

Consider the energy and fluctuation in energy of an arbitrary system in contact with a heat reservoir at absolute temperature $T = 1/k\beta$.

(a) Show that the average energy $\overline{E}$ of the system is

$$\overline{E} = -\frac{\partial \ln z}{\partial \beta}$$

where $z = \sum_n \exp(-\beta E_n)$ sums over all states of the system.

(b) Obtain an expression for $\overline{E^2}$ in terms of the derivatives of $\ln z$.

(c) Calculate the dispersion of the energy, $\overline{(\Delta E)^2} = \overline{E^2} - \overline{E}^2$.

(d) Show that the standard deviation $\widetilde{\Delta E} = \left(\overline{(\Delta E)^2}\right)^{1/2}$ can be expressed in terms of the heat capacity of the system and the absolute temperature.

(e) Use this result to derive an expression for $\widetilde{\Delta E}/\overline{E}$ for an ideal monatomic gas.

(*UC, Berkeley*)

Solution:

(a) $\overline{E} = \dfrac{\sum\limits_n E_n e^{-\beta E_n}}{\sum\limits_n e^{-\beta E_n}} = \dfrac{-\dfrac{\partial z}{\partial \beta}}{z} = -\dfrac{\partial}{\partial \beta} \ln z$.

(b) $\overline{E^2} = \dfrac{\sum\limits_n E_n^2 e^{-\beta E_n}}{\sum\limits_n e^{-\beta E_n}} = \dfrac{\dfrac{\partial^2 z}{\partial \beta^2}}{z} = \dfrac{\partial}{\partial \beta}\left(\dfrac{\partial}{\partial \beta} \ln z\right) + \left(\dfrac{\partial}{\partial \beta} \ln z\right)^2$.

(c) $\overline{(\Delta E)^2} = \overline{E^2} - (\overline{E})^2 = \dfrac{\partial^2}{\partial \beta^2} \ln z = -\dfrac{\partial}{\partial \beta}\overline{E} = kT^2 c_v$.

(d) $\widetilde{\Delta E} = \sqrt{\overline{(\Delta E)^2}} = \sqrt{kc_v} \, T$.

(e) For an ideal monatomic gas,

$$\overline{E} = \frac{3}{2}NkT, \quad c_v = \frac{3}{2}Nk$$

and thus

$$\frac{\widetilde{\Delta E}}{\overline{E}} = \sqrt{\frac{2}{3N}} \ .$$

2119

A useful way to cool He^3 is to apply pressure P at sufficiently low temperature T to a co-existing liquid-solid mixture. Describe qualitatively how this works on the basis of the following assumptions:

(a) The molar volume of the liquid V_L is greater than that of the solid V_S at all temperatures.

(b) The molar liquid entropy is given by

$$S_L = \gamma RT \quad \text{with} \quad \gamma \sim 4.6 \text{ K}^{-1} .$$

(c) The entropy of the solid S_S comes entirely from the disorder associated with the nuclear spins $(s = 1/2)$.

Note: Include in your answer a semi-quantitative graph of the $p-T$ diagram of He^3 at low temperatures (derived using the above information).

(*Chicago*)

Solution:

The Clausius-Clapeyron equation is

$$\frac{dp}{dT} = \frac{\Delta S}{\Delta V} = \frac{S_L - S_S}{V_L - V_S} .$$

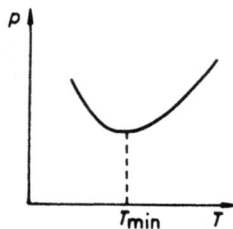

Fig. 2.24.

For particles of spin $\frac{1}{2}$, $S_S = kN_A \ln 2$. Thus

$$\frac{dp}{dT} = \frac{\gamma RT - kN_A \ln 2}{V_L - V_S} = \frac{\gamma RT - R \ln 2}{V_L - V_S} .$$

According to the problem, $V_L - V_S > 0$, thus when $T \to 0$, $\frac{dp}{dT} < 0$. Hence, when

$$T_{\min} = \frac{\ln 2}{\gamma} = \frac{\ln 2}{4.6} \text{ K},$$

the pressure reaches the minimum value. This means that at sufficiently low temperatures $(T < T_{\min})$, applying compression can lead to a decrease in temperature of the solid-liquid mixture.

A semi-quantitative $p - T$ diagram of He^3 at low temperatures is shown in Fig. 2.24.

2120

(a) Describe the third law of thermodynamics.

(b) Explain the physical meaning of negative absolute temperature. Does it violate the third law? Why?

(c) Suggest one example in which the negative temperature can actually be achieved.

(d) Discuss why the negative temperature does not make sense in classical thermodynamics.

(SUNY, Buffalo)

Solution:

(a) The third law or the Nernst heat theorem signifies that no system can have its absolute temperature reduced to zero.

(b) According to the Gibbs distribution, at equilibrium the ratio of the particle number of energy level E_n to that of E_m is $N_n/N_m = \exp[-(E_n - E_m)/kT]$. Hence, the particle number in the higher energy level is smaller than that in the lower energy level for $T > 0$. If the reverse is the case, i.e., under population inversion, the equation requires $T < 0$ and the system is said to be at negative temperature. This does not violate the third law for a system at negative temperature is further away from absolute zero than a system at positive temperature, from the point of view of energy.

(c) One such example is a localized system of spin $\frac{1}{2}$ particles. We can introduce a strong magnetic field to align all the spins in the same direction as, i.e., parallel to, the direction of the magnetic field. We then reverse the magnetic field quickly so that there is no time for most of the spins to change direction. Thus negative temperature is achieved.

(d) In classical thermodynamics, a negative temperature system is mechanically unstable. We divide a substance at rest into several parts. Let the internal energy and entropy of part i be U_i and $S_i(U_i)$ respectively. We have

$$U_i = E_i - p_i^2/2M_i$$

where E_i is the total energy of the part, M_i is its mass, and p_i is its momentum with $\sum_i \mathbf{p}_i = 0$. Mechanical equilibrium requires all $\mathbf{p}_i = 0$. As we have for a negative temperature system $dS_i(U_i)/dU_i = 1/T < 0$, S_i will increase when U_i decreases, i.e., p_i increases. Thus the entropies $S_i(U_i)$ are maximum when the $|\mathbf{p}_i|$'s reach maximum. This contradicts the mechanical equilibrium condition $\mathbf{p}_i = 0$.

2121

Consider a system of two atoms, each having ony 3 quantum states of energies 0, ε and 2ε. The system is in contact with a heat reservoir at temperature T. Write down the partition function Z for the system if the particles obey

(a) Classical statistics and are distinguishable.

(b) Classical statistics and are indistinguishable.

(c) Fermi-Dirac statistics.

(d) Bose-Einstein statistics.

(SUNY, Buffalo)

Solution:

(a) $Z_1 = A^2$, where $A = 1 + \exp(-\beta\varepsilon) + \exp(-2\beta\varepsilon)$.

(b) $Z_2 = \dfrac{Z_1}{2}$.

(c) $Z_3 = A\exp(-\beta\varepsilon)$.

(d) $Z_4 = A(1 + \exp(-2\beta\varepsilon))$.

2122

(a) You are given a system of two identical particles which may occupy any of the three energy levels

$$\varepsilon_n = n\varepsilon, \quad n = 0, 1, 2, \ .$$

The lowest energy state, $\varepsilon_0 = 0$, is doubly degenerate. The system is in thermal equilibrium at temperature T. For each of the following cases determine the partition function and the energy and carefully enumerate the configurations.

1) The particles obey Fermi statistics.

2) The particle obey Bose statistics.

3) The (now distinguishable) particles obey Boltzmann statistics.

(b) Discuss the conditions under which Fermions or Bosons may be treated as Boltzmann particles.

(SUNY, Buffalo)

Solution:

(a) Considering the systems as a canonical ensemble, the partition function is $z = \sum_n \omega_n \exp(-\beta E_n)$, where ω_n is the degeneracy of energy level n.

1) The particles obey Fermi statistics. We have

$$z = 1 + 2e^{-\beta\varepsilon} + e^{-3\beta\varepsilon}\left(1 + 2e^{\beta\varepsilon}\right) ,$$
$$E = -\frac{\partial}{\partial\beta}\ln z = -\frac{1}{z}\frac{\partial z}{\partial\beta}$$
$$= \frac{\varepsilon}{z}e^{-\beta\varepsilon}\left(2 + 4e^{-\beta\varepsilon} + 3e^{-2\beta\varepsilon}\right) .$$

The configurations are shown in Fig. 2.25(a)

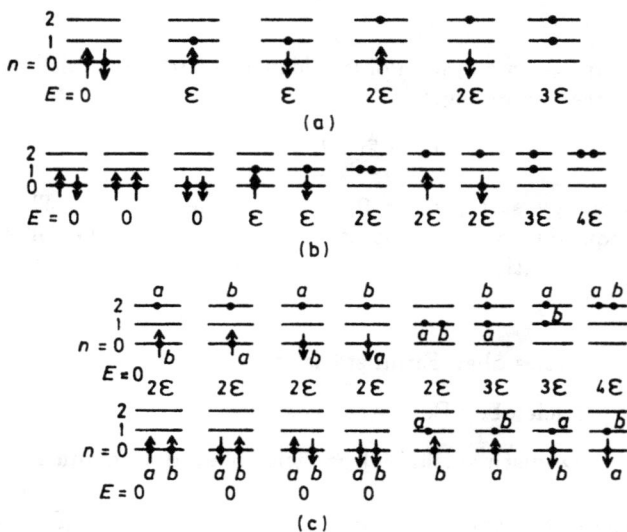

Fig. 2.25.

2) The particles obey Bose statistics. We have

$$z = 3 + 2e^{-\beta\epsilon} + 3e^{-2\beta\epsilon} + e^{-3\beta\epsilon} + e^{-4\beta\epsilon} .$$

$$E = -\frac{1}{z}\frac{\partial z}{\partial \beta}$$

$$= \frac{\epsilon}{z}e^{-\beta\epsilon}\left(2 + 6e^{-\beta\epsilon} + 3e^{-2\beta\epsilon} + 4e^{-3\beta\epsilon}\right) .$$

The configurations are shown in Fig. 2.25(b).

3) The particles obey Boltzmann statistics. We have

$$z = 4 + 4e^{-\beta\epsilon} + 5e^{-2\beta\epsilon} + 2e^{-3\beta\epsilon} + e^{-4\beta\epsilon} .$$

$$E = \frac{2\epsilon}{z}e^{-\beta\epsilon}\left(2 + 5e^{-\beta\epsilon} + 3e^{-2\beta\epsilon} + 2e^{-3\beta\epsilon}\right) .$$

The configurations are shown in Fig. 2.25(c).

(b) When the non-degeneracy condition is satisfied, i.e., when $e^{-\alpha} \approx \frac{N}{V}\left(\frac{h^2}{2\pi mkT}\right)^{3/2} \ll 1$, the indistinguishability of particles becomes unimportant and Fermions and Bosons can be treated as Boltzmann particles.

2123

(a) Give a definition of the partition function z for a statistical system.

(b) Find a relation between the heat capacity of a system and $\dfrac{\partial^2 \ln z}{\partial \beta^2}$, where $\beta = \dfrac{1}{kT}$.

(c) For a system with one excited state at energy Δ above the ground state, find an expression for the heat capacity in terms of Δ. Sketch the dependence on temperature and discuss the limiting behavior for high and low temperatures.

(UC, Berkeley)

Solution:

(a) The partition function is the sum of statistical probabilities.

For quantum statistics, $z = \sum_s \exp(-\beta E_s)$, summing over all the quantum states.

For classical statistics, $z = \int \exp(-\beta E)d\Gamma/h^\gamma$, integrating over the phase-space where γ is the number of degrees of freedom.

(b)
$$\overline{E} = -\frac{\partial}{\partial \beta}\ln z \ ,$$

$$c_v = \frac{\partial \overline{E}}{\partial T} = -\frac{1}{k\beta^2}\frac{\partial}{\partial \beta}\overline{E} = \frac{1}{k\beta^2}\frac{\partial^2}{\partial \beta^2}\ln z \ .$$

(c) Assume the two states are non-degenerate, then

$$\overline{E} = \frac{\Delta e^{-\Delta/kT}}{1 + e^{-\Delta/kT}} = \frac{\Delta}{e^{\Delta/kT} + 1}$$

$$c_v = \frac{d\overline{E}}{dT} = k\left(\frac{\Delta}{kT}\right)^2 \frac{e^{\Delta/kT}}{(1 + e^{\Delta/kT})^2} \ .$$

The variation of specific heat with temperature is shown in Fig. 2.26.

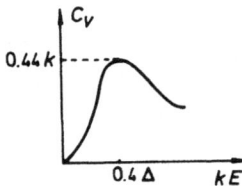

Fig. 2.26.

2124

Consider a collection of N two-level systems in thermal equilibrium at a temperature T. Each system has only two states: a ground state of energy 0 and an excited state of energy ε. Find each of the following quantities and make a sketch of the temperature dependence.

(a) The probability that a given system will be found in the excited state.

(b) The entropy of the entire collection.

(MIT)

Solution:

(a) The probability for a system to be in the excited state is $P = \frac{1}{z}e^{-\varepsilon/kT}$, where $z = 1 + e^{-\varepsilon/kT}$, i.e.,

$$P = \frac{1}{e^{\varepsilon/kT} + 1}.$$

The relation between probability and temperature is shown in Fig. 2.27.

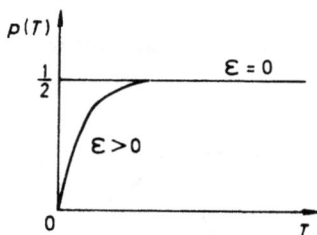

Fig. 2.27.

(b) $z_N = [1 + e^{-\varepsilon/kT}]^N$, $\quad F = -kT \ln z_N$,

$$S = -\left(\frac{\partial F}{\partial T}\right)_V = \frac{N\varepsilon}{T}(1 + e^{\varepsilon/kT})^{-1} + Nk\ln(1 + e^{-\varepsilon/kT}) .$$

The relation between entropy and temperature is shown in Fig. 2.28.

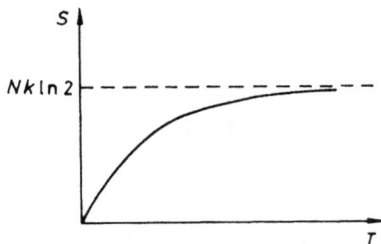

Fig. 2.28.

2125

N weakly coupled particles obeying Maxwell-Boltzmann statistics may each exist in one of the 3 non-degenerate energy levels of energies $-E$, 0, $+E$. The system is in contact with a thermal reservoir at temperature T.

(a) What is the entropy of the system at $T = 0$ K?

(b) What is the maximum possible entropy of the system?

(c) What is the minimum possible energy of the system?

(d) What is the partition function of the system?

(e) What is the most probable energy of the system?

(f) If $C(T)$ is the heat capacity of the system, what is the value of $\int_0^\infty \frac{C(T)}{T} dT$?

(*UC, Berkeley*)

Solution:

(a) At $T = 0$ K, the entropy of the system is $S(0) = 0$.

(b) The maximum entropy of the system is

$$S_{\max} = k\ln\Omega_{\max} = k\ln 3^N = Nk\ln 3 .$$

(c) The minimum energy of the system is $-NE$.

(d) The partition function of the system is

$$z = \left(e^{E/kT} + 1 + e^{-E/kT}\right)^N .$$

(e) When $N \gg 1$, the most probable energy is the average energy

$$NE_p \approx N\overline{E}$$

$$= -NE\frac{\left(a - \frac{1}{a}\right)}{\left(a + 1 + \frac{1}{a}\right)}$$

$$= -\frac{2NE \sinh\left(\dfrac{E}{kT}\right)}{1 + 2\cosh\left(\dfrac{E}{kT}\right)} \ ,$$

where $a = \exp(E/kT)$.

(f) $\displaystyle\int_0^\infty \frac{C(T)}{T}\,dT = \int_0^\infty dS = S(\infty) - S(0) = Nk\ln 3$.

2126

Find the pressure, entropy, and specific heat at constant volume of an ideal Boltzmann gas of indistinguishable particles in the extreme relativistic limit, in which the energy of a particle is related to its momentum by $\varepsilon = cp$. Express your answer as functions of the volume V, temperature T, and number of particle N.

(*Princeton*)

Solution:

Let z denote the partition function of a single particle, Z the total partition function, p the pressure, S the entropy, U the internal energy, and c the specific heat. We have

$$z = \frac{4\pi V}{h^3 c^3}\int_0^\infty \varepsilon^2 e^{-\varepsilon/kT}\,d\varepsilon = \frac{8\pi V}{h^3 c^3}(kT)^3 \ .$$

$$Z = \frac{z^N}{N!} = \frac{1}{N!}\left(\frac{8\pi V}{h^3 c^3}\right)^N (kT)^{3N} \ .$$

$$p = \frac{1}{\beta}\frac{\partial}{\partial V}\ln Z = \frac{N}{\beta V} = \frac{NkT}{V} \ .$$

$$S = k\left(\ln Z - \beta\frac{\partial}{\partial\beta}\ln Z\right)$$

$$= Nk\left(3\ln kT + \ln\frac{8\pi V}{Nh^3 c^3} + 4\right) \ .$$

$$u = -\frac{\partial}{\partial \beta} \ln Z = 3NkT \ .$$

$$c = 3Nk \ .$$

2127

A vessel of volume V contains N molecules of an ideal gas held at temperature T and pressure P_1. The energy of a molecule may be written in the form

$$E_k(p_x, p_y, p_z) = \frac{1}{2m}(p_x^2 + p_y^2 + p_z^2) + \varepsilon_k \ ,$$

where ε_k denotes the energy levels corresponding to the internal states of the molecules of the gas.

(a) Evaluate the free energy $F = -kT \ln Z$, where Z is the partition function and k is Boltzmann's constant. Explicitly display the dependence on the volume V_1.

Now consider another vessel, also at temperature T, containing the same number of molecules of an identical gas held at pressure P_2.

(b) Give an expression for the total entropy of the two gases in terms of P_1, P_2, T, N.

(c) The vessels are then connected to permit the gases to mix without doing work. Evaluate explicitly the change in entropy of the system. Check whether your answer makes sense by considering the special case $V_1 = V_2$ (*i.e.*, $P_1 = P_2$).

(*Princeton*)

Solution:

(a) The partition function of a single particle is

$$z = \frac{V}{h^3} \sum_k \iiint e^{-E_k/kT} d\mathbf{p}$$

$$= V \cdot \left(\frac{2\pi mkT}{h^2}\right)^{3/2} z_0 \ ,$$

where $z_0 = \sum_n \exp(-\varepsilon_n/kT)$ refers to the internal energy levels. Taking account of the indistinguishability of the particles, the partition function of the N particle system is

$$Z = \frac{z^N}{N!} = \frac{V^N}{N!} \left(\frac{2\pi mkT}{h^2}\right)^{3N/2} \cdot z_0^N .$$

So

$$F = -kT \ln Z$$
$$= -kT \left(N \ln V + N \ln z_0 + \frac{3N}{2} \ln \left(\frac{2\pi mkT}{h^2}\right) - \ln N!\right)$$

(b) $\quad S = k \left(\ln Z - \beta \frac{\partial}{\partial \beta} \ln Z\right)$
$$= Nk \left(\ln \frac{V}{N} + \frac{3}{2} \ln \left(\frac{2\pi mkT}{h^2}\right) + \frac{5}{2} + \ln z_0 - \beta \frac{\partial}{\partial \beta} \ln z_0\right) .$$

Thus

$$S_1 = Nk \left(\frac{V_1}{N} + \frac{3}{2} \ln \left(\frac{2\pi mkT}{h^2}\right) + \frac{5}{2} + S_0\right) ,$$
$$S_2 = Nk \left(\ln \frac{V_2}{N} + \frac{3}{2} \ln \left(\frac{2\pi mkT}{h^2}\right) + \frac{5}{2} + S_0\right) ,$$

where

$$S_0 = \ln z_0 - \beta \frac{\partial}{\partial \beta} \ln z_0 .$$

The total entropy is

$$S = S_1 + S_2$$
$$= 2Nk \left[\ln \frac{\sqrt{V_1 V_2}}{N} + \frac{3}{2} \ln \left(\frac{2\pi mkT}{h^2}\right) + \frac{5}{2} + S_0\right] .$$

(c) After mixing, the temperature of the ideal gas is the same as before, so that

$$S' = 2Nk \left[\ln \frac{V_1 + V_2}{2N} + \frac{3}{2} \ln \left(\frac{2\pi mkT}{h^2}\right) + \frac{5}{2} + S_0\right] .$$

Hence

$$\Delta S = S' - S = 2Nk \ln \frac{V_1 + V_2}{2\sqrt{V_1 V_2}} = 2Nk \ln \frac{P_1 + P_2}{2\sqrt{P_1 P_2}} .$$

When $V_1 = V_2$, $\Delta S = 0$ as expected.

2128

(a) Calculate the partition function z of one spinless atom of mass M moving freely in a cube of volume $V = L^3$. Express your result in terms of the quantum concentration

$$n_q = \left(\frac{MkT}{2\pi} \right)^{3/2} .$$

Explain the physical meaning of n_q.

(b) An ideal gas of N spinless atoms occupies a volume V at temperature T. Each atom has only two energy levels separated by an energy Δ. Find the chemical potential, free energy, entropy, pressure and heat capacity at constant pressure.

(*SUNY, Buffalo*)

Solution:

(a) The energy eigenvalues are given by

$$\varepsilon = \frac{h^2}{2mL^2} (n_x^2 + n_y^2 + n_z^2) ,$$

$$= \frac{1}{2M} (p_x^2 + p_y^2 + p_z^2) = \frac{p^2}{2M} ,$$

where $n_x, n_y, n_z = 0, \pm 1, \dots$.

The energy levels can be thought of as quasi-continuous, so that the number of quantum states in the range $p \to p + dp$ is $\frac{4\pi V}{h^3} p^2 dp$, whence the number of states in the energy interval $\varepsilon \to \varepsilon + d\varepsilon$ is $\frac{2\pi V}{h^3} (2M)^{3/2} \sqrt{\varepsilon} d\varepsilon$.

Hence

$$z = \int_0^\infty \frac{2\pi V}{h^3} (2M)^{3/2} \varepsilon^{1/2} e^{-\beta\varepsilon} d\varepsilon$$

$$= \frac{V}{h^3} \left(\frac{MkT}{2\pi} \right)^{3/2} \cdot (4\pi^2)^{3/2} = \frac{8\pi^3 V}{h^3} \cdot n_q$$

where $n_q = \left(\frac{MkT}{2\pi} \right)^{3/2}$ is the average number of quantum states in unit volume.

(b) The classical ideal gas satisfies the non-degeneracy condition. The partition function of a sub-system is $z = \exp(-\beta\varepsilon_1) + \exp(-\beta\varepsilon_2)$, $\varepsilon_2 =$

$\varepsilon_1 + \Delta$. Hence the partition function of the system is

$$Z = [z]^N = (e^{-\beta\varepsilon_1} + e^{-\beta\varepsilon_2})^N .$$

The free energy is

$$F = -kT \ln Z = -NkT \ln \left(e^{-\beta\varepsilon_1} + e^{-\beta\varepsilon_2}\right) .$$

The chemical potential is

$$\mu = \left(\frac{\partial F}{\partial N}\right)_{T,V} = -kT \ln \left(e^{-\beta\varepsilon_1} + e^{-\beta\varepsilon_2}\right) .$$

The pressure is

$$p = kT \frac{\partial}{\partial V} \ln Z = -N \frac{\frac{\partial \varepsilon_1}{\partial V} e^{-\beta\varepsilon_1} + \frac{\partial \varepsilon_2}{\partial V} e^{-\beta\varepsilon_2}}{e^{-\beta\varepsilon_1} + e^{-\beta\varepsilon_2}} .$$

The entropy is

$$S = Nk \left(\ln z - \beta \frac{\partial}{\partial \beta} \ln z\right) - k \ln N! = Nk \left[1 + \ln \left(\frac{e^{-\beta\varepsilon_1} + e^{-\beta\varepsilon_2}}{N}\right)\right]$$
$$+ \frac{N(\varepsilon_1 e^{-\beta\varepsilon_1} + \varepsilon_2 e^{-\beta\varepsilon_2})}{T(e^{-\beta\varepsilon_1} + e^{-\beta\varepsilon_2})} .$$

The heat capacity at constant pressure is

$$C_p = \left(\frac{\partial U}{\partial T}\right)_p = \frac{\partial}{\partial T} \left[NkT^2 \frac{\partial}{\partial T} \ln z\right]_p$$
$$= \frac{N\Delta^2}{2kT^2 \left(1 + \cosh \dfrac{\Delta}{kT}\right)} = \frac{N\Delta^2}{4kT^2 \cosh \left(\dfrac{\Delta}{2kT}\right)} .$$

2129

(a) Consider an ideal gas of N particles of mass m confined to a volume V at a temperature T. Using the classical approximation for the partition function and assuming the particles are indistinguishable, calculate the chemical potential μ of the gas.

(b) A gas of N particles, also of mass m, is absorbed on a surface of area A, forming a two-dimensional ideal gas at temperature T on the surface. The energy of an absorbed particle is $\varepsilon = |\mathbf{p}|^2/2m - \varepsilon_0$, where $\mathbf{p} = (p_x, p_y)$ and ε_0 is the surface binding energy per particle. Using the same approximations and assumptions as in part (a), calculate the chemical potential μ of the absorbed gas.

(c) At temperature T, the particles on the surface and in the surrounding three-dimensional gas are in equilibrium. This implies a relationship between the respective chemical potentials. Use this condition to find the mean number n of molecules absorbed per unit area when the mean pressure of the surrounding three-dimensional gas is p. (The total number of particles in absorbed gas plus surrounding vapor is N_0) .

(*Princeton*)

Solution:
(a) The classical partition function is

$$z = \frac{V^N}{N!} \left(\frac{2\pi mkT}{h^2} \right)^{\frac{3N}{2}} \, ,$$

Thus

$$G = F + pV = -kT \ln z + NkT$$
$$= NkT \left[\ln \frac{N}{V} - \frac{3}{2} \ln \left(\frac{2\pi mkT}{h^2} \right) \right] \, ,$$
$$\mu = -kT \left[\ln \frac{V}{N} + \frac{3}{2} \ln \left(\frac{2\pi mkT}{h^2} \right) \right] \, .$$

(b) The classical partition function for the two-dimensional ideal gas is

$$z = \frac{A^N}{N!} \left(\frac{2\pi mkT}{h^2} \right)^N \cdot e^{N\varepsilon_0/kT} \, .$$

Thus

$$G = F + pA = -NkT \left[\ln \frac{A}{N} + \ln \left(\frac{2\pi mkT}{h^2} \right) + \frac{\varepsilon_0}{kT} \right] \, ,$$
$$\mu = -kT \left[\ln \frac{A}{N} + \ln \left(\frac{2\pi mkT}{h^2} \right) + \frac{\varepsilon_0}{kT} \right] \, .$$

(c) The chemical potential of the three-dimensional gas is equal to that of the two-dimensional gas. Note that in the expression of the chemical potential for the three-dimensional gas, $\dfrac{V}{N} = \dfrac{kT}{p}$, and in that for the two-dimensional gas, $\dfrac{A}{N} = \dfrac{1}{n}$. Since the two chemical potentials have the same value, one obtains

$$n = \frac{p}{kT} \left(\frac{h^2}{2\pi m k T} \right)^{1/2} e^{\epsilon_0/kT} .$$

2130

A simple harmonic one-dimensional oscillator has energy levels $E_n = (n + 1/2)\hbar\omega$, where ω is the characteristic oscillator (angular) frequency and $n = 0, 1, 2, \ldots$.

(a) Suppose the oscillator is in thermal contact with a heat reservoir at temperature T, with $\dfrac{kT}{\hbar\omega} \ll 1$. Find the mean energy of the oscillator as a function of the temperature T.

(b) For a two-dimensional oscillator, $n = n_x + n_y$, where

$$E_{n_x} = \left(n_x + \frac{1}{2} \right) \hbar\omega_x, \quad E_{n_y} = \left(n_y + \frac{1}{2} \right) \hbar\omega_y ,$$

$n_x = 0, 1, 2, \ldots$ and $n_y = 0, 1, 2, \ldots$, what is the partition function for this case for any value of temperature? Reduce it to the degenerate case $\omega_x = \omega_y$.

(c) If a one-dimensional classical anharmonic oscillator has potential energy $V(x) = cx^2 - gx^3$, where $gx^3 \ll cx^2$, at equilibrium temperature T, carry out the calculations as far as you can and give expressions as functions of temperature for

1) the heat capacity per oscillator and
2) the mean value of the position x of the oscillator.

(UC, Berkeley)

Solution:

(a) Putting $\alpha = \dfrac{\hbar\omega}{kT} = \hbar\omega\beta$, one has

$$z = \sum_{n=0}^{\infty} e^{-(n+1/2)\alpha} = \frac{e^{\alpha/2}}{e^\alpha - 1} ,$$

$$\overline{E} = -\frac{1}{z} \frac{\partial z}{\partial \beta} = \frac{\hbar\omega}{2} + \frac{\hbar\omega}{e^\alpha - 1} .$$

(b) There is no difference between a two-dimensional oscillator and two independent one-dimensional oscillators, then the partition function is

$$z = \frac{e^{\alpha_z/2}}{e^{\alpha_z} - 1} \cdot \frac{e^{\alpha_y/2}}{e^{\alpha_y} - 1} .$$

When $\omega_x = \omega_y, \alpha_x = \alpha_y = \alpha$, we have

$$z = \frac{e^\alpha}{(e^\alpha - 1)^2} .$$

(c) 1) We calculate the partition function

$$z = \int \exp[-(cx^2 - gx^3)/kT]dx .$$

(Note that the kinetic energy term has not been included in the expression, this is done by adding $\dfrac{k}{2}$ in the heat capacity later.) The non-harmonic term $[\exp(gx^3/kT) - 1]$ is a small quantity in the region of motion. Using Taylor's expansion retaining only the lowest order terms, we get

$$z = \int_{-\infty}^{+\infty} \left[1 + \frac{1}{2}\left(\frac{gx^3}{kT}\right)^2 \right] e^{-(cx^2/kT)}dx$$

$$= \sqrt{\frac{\pi}{c}} \left(\frac{1}{\sqrt{\beta}} + \frac{15g^2}{16c^3} \cdot \frac{1}{\beta\sqrt{\beta}} \right) .$$

The mean value of the potential energy is

$$\overline{V} = -\frac{\partial}{\partial\beta}\ln z = \frac{kT}{2}\left(1 + \frac{15g^2}{8c^3}kT \right) .$$

The heat capacity per oscillator is

$$C = \frac{k}{2} + \frac{\partial\overline{V}}{\partial T} = k + \frac{15g^2}{8c^3}k^2T .$$

2) In the first-order approximation, the mean value of the position x of the oscillator is

$$\overline{x} = \frac{\int_{-\infty}^{+\infty} x\left[1 + \dfrac{gx^3}{kT} \right] e^{-(cx^2/kT)}dx}{\int_{-\infty}^{+\infty} e^{-(cx^2/kT)}dx} = \frac{3gkT}{4c^2} .$$

2131

Consider a dilute diatomic gas whose molecules consist of non-identical pairs of atoms. The moment of inertia about an axis through the molecular center of mass perpendicular to the line connecting the two atoms is I. Calculate the rotational contributions to the specific heat and to the absolute entropy per mole at temperature T for the following limiting cases:

(a) $kT \gg \hbar^2/I$,
(b) $kT \ll \hbar^2/I$.

Make your calculations sufficiently exact to obtain the lowest order non-zero contributions to the specific heat and entropy.

(*CUSPEA*)

Solution:

The contribution of rotation to the partition function is

$$z_R^T = (z_R)^N ,$$

where N is the total number of the molecules in one mole of gas, and

$$z_R = \sum_J (2J + 1) \exp(-\beta \varepsilon_J) \quad \text{with} \quad \varepsilon_J = \frac{\hbar^2}{2I} J(J+1) .$$

The contribution to energy is

$$E_R = -\frac{\partial}{\partial \beta} \ln z_R^T = -N \frac{\partial}{\partial \beta} \ln z_R ,$$

The contribution to specific heat is

$$C_R = \frac{\partial E_R}{\partial T} .$$

The contribution to entropy is

$$S_R = Nk \ln z_R + \frac{E_R}{T} .$$

(a) $kT \gg \hbar^2/I$, i.e., $\beta \hbar^2/2I \ll 1$. We have

$$z_R = \int_0^\infty (2J + 1) \exp[-\beta \hbar^2 J(J+1)/2I] dJ = \frac{2IkT}{\hbar^2} .$$
$$E_R = NkT ,$$
$$C_R = Nk .$$
$$S_R = Nk \left(1 + \ln \frac{2IkT}{\hbar^2}\right) .$$

(b) $kT \ll \hbar^2/I$. We have

$$z_R = 1 + 3e^{-\beta(h^2/4\pi^2 I)} + \ldots \approx 1 + 3e^{-h^2/4\pi^2 kTI} .$$

$$E_R = 3N \frac{\dfrac{\hbar^2}{I} e^{-h^2/4\pi^2 IkT}}{1 + 3e^{-h^2/4\pi^2 IkT}} \approx \frac{3N\hbar^2}{I} e^{-h^2/4\pi^2 IkT} .$$

$$C_R = \frac{3N\hbar^4}{I^2 kT^2} e^{-h^2/4\pi^2 IkT} .$$

$$S_R = kN \ln(1 + 3e^{-h^2/4\pi^2 IkT}) + \frac{3N\hbar^2}{IT} e^{-h^2/4\pi^2 IkT}$$

$$\approx 3kN \left(1 + \frac{\hbar^2}{IkT}\right) e^{-h^2/4\pi^2 IkT} \approx \frac{3N\hbar^2}{IT} e^{-h^2/4\pi^2 IkT} .$$

2132

An assembly of N fixed particles with spin $\dfrac{1}{2}$ and magnetic moment μ_0 is in a static uniform applied magnetic field. The spins interact with the applied field but are otherwise essentially free.

(a) Express the energy of the system as a function of its total magnetic moment and the applied field.

(b) Find the total magnetic moment and the energy, assuming that the system is in thermal equilibrium at temperature T.

(c) Find the heat capacity and the entropy of the system under these same conditions.

(*UC, Berkeley*)

Solution:

(a) $E = -MH$.

(b) Assume that $\bar{\mu}$ is the average magnetic moment per particle under the influence of the external field when equilibrium is reached, then $M = N\bar{\mu}$ and

$$\bar{\mu} = \mu_0 \frac{e^{\mu_0 H/kT} - e^{-\mu_0 H/kT}}{e^{\mu_0 H/kT} + e^{-\mu_0 H/kT}} = \mu_0 \tanh\left(\frac{\mu_0 H}{kT}\right) .$$

Thus $E = -N\mu_0 H \tanh(\mu_0 H/kT)$.

(c) $C_H = \left(\dfrac{\partial E}{\partial T}\right)_H = Nk \dfrac{\left(\dfrac{\mu_0 H}{kT}\right)^2}{\cosh^2\left(\dfrac{\mu_0 H}{kT}\right)}$.

The partition function of the system is

$$z = (a + 1/a)^N \quad \text{with } a = \exp(\mu_0 H/kT) .$$

Therefore

$$S = k\left(\ln z - \beta\frac{\partial}{\partial \beta}\ln z\right)$$
$$= Nk[\ln(e^{\mu_0 H\beta} + e^{-\mu_0 H\beta}) - \mu_0 H\beta \tanh(\mu_0 H\beta)] .$$

2133

Given a system of N identical non-interacting magnetic ions of spin $\dfrac{1}{2}$, magnetic moment μ_0 in a crystal at absolute temperature T in a magnetic field B. For this system calculate:

(a) The partition function Z.

(b) The entropy σ.

(c) The average energy $\overline{U}$.

(d) The average magnetic moment $\overline{M}$, and the fluctuation in the magnetic moment, $\Delta M = \sqrt{(M - \overline{M})^2}$.

(e) The crystal is initially in thermal equilibrium with a reservoir at $T = 1$ K, in a magnetic field $B_i = 10,000$ Gauss. The crystal is then thermally isolated from the reservoir and the field reduced to $B_f = 100$ Gauss. What happens?

(UC, Berkeley)

Solution:

(a) Since there is no interaction between the ions, the partition function of the system is

$$z = (e^\alpha + e^{-\alpha})^N ,$$

where $\alpha = \mu_0 B/kT$.

(b) The entropy is

$$\sigma = k \left(\ln z - \beta \frac{\partial}{\partial \beta} \ln z \right) = Nk \left[\ln(e^{\alpha} + e^{-\alpha}) - \alpha \tanh \alpha \right] .$$

(c) The average energy is

$$\overline{U} = -\frac{\partial}{\partial \beta} \ln z = -N \mu_0 B \tanh \left(\frac{\mu_0 B}{kT} \right) .$$

(d) The average magnetic moment is

$$\overline{M} = N \mu_0 \tanh \left(\frac{\mu_0 B}{kT} \right) .$$

For a single ion, we have

$$\overline{(\Delta m)^2} = \overline{m^2} - (\overline{m})^2 = \mu_0^2 - \mu_0^2 \tanh^2 \left(\frac{\mu_0 B}{kT} \right) = \frac{\mu_0^2}{\cosh^2 \left(\frac{\mu_0 B}{kT} \right)} .$$

For the whole system, we have

$$\overline{(\Delta M)^2} = N \overline{(\Delta m)^2}$$

$$\Delta M = \sqrt{N} \frac{\mu_0}{\cosh \left(\frac{\mu_0 B}{kT} \right)} .$$

(e) We see from (b) that the entropy σ is a function of $\mu_0 B/kT$. If the entropy of the spin states is maintained constant, i.e., $\mu_0 B/kT$ is kept constant, then the temperature of the spin states can be lowered by reducing the magnetic field adiabatically. In the crystal, the decrease in the temperature of the spin states can result in a decrease of the temperature of the crystal vibrations by "heat transfer". Therefore, the whole crystal is cooled by an "adiabatic reduction of the magnetic field". From

$$\frac{T_f}{B_f} = \frac{T_i}{B_i} ,$$

we have $T_f = 10^{-3}$ K.

2134

Consider N fixed non-interacting magnetic moments each of magnitude μ_0. The system is in thermal equilibrium at temperature T and is in a uniform external magnetic field B. Each magnetic moment can be oriented only parallel or antiparallel to B. Calculate:

(a) the partition function,

(b) the specific heat,

(c) the thermal average magnetic moment $(\overline{M})$.

Show that in the high temperature limit the Curie Law is satisfied (i.e., $\chi = d\overline{M}/dB \propto 1/T$).

(UC, Berkeley)

Solution:

(a) Since the magnetic moments are nearly independent of one another, we can consider a single magnetic moment. As its partition function is

$$z = a + \frac{1}{a} \,,$$

where $a = \exp(\mu_0 B/kT)$, the partition function for the entire system is

$$Z = z^N \,.$$

(b) $\overline{E} = -\mu_0 BN \left(a - \frac{1}{a} \right) / \left(a + \frac{1}{a} \right)$
$$= -\mu_0 BN \tanh(\mu_0 B/kT) \,,$$

giving the specific heat as

$$c = \frac{d\overline{E}}{dT} = kN \left(\frac{\mu_0 B}{kT} \right)^2 \frac{1}{\cosh^2 \left(\frac{\mu_0 B}{kT} \right)} \,.$$

(c) $\overline{M} = N\mu_0 \tanh \left(\frac{\mu_0 B}{kT} \right) \,,$

$$\chi \equiv \frac{d\overline{M}}{dB} = \frac{N\mu_0^2}{kT} \frac{1}{\cosh^2 \left(\frac{\mu_0 B}{kT} \right)} \,.$$

At the high temperature limit, $\chi \propto \dfrac{1}{T}$.

2135

Consider a system of non-interacting spins in an applied magnetic field H. Using $S = k(\ln Z + \beta E)$, where Z is the partition function, E is the energy, and $\beta = 1/kT$, argue that the dependence of S on H and T is of the form $S = f(H/T)$ where $f(x)$ is some function that need not be determined.

Show that if such a system is magnetized at constant T, then thermally isolated, and then demagnetized adiabatically, cooling will result.

Why is this technique of adiabatic demagnetization used for refrigeration only at very low temperatures?

How can we have $T < 0$ for this system? Can this give a means of achieving $T = 0$?

(SUNY, Buffalo)

Solution:

A single spin has two energy levels: μH and $-\mu H$, and its partition function is $z = \exp(-b) + \exp(b)$, where $b = \mu H/kT$. The partition function for the system is given by

$$Z = z^N = (2 \cosh b)^N ,$$

where N is the total number of spins.
So

$$E = -\frac{\partial}{\partial \beta} \ln z = -N \frac{\partial}{\partial \beta} \ln[2 \cosh(\mu H \beta)]$$

$$= -N\mu H \tanh \left(\frac{\mu H}{kT} \right) .$$

Hence

$$S = k(\ln z + \beta E)$$

$$= Nk \left\{ \ln \left[2 \cosh \left(\frac{\mu H}{kT} \right) \right] - \frac{\mu H}{kT} \tanh \left(\frac{\mu H}{kT} \right) \right\} = f \left(\frac{H}{T} \right) .$$

When the system is magnetized at constant T, the entropy of the final state is S. Because the entropy of the system does not change in an adiabatic process, T must decrease when the system is demagnetized adiabatically in order to keep H/T unchanged. The result is that the temperature decreases. This cooling is achieved by using the property of magnetic particles with spins in an external magnetic field. In reality, these magnetic particles are in lattice ions. For the effect to take place we require the entropy and specific heat of the lattice ions to be much smaller than those of the magnetic particles. Therefore, we require the temperature to be very low, $T \lesssim 1$ K.

When the external magnetic field is increased to a certain value, we can suddenly reverse the external field. During the time shorter than the relaxation time of the spins, the system is in a state of $T < 0$. However, it is not possible to achieve $T = 0$ since $T < 0$ corresponds to a state of higher energy.

2136

The Curie-Weiss model

Consider a crystal of N atoms ($N \sim 10^{23}$) with spin quantum numbers $s = \frac{1}{2}$ and $m_s = \pm\frac{1}{2}$. The magnetic moment of the i-th atom is $\boldsymbol{\mu}_i = g\mu_B \mathbf{s}_i$, where g is the Lande g-factor, and $\mu_B = e\hbar/2mc$ is the Bohr magneton. Assume that the atoms do not interact appreciably but are in equilibrium at temperature T and are placed in an external magnetic field $\mathbf{H} = H\mathbf{z}$.

(a) Show that the partition function is $z = (2\cosh\eta)^N$ where $\eta = g\mu_B H/2kT$.

(b) Find an expression for the entropy S of the crystal (you need only consider the contributions from the spin states). Evaluate S in the strong field ($\eta \gg 1$) and weak field ($\eta \ll 1$) limits.

(c) An important process for cooling substances below 1 K is adiabatic demagnetization. In this process the magnetic field on the sample is increased from 0 to H_0 while the sample is in contact with a heat bath at temperature T_0. Then the sample is thermally isolated and the magnetic field is reduced to $H_1 < H_0$. What is the final temperature of the sample?

(d) The magnetization M and susceptibility χ are defined by $M = \left\langle \sum_{i=1}^{N} (\mu_i)_z \right\rangle$ and $\chi = M/H$, respectively. Find expressions for M and χ, and evaluate these expressions in the weak field limit.

Now suppose each atom interacts with each of its nearest n neighbors. To include this interaction approximately we assume that the nearest n

neighbors generate a 'mean field' $\overline{H}$ at the site of each atom, where

$$g\mu_{\mathrm{B}}\,\overline{H} = 2\alpha\Big\langle \sum_{k=1}^{N}(S_k)_z \Big\rangle \ ,$$

α is a parameter which characterizes the strength of the interaction.

(e) Use the mean field approximation together with the results of part (d) to calculate the susceptibility χ in the weak field (i.e., the high temperature) limit. At what temperature, T_c, does χ become infinite?

(*MIT*)

Solution

(a) $z = \left(e^{-g\mu_B H/2kT} + e^{g\mu_B H/2kT}\right)^N = (2\cosh\eta)^N$,

where $\eta = g\mu_B H/2kT$.

(b) $F = -kT\ln z$, $\ S = -\left(\dfrac{\partial F}{\partial T}\right)_H = Nk[\ln(2\cosh\eta) - \eta\tanh\eta]$.

When $\eta \gg 1$, $\quad S \approx Nk(1+\eta)\exp(-2\eta)$;

When $\eta \ll 1$, $\quad S \approx Nk\ln 2$.

(c) During adiabatic demagnetization, the entropy of the system remains constant, i.e., $S_1 = S_0$. Thus $\eta_1 = \eta_0$, i.e., $T_1 = H_1 T_0/H_0$. Hence $T_1 < T_0$.

(d) $M = \left\langle \sum_{i=1}^{N}(\mu_i)_z \right\rangle = -\dfrac{\partial F}{\partial H} = kT\left(\dfrac{\partial}{\partial H}\ln z\right)_T$

$\qquad = \dfrac{Ng\mu_B}{2}\tanh\dfrac{g\mu_B H}{2kT}\ ,$

$\chi \qquad = M/H = \dfrac{Ng\mu_B}{2H}\tanh\dfrac{g\mu_B H}{2kT}\ .$

In the weak field limit

$$M \approx \frac{N}{4kT}(g\mu_B)^2 H\ ,$$

so that

$$\chi = \frac{N}{4kT}(g\mu_B)^2\ .$$

(e) From the definition given for the mean field, we have

$$\overline{H} = \frac{2\alpha}{g\mu_B}\cdot\frac{n}{N}\cdot\frac{M}{g\mu_B}\ ,$$

where M is total magnetic moment. Using the result in (d), we have

$$M \approx \frac{N}{4kT}(g\mu_B)^2(H + \overline{H})$$
$$= \frac{N}{4kT}(g\mu_B)^2 \left[H + \frac{2\alpha}{(g\mu_B)^2} \cdot \frac{n}{N}M\right] \ .$$

Hence

$$M = H\frac{N}{4kT}(g\mu_B)^2 / \left(1 - \frac{\alpha n}{2kT}\right) \ ,$$
$$\chi = \frac{N}{4kT}(g\mu_B)^2 / \left(1 - \frac{\alpha n}{2kT}\right) \ .$$

When

$$T = T_c = \alpha n/2k, \chi \to \infty \ .$$

2137

Consider a system of free electrons in a uniform magnetic field $B = B_z$, with the electron spin ignored. Show that the quantization of orbits, in contrast to classical orbits, affect the calculation of diamagnetism in the high temperature limit by calculating:

(a) the degeneracy of the quantized energy levels,

(b) the grand partition function,

(c) the magnetic susceptibility in the high temperature limit.

(SUNY, Buffalo)

Solution:

(a) Assume that the electrons are held in a cubic box of volume L^3. The number of energy levels in the interval p_x to $p_x + dp_x, p_y$ to $p_y + dp_y$ without the external magnetic field is

$$L^2 dp_x dp_y / h^2 \ .$$

When the external magnetic field is applied, the electrons move in circular orbits in the x-y plane with angular frequency eB/mc. The energy levels are given by

$$\hbar\frac{eB}{mc}\left(l + \frac{1}{2}\right) + \frac{1}{2m}p_z^2, \ l = 0, 1, 2, \ldots \ .$$

The degeneracy of the quantized energy levels is given by

$$\frac{L^2}{h^2} \iint_{A_1} dp_x dp_y = \frac{L^2}{h^2} \int_{A_2} 2\pi p\, dp = \frac{L^2 eB}{hc} ,$$

where the integral limits A_1 represents $2\mu_B Bl < (p_x^2 + p_y^2)/2m < 2\mu_B B(l+1)$, and A_2 represents $2\mu_B Bl < p^2/2m < 2\mu_B B(l+1)$ with $\mu_B = e\hbar/2mc$.

(b) $\ln \Xi = \sum_u \ln(1 + e^{\beta\mu} \cdot e^{-\beta\epsilon})$

$$= \frac{L}{h} \int_{-\infty}^{\infty} dp_z \sum_{l=0}^{\infty} \frac{L^2 eB}{hc}$$

$$\times \ln\{1 + \lambda e^{-\beta[2\mu_B B(l+1/2)+p_z^2/2m]}\}$$

where $\lambda = \exp(\beta\mu)$.

In the high temperature limit, $\lambda \ll 1$, hence

$$\ln \Xi = \frac{eBV}{h^2 c} \lambda \int_{-\infty}^{\infty} dp_z \sum_{l=0}^{\infty} e^{-\beta[2\mu_B B(l+1/2)+p_z^2/2m]}$$

$$= \frac{\lambda V}{\lambda_T^3} \cdot \frac{\mu_B B}{kT \sinh \chi}$$

where $\lambda_T = h/\sqrt{2\pi mkT}$ and $\chi = \mu_B B/kT$.

(c) $M = -\dfrac{\partial F}{\partial B} = \dfrac{1}{\beta} \left(\dfrac{\partial \ln \Xi}{\partial B} \right)_{\mu,T,V} ,$

where F is the free energy of the system.

Hence

$$M = \frac{\lambda V}{\lambda_T^3} \mu_B \left[\frac{1}{\sinh x} - \frac{x \cosh x}{\sinh^2 x} \right] .$$

By

$$\overline{N} = \left(\lambda \frac{\partial}{\partial \lambda} \ln \Xi \right)_{B,T,V} = \frac{\lambda V}{\lambda_T^3} \frac{x}{\sinh x} ,$$

we have $M = -\overline{N}\mu_B L(x)$, where $L(x) = \coth x - 1/x$.

At high temperatures, $kT \gg \mu_B B$ or $x \ll 1$. Therefore,

$$L(x) = \frac{1}{3} x - \frac{1}{45} x^3 + \cdots ,$$

$$\overline{N} \approx \frac{\lambda V}{\lambda_T^3} ,$$

$$M \approx -\overline{N} \mu_B^2 B/3kT ,$$

$$\chi_\infty = \frac{M}{VB} = -\bar{n}\mu_B^2/3kT ,$$

where $\bar{n} = \overline{N}/V$ is the particle number density.

2138

A certain insulating solid contains N_A non-magnetic atoms and N_I magnetic impurities each of which has spin $\frac{3}{2}$. Each impurity spin is free to rotate independently of all the rest. There is a very weak spin-phonon interaction which we can for most purposes neglect completely. Thus the solid and the impurities are very weakly interacting.

(a) A magnetic field is applied to the system while it is held at constant temperature, T. The field is strong enough to line up the spins completely. What is the magnitude and sign of the change in entropy in the system as the field is applied?

(b) Now the system is held in thermal isolation, no heat is allowed to enter or leave. The magnetic field is reduced to zero. Will the temperature of the solid increase or decrease? Justify your answer.

(c) Assume the heat capacity of the solid is given by $C = 3N_A k$, where k is the Boltzmann constant. What is the temperature change produced by the demagnetization of Part (b)? (Neglect all effects of possible volume changes in the solid.)

(*UC, Berkeley*)

Solution:

(a) Entropy given by $S = k \ln$ (number of micro-states).

Before the external magnetic field is applied, $S = kN_I \ln 4$; after the external magnetic field is applied $S = 0$. Thus the decrease of the entropy is $kN_I \ln 4$.

(b) During adiabatic demagnetization part of the energy of the atomic system is transferred into the spin system. The energy of atomic system decreases and the temperature becomes lower and lower.

(c) With the magnetic field, $S = 3N_A k \ln T$. After the magnetic field is removed, $S = 3N_A k \ln T' + N_I k \ln 4$. During the process, the entropy is constant, giving

$$T' = T \exp(-N_I \ln 4/3N_A) .$$

2139

What is the root-mean-square fluctuation in the number of photons of mode frequency ω in a conducting rectangular cavity? Is it always smaller than the average number of photons in the mode?

(UC, Berkeley)

Solution:

Consider a photon mode (or state) of frequency ω. It can be occupied by 0, 1, 2,... photons.

Denote $\lambda = \hbar\omega/kT$, then

$$\langle n \rangle = \frac{\sum\limits_{n=0}^{\infty} n e^{-n\lambda}}{\sum\limits_{n=0}^{\infty} e^{-n\lambda}} = -\frac{1}{z}\frac{\partial z}{\partial \lambda} \ ,$$

where

$$z = \sum_{n=0}^{\infty} e^{-n\lambda} = \frac{1}{1 - e^{-\lambda}} \ .$$

Hence

$$\langle n \rangle = \frac{1}{e^{\lambda} - 1} \ ,$$

$$\langle n^2 \rangle = \frac{1}{z}\frac{\partial^2 z}{\partial \lambda^2} = \frac{\partial}{\partial \lambda}\left(\frac{1}{z}\frac{\partial z}{\partial \lambda}\right) + \left(\frac{1}{z}\frac{\partial z}{\partial \lambda}\right)^2 \ ,$$

$$\langle (\Delta n)^2 \rangle = \langle n^2 \rangle - \langle n \rangle^2 = -\frac{\partial}{\partial \lambda}\langle n \rangle = \frac{e^{\lambda}}{(e^{\lambda} - 1)^2} \ ,$$

$$\sqrt{\langle (\Delta n)^2 \rangle} = \frac{e^{\lambda/2}}{e^{\lambda} - 1} = \langle n \rangle e^{\lambda/2} > \langle n \rangle \ .$$

Thus root-mean-square fluctuation is always greater than the average number of photons.

2140

Consider an adsorbent surface having N sites, each of which can adsorb one gas molecule. This surface is in contact with an ideal gas with chemical potential μ (determined by the pressure p and the temperature T). Assuming that the adsorbed molecule has energy $-\varepsilon_0$ compared to one in a free state.

(a) Find the grand canonical partition function (sometimes called the grand sum) and

(b) calculate the covering ratio θ, i.e., the ratio of adsorbed molecules to adsorbing sites on the surface.
[A useful relation is $(1+x)^N = \sum_{N_1} N! x^{N_1}/N_1!(N-N_1)!$].

(UC, Berkeley)

Solution:

(a) With N_1 molecules adsorbed on the surface, there are

$$C_{N_1}^N = \frac{N!}{N_1!(N-N_1)!}$$

different configurations. The grand partition function is therefore

$$\Xi = \sum_{N_1=0}^{N} C_{N_1}^N \cdot e^{N_1(\mu+\varepsilon_0)/kT} = (1 + e^{(\mu+\varepsilon_0)/kT})^N .$$

(b) $\overline{N} = -\dfrac{\partial}{\partial \alpha} \ln \Xi = Ne^{(\mu+\varepsilon_0)/kT}/(1 + e^{(\mu+\varepsilon_0)/kT})$,

where $\alpha = -\dfrac{\mu}{kT}$, so that the covering ratio is

$$\theta = \frac{\overline{N}}{N} = \frac{1}{1 + e^{-(\mu+\varepsilon_0)/kT}} .$$

The chemical potential of the adsorbed molecules is equal to that of gas molecules. For an ideal gas,

$$\frac{p}{kT} = n = \int_0^\infty \frac{2\pi(2m)^{3/2}}{h^3} \sqrt{\varepsilon} e^{\mu/kT} e^{-\varepsilon/kT} d\varepsilon .$$

Hence

$$e^{\mu/kT} = \frac{p}{kT} \left(\frac{h^2}{2\pi mkT}\right)^{3/2} ,$$

$$\theta = \frac{1}{1 + \dfrac{kT}{p} \left(\dfrac{2\pi mkT}{h^2}\right)^{3/2} e^{-\varepsilon_0/kT}} .$$

2141

A zipper has N links. Each link has two states: state 1 means it is closed and has energy 0 and state 2 means it is open with energy ε. The zipper can only unzip from the left end and the sth link cannot open unless all the links to its left $(1,2,\ldots,s-1)$ are already open.

(a) Find the partition function for the zipper.

(b) In the low temperature limit, $\varepsilon \gg kT$, find the mean number of open links.

(c) There are actually an infinite number of states corresponding to the same energy when the link is open because the two parts of an open link may have arbitrary orientations. Assume the number of open states is g. Write down the partition function and discuss if there is a phase transition.

(SUNY, Buffalo)

Solution:

(a) The possible states of the zipper are determined by the open link number s. The partition function is

$$z = \sum_{s=0}^{N} e^{-s\varepsilon/kT} = \frac{1 - e^{-(N+1)\varepsilon/kT}}{1 - e^{-\varepsilon/kT}} .$$

(b) The average number of open links is

$$\bar{s} = \frac{kT^2}{\varepsilon} \frac{\partial}{\partial T} \ln z \approx e^{-\varepsilon/kT} , \qquad \varepsilon \gg kT ,$$

(c) $$z = \sum_{s=0}^{N} g^s e^{-s\varepsilon/kT} = \frac{[1 - (ge^{-\varepsilon/kT})^{N+1}]}{[1 - ge^{-\varepsilon/kT}]} .$$

Whether or not there is phase transition is determined by the continuity of the derivatives of the chemical potential $\mu = G/N$, where $G = F+pV$, with $F = -kT \ln z, p = -N(\partial \ln z/\partial V)/\beta$. Since z has no zero value, $\partial \mu/\partial T$ and $\partial \mu/\partial V$ are continuous, so that there is no first-order phase transition. Similarly, there is no second-order phase transition.

2142

A system consisting of three spins in a line, each having $s = \dfrac{1}{2}$, is coupled by nearest neighbor interactions (see Fig. 2.29).

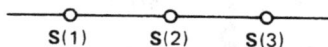

Fig. 2.29.

Each spin has a magnetic moment pointing in the same direction as the spin, $\mu = 2\mu s$. The system is placed in an external magnetic field H in the z direction and is in thermal equilibrium at temperature T. The Hamiltonian for the system is approximated by an Ising model, where the true spin-spin interaction is replaced by a term of the form $J S_z(i) S_z(i+1)$:

$$H = J S_z(1) S_z(2) + J S_z(2) S_z(3) - 2\mu H [S_z(1) + S_z(2) + S_z(3)] \ ,$$

where J and μ are positive constants.

(a) List each of the possible microscopic states of the system and its energy. Sketch the energy level diagram as a function of H. Indicate any degeneracies.

(b) For each of the following conditions, write down the limiting values of the internal energy $U(T, H)$, the entropy $S(T, H)$, and the magnetization $M(T, H)$.

1) $T = 0$ and $H = 0$,
2) $T = 0$ and $0 < H \ll J/\mu$,
3) $T = 0$ and $J/\mu \ll H$,
4) $J \ll kT$ and $H = 0$.

(c) On the basis of simple physical considerations, without doing any calculations, sketch the specific heat at constant field, $C_H(T, H)$ when $H = 0$. What is the primary temperature dependence at very high and very low T?

(d) Find a closed form expression for the partition function $Q(T, H)$.

(e) Find the magnetization $M(T, H)$. Find an approximate expression for $M(T, H)$ which is valid when $kT \gg \mu H$ or J.

(MIT)

Solution:

(a) Let $(S_z(1), S_z(2), S_z(3))$ stand for a microstate, and $E(S_z(1), S_z(2), S_z(3))$ stand for its energy. We have

$$E\left(\frac{1}{2}, \frac{1}{2}, \frac{1}{2}\right) = \frac{J}{2} - 3\mu H \ ,$$

$$E\left(\frac{1}{2}, -\frac{1}{2}, \frac{1}{2}\right) = -\frac{J}{2} - \mu H \ ,$$

$$E\left(\frac{1}{2}, \frac{1}{2}, -\frac{1}{2}\right) = E\left(-\frac{1}{2}, \frac{1}{2}, \frac{1}{2}\right) = -\mu H \ ,$$

$$E\left(-\frac{1}{2}, \frac{1}{2}, -\frac{1}{2}\right) = -\frac{J}{2} + \mu H \ ,$$

$$E\left(-\frac{1}{2}, -\frac{1}{2}, \frac{1}{2}\right) = E\left(\frac{1}{2}, -\frac{1}{2}, -\frac{1}{2}\right) = \mu H \ ,$$

$$E\left(-\frac{1}{2}, -\frac{1}{2}, -\frac{1}{2}\right) = \frac{J}{2} + 3\mu H \ .$$

These energy levels are sketched in Fig. 2.30, where we have assumed $2\mu H > \frac{J}{2} > \mu H$.

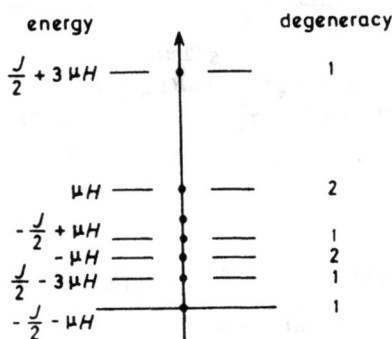

Fig. 2.30.

(b) 1) $U = -\dfrac{J}{2}$, $M = \pm\mu$, $S = k\ln 2$;

    2) $U = -\dfrac{J}{2} - \mu H$, $M = \mu$, $S = 0$;

    3) $U = \dfrac{J}{2} - 3\mu H$, $M = 3\mu$, $S = 0$;

    4) $U = 0$, $M = 0$, $S = 3k\ln 2$.

(c) When $H = 0$, the system has three energy levels $\left(-\dfrac{J}{2}, 0, \dfrac{J}{2}\right)$. At $T = 0$ K, the system is at ground state; when $0 < kT \ll \dfrac{J}{2}$, the energy of the system is enhanced by

$$\Delta E \propto e^{-J/2kT},$$

and

$$C_H \propto \frac{1}{T^2} e^{-J/2kT}.$$

As temperature increases, E and C_H increase rapidly. When the system is near the state where all the energy levels are uniformly occupied, the increase of energy becomes slower and C_H drops. When $kT \gg J$, $\Delta E \propto 1/\exp(J/kT)$ and $C_H \propto \dfrac{1}{T^2}$. Finally the energy becomes constant and $C_H = 0$. The $C_H(T, H = 0)$ vs T curve is sketched in Fig. 2.31.

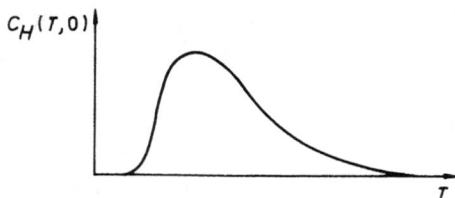

Fig. 2.31.

(d) $Q = e^{-\beta(J/2-3\mu H)} + e^{\beta(J/2+\mu H)} + 2e^{\beta\mu H}$
$\qquad + e^{-\beta(-J/2+\mu H)} + 2e^{-\beta\mu H} + e^{-\beta(J/2+3\mu H)}$
$\qquad = 2e^{-\beta J/2}\cosh(3\beta\mu H) + 2(e^{\beta J/2} + 2)\cosh(\beta\mu H)$,

where $\beta = \dfrac{1}{kT}$.

(e) $M = \dfrac{1}{\beta}\dfrac{\partial}{\partial H}\ln Q = \dfrac{2\mu}{Q}[3e^{-\beta J/2}\sinh(3\beta\mu H)$
$\qquad + (e^{\beta J/2} + 2)\sinh(\beta\mu H)]$.

When $kT \gg \mu H$ or J, $Q \approx 6$, $M \approx 4\beta\mu^2 H$.

2143

Consider a crystalline lattice with Ising spins $s_\ell = \ell$ at each site ℓ. In the presence of an external field $\mathbf{H} = (0, 0, H_0)$, the Hamiltonian of the system may be written as

$$H = -J \sum_{\ell,\ell'} s_\ell s_{\ell'} - \mu_0 H_0 \sum_\ell s_\ell ,$$

where $J > 0$ is a constant and the sum $\sum_{\ell,\ell'}$ is over all nearest-neighbor sites only (each site has p nearest neighbors).

(a) Write an expression for the free energy of the system at temperature T (do not try to evaluate it).

(b) Using the mean-field approximation, derive an equation for the spontaneous magnetization $m = \langle s_0 \rangle$ for $H_0 = 0$ and calculate the critical temperature T_c below which $m \neq 0$.

(c) Calculate the critical exponent β defined by $m(T, H_0 = 0) \sim$ const. $(1 - T/T_c)^\beta$ as $T \to T_c$.

(d) Describe the behavior of the specific heat at constant $\mathbf{H}_0$, $C(\mathbf{H}_0 = 0)$, near $T = T_c$.

(Princeton)

Solution:

Denote by N_A and N_B the total numbers of particles of $s_\ell = +1$ and $s_\ell = -1$ respectively. Also denote by N_{AA}, N_{BB}, and N_{AB} the total number of pairs of the nearest-neighbor particles that both have $s_\ell = 1$, that both have $s_\ell = -1$, and that have spins antiparallel to each other respectively. The Hamiltonian can be written as

$$H = -J(N_{AA} + N_{BB} - N_{AB}) - \mu_0 H_0 (N_A - N_B) .$$

Considering the number of nearest-neighbor pairs with at least one $s_\ell = +1$, we have

$$PN_A = 2N_{AA} + N_{AB} .$$

Similarly, $PN_B = 2N_{BB} + N_{AB}$. As $N = N_A + N_B$, among N_A, N_B, N_{AB}, N_{AA} and N_{BB} only two are independent. We can therefore write in terms of N_A and N_{AA}

$$N_B = N - N_A ,$$
$$N_{AB} = PN_A - 2N_{AA} ,$$
$$N_{BB} = PN/2 - PN_A + N_{AA} .$$

Hence $H = -4JN_{AA} - 2(-PJ + \mu_0 H_0)N_A + (-PJ/2 + \mu_0 H_0)N$.

(a) The partition function is

$$z = \sum_{\text{all states}} \exp\left(\frac{-H}{kT}\right).$$

The free energy is $F = -kT \ln z$.

(b) Using the mean-field approximation, the ratio of the number of the nearest-neighbor pairs with spins upward to the total number of pairs equals the probability that the spins are all upward in the nearest-neighbor sites, i.e.,

$$\frac{2N_{AA}}{PN} = \left(\frac{N_A}{N}\right)^2.$$

Thus,

$$H = -2PJ\left(\frac{N_A}{N}\right)^2 N + 2(PJ - \mu_0 H_0)N_A + \left(-\frac{PJ}{2} + \mu_0 H_0\right)N.$$

The partition function is then

$$z = \sum_{N_A=0}^{N} C_{N_A}^{N} \exp\left(-\frac{H}{kT}\right).$$

Defining the magnetization by

$$m = \frac{N_A - N_B}{N_A + N_B},$$

we have

$$\frac{1}{N}\ln z \approx -\frac{H}{NkT} - \frac{1}{2}(1 + m)\ln\frac{1 + m}{2}$$
$$- \frac{1}{2}(1 - m)\ln\frac{1 - m}{2}.$$

For $\partial \ln z/\partial m = 0$, we obtain

$$\frac{\mu_0 H_0}{kT} + \frac{PJ}{kT}m - \frac{1}{2}\ln\frac{1 + m}{1 - m} = 0,$$

or

$$\frac{\mu_0 H_0}{kT} + \frac{PJm}{kT} = \tanh^{-1} m \; .$$

With $H_0 = 0, m = \tanh\left(\frac{T_c}{T}m\right)$. Therefore, only when $T < T_c = PJ/k$, has the above equation a solution $m \neq 0$. Thus, the critical temperature is $T_c = PJ/k$.

(c) When $H_0 = 0$, we have $m = \tanh\left(\frac{T_c}{T}m\right)$. For $T \to T_c$ we can use the Taylor expansion and write

$$m \sim \text{const.} \left(1 - \frac{T}{T_c}\right)^{1/2} \; .$$

Hence $\beta = \frac{1}{2}$.

(d) From $E = -\frac{\partial}{\partial \beta} \ln z$, we obtain

$$\frac{E}{N} = -\mu_0 H_0 m - \frac{1}{2} PJm^2 \; ,$$

and

$$C = \frac{1}{N}\frac{\partial E}{\partial T} = (-\mu_0 H_0 - PJm)\frac{\partial m}{\partial T} \; .$$

When $H_0 = 0, C = -PJm\frac{\partial m}{\partial T}$. When $T \geq T_c, m = 0, C = 0$. When $T < T_c$, we have near T_c,

$$m = \text{const.} \left(1 - \frac{T}{T_c}\right)^{1/2} \; ,$$

$$m^2 \propto 1 - \frac{T}{T_c} \; ,$$

$$C \propto \frac{PJ}{T_c} \; .$$

2144

Consider a gas of hard spheres with the 2-body interaction

$$V(|\mathbf{r}_i - \mathbf{r}_j|) = 0 , \qquad |\mathbf{r}_i - \mathbf{r}_j| > a ,$$
$$= \infty , \qquad |\mathbf{r}_i - \mathbf{r}_j| < a .$$

Using the classical partition function, calculate the average energy at a given temperature and density (thermodynamics: the internal energy).

On the basis of simple physical arguments, would you expect this same simple answer to also result from a calculation with the quantum mechanical partition function?

(Wisconsin)

Solution:

The partition function of the whole system is

$$Z = Z_T Z_V ,$$

where Z_T is that of the thermal motion of the particles and Z_V is that of the interactions between particles:

$$Z_T = \left(\frac{2\pi m k T}{h^2}\right)^{\frac{3N}{2}} ,$$

$$Z_V = \int \ldots \int d\mathbf{r}_1 \ldots d\mathbf{r}_N \exp\left[-\beta \sum_{j \neq i} V(|\mathbf{r}_i - \mathbf{r}_j|)\right]$$
$$= [V - (N-1)\alpha][V - (N-2)\alpha] \ldots [V - \alpha]V$$
$$= V^N \left[1 - \frac{N(N-1)}{2V}\alpha + \ldots\right] ,$$

where $\alpha = \frac{4}{3}\pi a^3$. The average energy is

$$\langle E \rangle = kT^2 \frac{\partial \ln Z}{\partial T} = \frac{3N}{2}kT .$$

That is, in this model, the average energy of the system (the internal energy of thermodynamics) is equal to the sum of the energies of thermal motion of the particles and is independent of the interactions between particles. As the interactions between particles do not come in the result, we expect to obtain the same result from the quantum partition function.

2145

A classical gas of N point particles occupies volume V at temperature T. The particles interact pairwise, $\phi(r_{ij})$ being the potential between particles i and j, $r_{ij} = |\mathbf{r}_i - \mathbf{r}_j|$. Suppose this is a "hard sphere" potential

$$\phi(r_{ij}) = \begin{cases} \infty, & r_{ij} < a, \\ 0, & r_{ij} > a. \end{cases}$$

(a) Compute the constant volume specific heat as a function of temperature and specific volume $v = \dfrac{V}{N}$.

(b) The virial expansion for the equation of state is an expansion of $\dfrac{pV}{RT}$ in inverse powers of V:

$$\frac{pV}{RT} = 1 + \frac{A_1(T)}{V} + \frac{A_2(T)}{V^2} + \cdots .$$

Compute the virial coefficient A_1.

(*Princeton*)

Solution

For the canonical distribution, the partition function is

$$z = \frac{1}{N!} \cdot \frac{1}{h^{3N}} \int e^{-\beta E} dq dp$$

$$= \frac{1}{N!} \left(\frac{2\pi m}{\beta h^2} \right)^{3N/2} \cdot Q ,$$

with

$$Q = \int \cdots \int e^{-\beta \sum_{i<j} \phi(r_{ij})} dq_1 \ldots dq_{3N} ,$$

where q_i represents the coordinates and p_i the momentum of the ith particle. Defining the function $f_{ij} = \exp[-\beta\phi(r_{ij})] - 1$ with $f_{ij} = 0$ for $r_{ij} > a$, we can write

$$Q = \int \cdots \int \left(1 + \sum_{i<j} f_{ij} + \sum_{i<j} f_{ij} \cdot \sum_{i'<j'} f_{i'j'} + \cdots \right) dq_1 \ldots dq_{3N} .$$

Keeping the first two terms, we have

$$Q = \int \ldots \int \left(1 + \sum_{i<j} f_{ij}\right) dq_1 \ldots dq_{3N} \ ,$$

$$Q \approx V^N + \frac{N^2}{2} V^{N-1} \int f_{12} dq_1 dq_2 dq_3$$

$$= V^N \left(1 + \frac{N^2}{2V} \cdot (-\tau)\right) \ ,$$

where $\tau = \dfrac{4\pi a^3}{3}$.

Hence

$$\ln Q = N \ln V + \ln\left[1 - \frac{\tau}{2V} N^2\right] \approx N \ln V - \frac{\tau}{2V} N^2 \ ,$$

so that

$$u = -\frac{\partial}{\partial \beta} \ln z = \frac{3N}{2} kT \ ,$$

$$c_v = \frac{3}{2} Nk \ ,$$

and

$$p = \frac{1}{\beta} \frac{\partial}{\partial V} \ln z = \frac{1}{\beta} \frac{\partial}{\partial V} \ln Q = \frac{1}{\beta} \left(\frac{N}{V} + \frac{\tau N^2}{2V^2}\right) \ ,$$

giving

$$\frac{pV}{NkT} = 1 + \frac{\tau N}{2V} \ .$$

Thus

$$A_1(T) = \frac{\tau N}{2} = \frac{2\pi a^3}{3} N \ .$$

2146

Consider a classical system of N point particles of mass m in a volume V at temperature T. Let U be the total energy of the system, p the pressure. The particles interact through a two-body central potential

$$\phi(r_{ij}) = \frac{A}{r_{ij}^n} \ , \quad A > 0 \ , \quad n > 0 \ , \quad r_{ij} = |\mathbf{r}_i - \mathbf{r}_j| \ .$$

Notice the scaling property $\phi(\gamma r) = \gamma^{-n}\phi(r)$ for any γ. From this, and from scaling arguments (e.g. applied to the partition function) show that

$$U = apV + bNkT, \quad k = \text{Boltzmann's const.},$$

where the constants a and b depend on the exponent n in the pairwise potential. Express a and b in terms of n.

(*Princeton*)

Solution:

The partition function of the system is

$$z(T,V) = \frac{1}{h^{3N}} \int e^{-\beta E} dp\, dr$$

$$= \left(\frac{2\pi m kT}{h^2}\right)^{3N/2} \int_V e^{-\beta \Sigma \phi(r_{ij})} d\mathbf{r} .$$

Replacing T with λT, and noticing that $\phi(r_{ij}) = A/r_{ij}^n$, we have

$$z(\lambda T,V) = \left(\frac{2\pi m k \lambda T}{h^2}\right)^{3N/2} \int_V e^{-(\beta/\lambda)\Sigma \phi(r_{ij})} d\mathbf{r}$$

$$= \left(\frac{2\pi m k \lambda T}{h^2}\right)^{3N/2} \int_V \exp\left(-\beta \sum \phi(\lambda^{\frac{1}{n}} r_{ij})\right) d\mathbf{r}$$

$$= \left(\frac{2\pi m k \lambda T}{h^2}\right)^{3N/2} \int_{\lambda^{3/n}V} \lambda^{-3N/n} e^{-\beta \Sigma \phi(r_{ij})} d\mathbf{r}$$

$$= \lambda^{3N(\frac{1}{2}-\frac{1}{n})} z(T, \lambda^{3/n}V) .$$

This can be rewritten as

$$z(\lambda T, \lambda^{-3/n}V) = \lambda^{3N(\frac{1}{2}-\frac{1}{n})} z(T,V) .$$

The free energy

$$F(\lambda T, \lambda^{-3/n}V) = -kT\lambda \ln z(\lambda T, \lambda^{-3/n}V)$$

$$= -3N\left(\frac{1}{2}-\frac{1}{n}\right) kT\lambda \ln\lambda + \lambda F(T,V) .$$

We differentiate it with respect to λ, take $\lambda = 1$, and get

$$T\left(\frac{\partial F}{\partial T}\right)_V - \frac{3}{n}V\left(\frac{\partial F}{\partial V}\right)_T = -3N\left(\frac{1}{2}-\frac{1}{n}\right)kT + F .$$

On the other hand, from

$$U = F - T \left(\frac{\partial F}{\partial T} \right)_V \;, \quad p = - \left(\frac{\partial F}{\partial V} \right)_T \;,$$

we have $U = 3 \left(\frac{1}{2} - \frac{1}{n} \right) NkT + \frac{3}{n} pV = apV + bNkT$ giving $a = \frac{3}{n}, b = 3 \left(\frac{1}{2} - \frac{1}{n} \right)$.

2147

(a) Given $\int_{-\infty}^{+\infty} \exp(-\alpha x^2) dx = \sqrt{\pi/\alpha}$, show that

$$\int_{-\infty}^{\infty} x^2 e^{-\alpha x^2} dx = \frac{\sqrt{\pi}}{2} \alpha^{-3/2} \;, \quad \int_{-\infty}^{\infty} x^4 e^{-\alpha x^2} dx = \frac{3}{4} \sqrt{\pi} \alpha^{-5/2} \;.$$

(b) Given that $\frac{\beta}{\sqrt{\alpha}} \ll 1$ and that a is of the order of $\frac{1}{\sqrt{\alpha}}$, show

$$\int_{-a}^{a} f(x) e^{-\alpha[x^2 + \beta x^3]} dx \approx \int_{-a}^{a} f(x)(1 - \alpha\beta x^3) e^{-\alpha x^2} dx \;.$$

(c) Two atoms interact through a potential

$$U(x) = U_0 \left[\left(\frac{a}{x} \right)^{12} - 2 \left(\frac{a}{x} \right)^6 \right] \;,$$

where x is their separation. Sketch this potential. Calculate the value of x for which $U(x)$ is minimum.

(d) Given a row of such atoms constrained to move only on the x axis, each assumed to interact only with its nearest neighbors, use classical statistical mechanics to calculate the mean interatomic separation $\bar{x}(T)$.

To do this, expand U about its minimum, keeping as many terms as necessary to obtain the lowest order temperature dependence of $\bar{x}(T)$. Assume that $kT \ll U_0$, and in the relevant integrals extend the limits of integration to $\pm\infty$ where appropriate. Explain clearly the justification for extending the limits. Also calculate

$$\lambda = \frac{\left(\dfrac{d\bar{x}}{dT}\right)}{a} .$$

(*CUSPEA*)

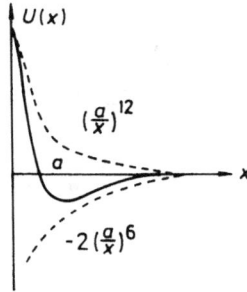

Fig. 2.32.

Solution:

(a) $\displaystyle\int_{-\infty}^{\infty} x^2 e^{-\alpha x^2} dx = -\frac{d}{d\alpha} \int_{-\infty}^{\infty} e^{-\alpha x^2} dx = -\frac{d}{d\alpha} \cdot \sqrt{\frac{\pi}{\alpha}}$

$$= \frac{\sqrt{\pi}}{2} \alpha^{-3/2} ,$$

$\displaystyle\int_{-\infty}^{\infty} x^4 e^{-\alpha x^2} dx = \frac{d^2}{d\alpha^2} \int_{-\infty}^{\infty} e^{-\alpha x^2} dx$

$$= \frac{d^2}{d\alpha^2} \sqrt{\frac{\pi}{\alpha}} = \frac{3}{4} \sqrt{\pi} \alpha^{-5/2} .$$

(b) $\displaystyle\int_{-a}^{a} f(x) e^{-\alpha[x^2 + \beta x^3]} dx = \int_{-a}^{a} f(x) e^{-\alpha\beta x^3} e^{-\alpha x^2} dx .$

Since

$$|\alpha\beta x^3| = \left|\alpha x^2 \cdot \sqrt{\alpha} x \cdot \frac{\beta}{\sqrt{\alpha}}\right| = \left|(\sqrt{\alpha} x)^3 \frac{\beta}{\sqrt{\alpha}}\right|$$

$$< |\sqrt{\alpha} a|^3 \frac{\beta}{\sqrt{\alpha}} \sim \frac{\beta}{\sqrt{\alpha}} \ll 1 ,$$

we have

$$e^{-\alpha\beta x^3} = 1 - \alpha\beta x^3 + \ldots \approx 1 - \alpha\beta x^3 ,$$

and hence

$$\int_{-a}^{a} f(x) e^{-\alpha[x^2 + \beta x^2]} dx \approx \int_{-a}^{a} f(x)(1 - \alpha\beta x^3) e^{-\alpha x^2} dx .$$

(c) The given potential is as shown in Fig. 2.32. Letting

$$\frac{dU}{dx} = U_0 \left[-12 \frac{a^{12}}{x^{13}} - 2(-6) \frac{a^6}{x^7} \right] = 0 ,$$

we find that $U(x)$ is minimum at $x = a$.

(d) According to classical mechanics, atoms are at rest at their equilibrium positions when $T = 0$. The distance between neighboring atoms is a. If $T \neq 0$, the interacting potential is

$$\begin{aligned} U_T(x) &= U(x + a) + U(a - x) \\ &= U_T(a) + U_{T_2} x^2 + U_{T_3} x^3 + \dots \end{aligned}$$

where x is the displacement from equilibrium position, and $U_{T_2} = 72 U_0 / a^2$, $U_{T_3} = -504 U_0 / a^3$ etc.

Using classical statistical mechanics, we obtain

$$\bar{x}(T) = \int_{-\infty}^{\infty} dx \cdot x e^{-U_T(x)/kT} \Big/ \int_{-\infty}^{\infty} dx \cdot e^{-U_T(x)/kT}$$

$$\approx \int_{-\infty}^{\infty} dx \cdot x e^{-(U_{T_2} x^2 + U_{T_3} x^3)/kT} \Big/ \int_{-\infty}^{\infty} dx \cdot e^{-(1/kt)U_{T_2} x^2} .$$

Since $kT \ll U_0$, we obtain

$$\bar{x}(T) \approx \int_{-\infty}^{\infty} dx \cdot x \left(1 - \frac{U_{T_3}}{kT} x^3 \right) e^{-(U_{T_2}/kT)x^2} \Big/ \int_{-\infty}^{\infty} dx \cdot e^{-(U_{T_2}/kT)x^2}$$

$$= 7akT/96U_0 ,$$

and $\lambda = \dfrac{7k}{96 U_0}$.

2148

A classical system is described by its Hamiltonian H, which is a function of a set of generalized co-ordinates q_i and momenta p_i. The canonical equations of motion are

$$\dot{p}_i = -\frac{\partial H}{\partial q_i} , \quad \dot{q}_i = \frac{\partial H}{\partial p_i} .$$

Write the equation of continuity for ρ, the phase space density, and use it to show that the entropy of this system is constant in time.

Now consider a system whose motion is damped by frictional force. For simplicity, consider a damped harmonic oscillator in one dimension. The equations of motion are

$$\dot{p} = -kq - \frac{\gamma p}{m} , \quad \dot{q} = \frac{p}{m} ,$$

where m is the mass, k is the spring constant, and γ is related to the friction, m, k and γ being all positive.)

What is the equation of motion for the phase space density ρ? Show that the entropy is now a decreasing function of time.

Can the last result be reconciled with the second law of thermodynamics?

(*Princeton*)

Solution:

The conservation of ρ is described by

$$\frac{d\rho}{dt} + \sum_i \rho \left[\frac{\partial}{\partial q_i}(\dot{q}_i) + \frac{\partial}{\partial p_i}(\dot{p}_i) \right] = 0 .$$

With the canonical equations of motion we have

$$\frac{d\rho}{dt} = 0 .$$

That is, along the phase orbit ρ is a constant. Since the phase orbits are on the surface of constant energy, $d\rho/dt = 0$ implies that there is no transition among the energy levels. Thus the entropy of the system is constant in time.

For a damped one-dimensional harmonic oscillator, the equations of motion give

$$\frac{d\rho}{dt} - \frac{\gamma \rho}{m} = 0 .$$

Hence along the phase orbit (flow line), we have state density

$$\rho = \rho_0 \exp(\gamma t/m) ,$$

which is always increasing. In addition, energy consideration gives

$$\frac{d}{dt}\left(\frac{p^2}{2m} + \frac{k}{2}q^2 \right) = -\gamma \left(\frac{dq}{dt} \right)^2 < 0 ,$$

i.e., along the phase orbit the energy decreases. In other words, the flow line points to region of lower energies. Thus, combining these two results, we would expect transitions from high-energy states to low-energy states. Hence the entropy of the system is a decreasing function of time. As the oscillator is not an isolated system, the decrease of entropy is not in contradiction to the second law of thermodynamics.

5. KINETIC THEORY OF GASES (2149-2208)

2149

Estimate

(a) The number of molecules in the air in a room.

(b) Their energy, in joules or in ergs, per mole.

(c) What quantity of heat (in joules or in ergs) must be added to warm one mole of air at 1 atm from 0°C to 20°C?

(d) What is the minimum energy that must be supplied to a refrigerator to cool 1 mole of air at 1 atm from 20°C to 18°C? The refrigerator acts in a cyclic process and gives out heat at 40°C.

$$(UC, Berkeley)$$

Solution:

(a) 1 mole of gas occupies about 23 l and an average-sized room has a volume of 50m^3. Then the number of molecules therein is about

$$N \simeq \frac{50 \times 1000}{23} \times 6.02 \times 10^{23} \sim 10^{27} \ .$$

(b) The energy per mole of gas is

$$E = \frac{5}{2} RT = \frac{5}{2} \times 8.31 \times 300 = 6.2 \times 10^3 \ \text{J} \ .$$

(c) The heat to be added is

$$Q = C_p \Delta T = \frac{7}{2} R \cdot \Delta T = 5.8 \times 10^2 \ \text{J} \ .$$

(d) $dW = \dfrac{T_1 - T_2}{T_2} dQ.$

With $T_1 = 313$ K, $T_2 = 293$ K, we require

$$\Delta W = \frac{T_1 - T_2}{T_2} C_p \Delta T = 4 \ \text{J} \ ,$$

where we have taken $\Delta T = 2$ K.

2150

Consider a cube, 10 cm on a side, of He gas at STP. Estimate (order of magnitude) the number of times one wall is struck by molecules in one second.

(*Columbia*)

Solution:

Under STP, pressure $p \approx 10^6$ dyn/cm^2, temperature $T \approx 300$ K, thus the number of times of collisions is

$$N \approx \frac{1}{6}\bar{v}nS = \frac{\sqrt{2}pS}{3\sqrt{\pi mkT}} \approx 5 \times 10^{25} \text{ s}^{-1} ,$$

where $\bar{v} = (8kT/\pi m)^{\frac{1}{2}}$ is the average velocity, n is the number density of the gas molecules, S is the area of one wall.

2151

Estimate the mean free path of a cosmic ray proton in the atmosphere at sea level.

(*Columbia*)

Solution:

The proton is scattered by interacting with the nuclei of the molecules of the atmosphere. The density of the atmosphere near sea level is

$$n = \frac{N}{V} = \frac{p}{kT} .$$

The mean free path is

$$l = \frac{1}{\pi n\sigma} = \frac{1}{\pi\sigma} \cdot \frac{kT}{p} = 10^6 \text{ cm} ,$$

where we have taken $\sigma = 10^{-26}$ cm^2.

2152

Even though there is a high density of electrons in a metal (mean separation $r \sim 1 - 3$ Å), electron-electron mean free paths are long. ($\lambda_{ee} \sim 10^4$Å at room temperature.) State reasons for the long electron-electron collision mean free path and give a qualitative argument for its temperature dependence.

(*Wisconsin*)

Solution:

The mean free path $\lambda \propto \frac{1}{n_{\text{eff}}}$, where n_{eff} is the effective number density of electrons. For the electron gas in a metal at temperature T, only the electrons near the Fermi surfaces are excited and able to take part in collisions with one another. The effective number density of electrons near the Fermi surface is

$$n_{\text{eff}} = \frac{nkT}{\varepsilon_F} .$$

Hence λ is very long even though the electron density is high quantitatively, we have from the above

$$\lambda_{ee} \propto \frac{\varepsilon_F}{kTn\sigma} \propto \frac{1}{T} .$$

That is, when temperature increases more electrons are excited and able to collide with one another. This reduces the mean free path.

2153

Estimate the following:

(a) The mean time between collisions for a nitrogen molecule in air at room temperature and atmospheric pressure.

(b) The number density of electrons in a degenerate Fermi electron gas at $T = 0$ K and with a Fermi momentum $p_F = m_e c$.

(*UC, Berkeley*)

Solution:

(a) Assume that the mean free path of a molecule is l, its average velocity is v, and the mean collision time is τ. We have

$$l = \frac{1}{n\sigma} = \frac{kT}{p\sigma} \approx 4 \times 10^{-6} \text{ m},$$

$$v = \sqrt{\frac{3kT}{m}} \approx 9 \times 10^2 \text{ m/s}.$$

$$\tau = \frac{l}{v} \approx 4 \times 10^{-9} \text{ s}.$$

(b) The electron number density is

$$n = 2 \iiint_{p \le p_F} \frac{d\mathbf{p}}{h^3} = \frac{8\pi}{3h^3} p_F^3 = \left(\frac{8\pi}{3}\right)\left(\frac{m_e c}{h}\right)^3 = 6 \times 10^{35} \text{ m}^{-3}.$$

2154

A container is divided into two parts by a partition containing a small hole of diameter D. Helium gas in the two parts is held at temperature $T_1 = 150$ K and $T_2 = 300$ K respectively through heating of the walls.

(a) How does the diameter D determine the physical process by which the gases come into a steady state?

(b) What is the ratio of the mean free paths l_1/l_2 between the two parts when $D \ll l_1, D \ll l_2$, and the system has reached a steady state?

(c) What is the ratio l_1/l_2 when $D \gg l_1, D \gg l_2$?

(*Princeton*)

Solution:

(a) At the steady state the number of molecules in each part is fixed. If $D \gg l_1$ and $D \gg l_2$, the molecules are exchanged by macroscopic gas flow. If $D \ll l_1, D \ll l_2$, the molecules are exchanged by leakage gas flowing through the pinhole.

(b) When $l_1 \gg D$ and $l_2 \gg D$, the steady state occurs under the condition

$$\frac{n_1 v_1}{4} = \frac{n_2 v_2}{4},$$

i.e., the numbers of collision are equal. Hence

$$\frac{l_1}{l_2} = \frac{n_2}{n_1} = \frac{v_1}{v_2} = \sqrt{\frac{T_1}{T_2}} = 0.707 \ .$$

(c) When $l_1 \ll D$ and $l_2 \ll D$, the steady state occurs under the condition $p_1 = p_2$, i.e., the pressures are equal. Hence

$$\frac{l_1}{l_2} = \frac{n_2}{n_1} = \frac{T_1}{T_2} = 0.5 \ .$$

2155

Consider the orthogonalized drunk who starts out at the proverbial lamp-post: Each step he takes is either due north, due south, due east or due west, but which of the four directions he steps in is chosen purely randomly at each step. Each step is of fixed length L. What is the probability that he will be within a circle of radius $2L$ of the lamp-post after 3 steps?

(*Columbia*)

Solution:

The number of ways of walking three steps is $4 \times 4 \times 4 = 64$. The drunk has two ways of walking out from the circle:

     i) Walk along a straight line

     ii) Two steps forward, one step to right (or left).

Corresponding to these the numbers of ways are $C_1^4 = 4$ and $C_1^4 \cdot C_1^3 \cdot C_1^2 = 24$ respectively. Hence the propability that he will remain within the circle after 3 steps is

$$P = 1 - \frac{4 + 24}{64} = \frac{9}{16} \ .$$

2156

Estimate how long it would take a molecule of air in a room, in which the air is macroscopically 'motionless' and of perfectly uniform temperature and pressure, to move to a position of distance 5 meters away.

(*Columbia*)

Solution:

As molecular diffusion is a random process, we have

$$L^2 = nl^2 ,$$

where L is the total displacement of the molecule, l is its mean free path, n is the number of collisions it suffers as it moves through the displacement L. Therefore, the required time is

$$t = n\frac{l}{v} = \frac{L^2}{lv} = 10^4 \text{ s}$$

where we have taken $l = 5 \times 10^{-6}$ m and $v = 5 \times 10^2$ m/s.

2157

You have just very gently exhaled a helium atom in this room. Calculate how long (t in seconds) it will take to diffuse with a reasonable probability to some point on a spherical surface of radius $R = 1$ meter surrounding your head.

(*UC, Berkeley*)

Solution:

First we estimate the mean free path l and mean time interval τ for molecular collisions:

$$l = \frac{1}{n\sigma} = \frac{kT}{p\sigma} = \frac{1.38 \times 10^{-23} \times 300}{1 \times 10^5 \times 10^{-20}} = 4 \times 10^{-6} \text{ m} ,$$

$$\tau = \frac{l}{v} = \frac{4 \times 10^{-6}}{300} = 1.4 \times 10^{-8} \text{ s} .$$

Since $R^2 = Nl^2$, where N is the number of collisions it suffers in traversing the displacement R, we have

$$t = N\tau = \left(\frac{R}{l}\right)^2 \tau = \frac{R^2}{lv} = 8.6 \times 10^2 \text{ s} \approx 14 \text{ min} .$$

2158

In an experiment a beam of silver atoms emerges from an oven, which contains silver vapor at $T = 1200$ K. The beam is collimated by being passed through a small circular aperture.

(a) Give an argument to show that it is not possible, by narrowing the aperture a, to decrease indefinitely the diameter of the spot, D, on the screen.

(b) If the screen is at $L = 1$ meter from the aperture, estimate numerically the smallest D that can be obtained by varying a.

(You may assume for simplicity that all atoms have the same momentum along the direction of the beam and have a mass of $M_{Ag} = 1.8 \times 10^{-22}$ g).

(UC, Berkeley)

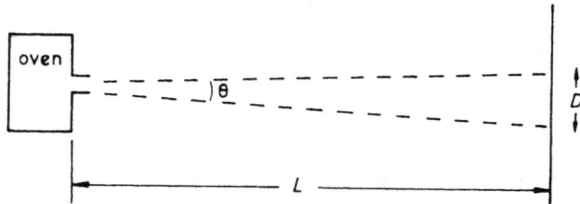

Fig. 2.33.

Solution:

(a) According to the uncertainty principle, the smaller a is, the greater is the uncertainty in the y-component of the momentum of the silver atoms that pass through the aperture and the larger is the spot.

(b) Using the uncertainty principle, we obtain the angle of deflection of the outgoing atoms

$$\theta \approx \frac{\lambda}{a} \approx \frac{h}{pa} = \frac{h}{a\sqrt{3mkT}} \; .$$

Thus, $D = a + 2\theta L = a + \dfrac{2hL}{a\sqrt{3mkT}} \geq 2\sqrt{\dfrac{2hL}{\sqrt{3mkT}}}.$

$$D_{\min} = 2\frac{(2hL)^{1/2}}{(3mkT)^{1/4}} = 8.0 \times 10^{-6} \text{ m} \; .$$

That is, the smallest diameter is about 80×10^3 Å.

2159

Scattering.

The range of the potential between two hydrogen atoms is approximately 4Å. For a gas in thermal equilibrium obtain a numerical estimate of the temperature below which the atom-atom scattering is essentially S-wave.

(*MIT*)

Solution:

For S-wave scattering, $Ka < 2\pi$, where $a = 4$Å is the range of the potential. As

$$\frac{1}{2}kT = \frac{h^2 K^2}{8\pi^2 m} \, ,$$

we get

$$T < \frac{h^2}{mka^2} \approx 1 \text{ K} \, .$$

2160

Show that a small object immersed in a fluid at temperature T will undergo a random motion, due to collisions with the molecules of the fluid, such that the mean-square displacement in any direction satisfies

$$\langle (\Delta x)^2 \rangle = Tt/\lambda \, ,$$

where t is the elapsed time, and λ is a constant proportional to the viscosity of the fluid.

(*Columbia*)

Solution:

Because of the thermal motion of the molecules of the fluid, the small object is continually struck by them. The forces acting on the object are the damping force $-\gamma v$ (γ is the viscosity of the fluid) and a random force $f(t)$ for which

$$\langle f(t) \rangle = 0 \, , \quad \langle f(t)f(t') \rangle = a\delta(t - t') \, , \quad \text{where } a \text{ is to be determined.}$$

The equation of the motion of the object is

$$m\frac{dv}{dt} = -\gamma v + f(t) \, .$$

Assuming $v(0) = 0$, we have

$$v(t) = \int_0^t F(s) \exp[(s - t)\beta] ds \ ,$$

where $t > 0, \beta = \gamma/m, F(t) = f(t)/m$. Consider

$$
\begin{aligned}
\langle v(t)v(t')\rangle &= \int_0^t ds \int_0^{t'} ds' e^{-(t-s)\beta - (t'-s')\beta} \langle F(s)F(s')\rangle \\
&= \frac{a}{m^2} \int_0^t ds \int_0^{t'} e^{-(t+t')\beta + (s+s')\beta} \delta(s - s') ds' \\
&= \frac{a}{m^2} e^{-(t+t')\beta} \int_0^t ds \int_0^{t'} e^{2\beta s} \delta(s - s') ds' \\
&= \frac{a}{m^2} e^{-(t+t')\beta} \int_0^t e^{2\beta s} \theta(s)\theta(t' - s) ds \\
&= \frac{a}{m^2} e^{-(t+t')\beta} \frac{1}{2\beta} \exp[2\beta \min(t, t')] \\
&= \frac{a}{2\beta m^2} \exp(-\beta|t - t'|)
\end{aligned}
$$

where, $\theta(x) = \begin{cases} 1, & x > 0 \ , \\ 0, & x < 0 \ . \end{cases}$

At thermal equilibrium, $\langle v^2(t)\rangle = \dfrac{kT}{m}$. Comparing this with the above result gives

$$a = 2\beta mkT = 2\gamma kT \ .$$

Next, consider

$$\Delta x = x(t) - x(0) = \int_0^t v(s) ds \ ,$$

and

$$
\begin{aligned}
\langle (\Delta x)^2\rangle &= \int_0^t ds \int_0^{t'} \langle v(s)v(s')\rangle ds' \\
&= 2kT \int_0^t ds \int_0^s \exp[-\gamma(s - s')/m]/m ds' \\
&= \frac{2kT}{m}\left[\frac{m}{\gamma}t - \frac{m^2}{\gamma^2}\left(1 - e^{-\gamma t/m}\right)\right] \underset{t\to\infty}{\longrightarrow} \frac{2kT}{\gamma}t \ .
\end{aligned}
$$

This can be written as

$$\langle (\Delta x)^2 \rangle = \frac{Tt}{\lambda}$$

with

$$\lambda = \gamma/2k \propto \gamma .$$

2161

A box of volume $2V$ is divided into halves by a thin partition. The left side contains a perfect gas at pressure p_0 and the right side is initially vacuum. A small hole of area A is punched in the partition. What is the pressure p_1 in the left hand side as a function of time? Assume the temperature is constant on both sides. Express your answer in terms of the average velocity v.

(*Wisconsin*)

Solution:

Because the hole is small, we can assume the gases of the two sides are at thermal equilibrium at any moment. If the number of particles of the left side per unit volume at $t = 0$ is n_0, the numbers of particles of the left and right sides per unit volume at the time t are $n_1(t)$ and $n_0 - n_1(t)$ respectively. We have

$$V \frac{dn_1(t)}{dt} = -\frac{A}{4} n_1 v + \frac{A}{4}(n_0 - n_1)v ,$$

where $v = \sqrt{\frac{8kT}{\pi m}}$ is the average velocity of the particles. The first term is the rate of decrease of particles of the left side due to the particles moving to the right side, the second term is the rate of increase of particles of the left side due to the particles moving to the left side. The equation is simplified to

$$\frac{dn_1(t)}{dt} + \frac{A}{2V} n_1 v = \frac{A}{4V} n_0 v .$$

With the initial condition $n_1(0) = n_0$, we have

$$n_1(t) = \frac{n_0}{2}\left(1 + e^{\frac{-Avt}{2V}}\right) ,$$

and

$$p_1(t) = \frac{p_0}{2}\left(1 + e^{-\frac{Avt}{2V}}\right) .$$

2162

Starting with the virial theorem for an equilibrium configuration show that:

(a) the total kinetic energy of a finite gaseous configuration is equal to the total internal energy if $\gamma = C_p/C_v = 5/3$, where C_p and C_v are the molar specific heats of the gas at constant pressure and at constant volume, respectively,

(b) the finite gaseous configuration can be in Newtonian gravitational equilibrium only if $C_p/C_v > 4/3$.

(*Columbia*)

Solution:

For a finite gaseous configuration, the virial theorem gives

$$2\overline{K} + \overline{\sum_i \mathbf{r}_i \cdot \mathbf{F}_i} = 0$$

$\overline{K}$ is the average total kinetic energy, $\mathbf{F}_i$ is the total force acting on molecule i by all the other molecules of the gas. If the interactions are Newtonian gravitational of potentials $V(r_{ij}) \sim \dfrac{1}{r_{ij}}$, we have

$$\sum_i \mathbf{r}_i \cdot \mathbf{F}_i = -\sum_{j \neq i} r_{ij} \frac{\partial V(r_{ij})}{\partial r_{ij}} = \sum_{j<i} V(r_{ij}) \ ,$$

$$r_{ij} = |\mathbf{r}_i - \mathbf{r}_j| \ .$$

Hence $2\overline{K} + \overline{V} = 0$, where $\overline{V}$ is the average total potential energy.

We can consider the gas in each small region of the configuration as ideal, for which the internal energy density $\overline{u}(\mathbf{r})$ and the kinetic energy density $\overline{K}(\mathbf{r})$ satisfy

$$\overline{u}(\mathbf{r}) = \frac{2}{3}\frac{1}{\gamma - 1}\overline{K}(\mathbf{r}) \quad \text{with} \quad \overline{K}(\mathbf{r}) = \frac{3}{2}kT(\mathbf{r}) \ .$$

Hence the total internal energy is $\overline{U} = 2\overline{K}/3(\gamma - 1)$.

When $\gamma = \frac{5}{3}$, $\overline{U} = \overline{K}$. In general the Virial theorem gives $3(\gamma - 1)\overline{U} + \overline{V} = 0$, so the total energy of the system is

$$E = \overline{U} + \overline{V} = (4 - 3\gamma)\overline{U} \ .$$

For the system to be in stable equilibrium and not to diverge infinitely, we require $E < 0$. Since $\overline{U} > 0$, we must have

$$\gamma > \frac{4}{3} \ .$$

2163

A system consists of N very weakly interacting particles at a temperature sufficiently high such that classical statistics are applicable. Each particle has mass m and oscillates in one direction about its equilibrium position. Calculate the heat capacity at temperature T in each of the following cases:

(a) The restoring force is proportional to the displacement x from the equilibrium position.

(b) The restoring force is proportional to x^3.

The results may be obtained without explicitly evaluating integrals.

(UC, Berkeley)

Solution:
According to the virial theorem, if the potential energy of each particle is $V \propto x^n$, then the average kinetic energy $\overline{T}$ and the average potential energy $\overline{V}$ satisfy the relation $2\overline{T} = n\overline{V}$. According to the theorem of equipartition of energy. $\overline{T} = \frac{1}{2}kT$ for a one-dimensional motion. Hence we can state the following:

(a) As $f \propto x, V \propto x^2$, and $n = 2$. Then $\overline{V} = \overline{T} = \frac{1}{2}kT$, $E = \overline{V} + \overline{T} = kT$. Thus the heat capacity per particle is k and $C_v = Nk$.

(b) As $f \propto x^3$, $V \propto x^4$ and $n = 4$. Then $\overline{V} = \frac{1}{2}\overline{T} = \frac{1}{4}kT$, $E = \frac{3}{4}kT$. Thus the heat capacity per particle is $\frac{3}{4}k$ and $C_v = \frac{3}{4}Nk$ for the whole system.

2164

By treating radiation in a cavity as a gas of photons whose energy ε and momentum k are related by the expression $\varepsilon = ck$, where c is the velocity of light, show that the pressure p exerted on the walls of the cavity is one-third of the energy density.

With the above result prove that when radiation contained in a vessel with perfectly reflecting walls is compressed adiabatically it obeys the equation

$$PV^\gamma = \text{constant} .$$

Determine the value of γ.

(UC, Berkeley)

Solution:

Let $n(\omega)d\omega$ denote the number of photons in the angular frequency interval $\omega \sim \omega + d\omega$. Consider the pressure exerted on the walls by such photons in the volume element dV at (r, θ, φ) (Fig. 2.34). The probability that they collide with an area dA of the wall is $dA \cdot \cos\theta/4\pi r^2$, each collision contributing an impulse $2k\cos\theta$ perpendicular to dA. Therefore, we have

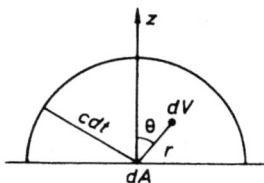

Fig. 2.34.

$$dp_\omega = \frac{df}{dAdt}$$

$$= \frac{nd\omega \cdot dV \cdot \dfrac{dA \cdot \cos\theta}{4\pi r^2} \cdot 2k\cos\theta}{dAdt}$$

$$= \frac{d\omega}{4\pi dt} 2nk\cos^2\theta \sin\theta \, dr \, d\theta \, d\varphi \ ,$$

$$p = \int_{r \le cdt} dp_\omega = \int \frac{n}{3} kc d\omega = \int \frac{u(\omega)}{3} d\omega \ .$$

Integrating we get $p = \dfrac{u}{3} = \dfrac{U}{3V}$, where u is the energy density and U is the total energy. From the thermodynamic equation

$$dU = TdS - pdV \ ,$$

and $p = U/3V$, we obtain $dU = 3pdV + 3Vdp$. Hence $4pdV + 3Vdp = TdS$. For an adiabatic process $dS = 0$, $4\dfrac{dV}{V} + 3\dfrac{dp}{p} = 0$. Integrating we have

$$pV^{4/3} = \text{const.}, \quad \gamma = \frac{4}{3} \ .$$

2165

Radiation pressure.

One may think of radiation as a gas of photons and apply many of the results from kinetic theory and thermodynamics to the radiation gas.

(a) Prove that the pressure exerted by an isotropic radiation field of energy density u on a perfectly reflecting wall is $p = u/3$.

(b) Blackbody radiation is radiation contained in, and in equilibrium with, a cavity whose walls are at a fixed temperature T. Use thermodynamic arguments to show that the energy density of blackbody radiation depends only on T and is independent of the size of the cavity and the material making up the walls.

(c) From (a) and (b) one concludes that for blackbody radiation the pressure depends only on the temperature, $p = p(T)$, and the internal energy U is given by $U = 3p(T)V$ where V is the volume of the cavity. Using these two facts about the gas, derive the functional form of $p(T)$, up to an unspecified multiplicative constant, from purely thermodynamic reasoning.

(*MIT*)

Solution:

(a) Consider an area element dS of the perfectly reflecting wall and the photons impinging on dS from the solid angle $d\Omega = \sin\theta d\theta d\varphi$. The change of momentum per unit time in the direction perpendicular to dS is $u \cdot \sin\theta d\theta d\varphi \cdot dS \cos\theta \cdot 2\cos\theta/4\pi$. Hence the pressure on the wall is

$$p = \left(\frac{u}{2\pi}\right) \int_0^{\pi/2} d\theta \int_0^{2\pi} d\varphi \cos^2\theta \sin\theta = \frac{u}{3} .$$

(b) Consider the cavity as consisting of two arbitrary halves separated by a wall. The volumes and the materials making up the sub-cavities are different but the walls are at the same temperature T. Then in thermal equilibrium, the radiations in the sub-cavities have temperature T but different energy densities if these depend also on factor other than temperature. If a small hole is opened between the sub-cavities, there will be a net flow of radiation from the sub-cavity of higher u because of the pressure difference. A heat engine can then absorb this flow of heat radiation and produce mechanical work. This contradicts the second law of thermodynamics if no other external effect is involved. Hence the energy density of black body radiation depends only on temperature.

(c) Since the free energy F is an extensive quantity and

$$\left(\frac{\partial F}{\partial V}\right)_T = -p = -\frac{1}{3}u(T) ,$$

we have

$$F = -\frac{1}{3}u(T)V .$$

From thermodynamics we also have $F = U - TS$, where $U = uV$ is the internal energy, S is the entropy, and

$$S = -\left(\frac{\partial F}{\partial T}\right)_V$$
$$= \frac{1}{3}\frac{du(T)}{dT}V .$$

Hence

$$\frac{du}{u} = 4\frac{dT}{T} ,$$

giving $u = aT^4, p = \frac{1}{3}aT^4$, where a is a constant.

2166

A gas of interacting atoms has an equation of state and heat capacity at constant volume given by the expressions

$$p(T,V) = aT^{1/2} + bT^3 + cV^{-2} ,$$
$$C_v(T,V) = dT^{1/2}V + eT^2V + fT^{1/2} ,$$

where a through f are constants which are independent of T and V.

(a) Find the differential of the internal energy $dU(T,V)$ in terms of dT and dV.

(b) Find the relationships among a through f due to the fact that $U(T,V)$ is a state variable.

(c) Find $U(T,V)$ as a function of T and V.

(d) Use kinetic arguments to derive a simple relation between p and U for an ideal monatomic gas (a gas with no interactions between the atoms, but whose velocity distribution is arbitrary). If the gas discussed in the previous parts were to be made ideal, what would be the restrictions on the constants a through f?

(MIT)

Solution:

(a) We have $dU(T,V) = C_v\, dT + \left(\dfrac{\partial U}{\partial V}\right)_T dV,$

$$p = -\left(\frac{\partial F}{\partial V}\right)_T = -\left(\frac{\partial U}{\partial V}\right)_T + T\left(\frac{\partial p}{\partial T}\right)_V .$$

Hence $dU = (dT^{1/2}V + eT^2V + fT^{1/2})dT - \left(\dfrac{a}{2}T^{1/2} - 2bT^3 + cV^{-2}\right)dV,$

(b) Since $U(T,V)$ is a state variable $dU(T,V)$ is a total differential, which requires

$$\frac{\partial}{\partial V}\left(\frac{\partial U}{\partial T}\right)_V = \frac{\partial}{\partial T}\left(\frac{\partial U}{\partial V}\right)_T ,$$

that is,

$$dT^{1/2} + eT^2 = -\left(\frac{1}{4}aT^{-1/2} - 6bT^2\right) .$$

Hence $a = 0, d = 0, e = 6b$.

(c) Using the result in (b) we can write

$$dU(T,V) = d(2bT^3V) + fT^{1/2}dT - cV^{-2}dV .$$

Hence

$$U(T,V) = 2bT^3V + 2fT^{3/2}/3 + cV^{-1} + \text{const.}$$

(d) Imagine that an ideal reflecting plane surface is placed in the gas. The pressure exterted on it by atoms of velocity v is

$$p_v = \int_0^{\frac{\pi}{2}}\int_0^{2\pi} \frac{N}{V}\frac{\sin\theta\, d\theta\, d\varphi}{4\pi} v\cos\theta \cdot 2mv\cos\theta = \frac{1}{3}\frac{N}{V}mv^2 .$$

The mean internal energy density of an ideal gas is just its mean kinetic energy density, i.e.,

$$u = \frac{1}{2}mv^2 \cdot \frac{N}{V} .$$

The average pressure is $p = \bar{p}_v = 2u/3$, giving $pV = \frac{2}{3}U$. For the gas discussed above to be made ideal, we require the last equation to be satisfied:

$$\left(bT^3 + \frac{c}{V^2}\right)V = \frac{2}{3}\left(2bT^3V + \frac{2}{3}fT^{3/2} + \frac{c}{V} + \text{const.}\right)$$

i.e., $3bT^3V + \frac{4fT^{\frac{3}{2}}}{3} - \frac{c}{V} = \text{const.}$.

It follows that b and f cannot be zero at the same time. The expression for p means that b and c cannot be zero at the same time.

2167

(a) From simplest kinetic theory derive an approximate expression for the diffusion coefficient of a gas, D. For purposes of this problem you need not be concerned about small numerical factors and so need not integrate over distribution functions etc.

(b) From numbers you know evaluate D for air at STP.

(Wisconsin)

Solution:

(a) We take an area element dS of an imaginary plane at $z = z_0$ which divides the gas into two parts A and B as shown in Fig. 2.35. For a uniform gas the fraction of particles moving parallel to the z-axis (upward or downward) is $\frac{1}{6}$. Therefore the mass of the gas traveling along the positive direction of the z-axis through the area element dS in time interval dt is

$$dM = m\left(\frac{1}{6}n_A\bar{v}dSdt - \frac{1}{6}n_B\bar{v}dSdt\right)$$

$$= \frac{1}{6}\bar{v}dSdt(\rho_A - \rho_B)$$

$$= -\frac{1}{6}\bar{v}dSdt \cdot 2\bar{\lambda}\left(\frac{d\rho}{dz}\right)_{z_0}$$

$$= -\frac{1}{3}\bar{v}\bar{\lambda}\left(\frac{d\rho}{dz}\right)_{z_0}dSdt \ .$$

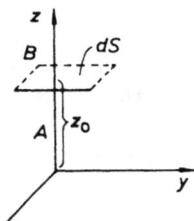

Fig. 2.35.

where $\bar{v}$ is the average velocity and $\bar{\lambda}$ the mean free path of the particles of the gas. By definition the diffusion coefficient is

$$D = \frac{-dM}{dS\,dt} \Big/ \left(\frac{d\rho}{dz}\right)_{z_0} = \frac{1}{3}\bar{v}\bar{\lambda} \ .$$

(b) At STP the average speed of air molecules is

$$\bar{v} = \sqrt{\frac{8kT}{\pi m}} \approx 448 \text{ m/s} \ ,$$

and the mean free path length is

$$\bar{\lambda} \approx 6.9 \times 10^{-8} \text{ m} \ .$$

Thus the diffusion coefficient is

$$D = \frac{1}{3}\bar{v}\bar{\lambda} \approx 3.1 \times 10^{-5} \text{ m}^2/\text{s} \ .$$

2168

(a) Show that the ratio of the pressure to the viscosity coefficient gives approximately the number of collisions per unit time for a molecule in a gas.

(b) Calculate the number of collisions per unit time for a molecule in a gas at STP using the result of (a) above or by calculating it from the mean velocity, molecular diameter, and number density.

The coefficient of viscosity for air at STP is 1.8×10^{-4} in cgs units. Use values you know for other constants you need.

(*Wisconsin*)

Solution:

(a) The coefficient of viscosity is $\eta = \frac{n}{3} m \bar{v} \lambda$, where n is the particle number density. The pressure of the gas is

$$p = nkT .$$

Hence

$$\frac{p}{\eta} = \frac{3kT}{m \bar{v} \lambda} .$$

The mean-square speed of the molecues is $\overline{v^2} = \frac{3kT}{m}$. Neglecting the difference between the average speed and the rms speed, we have

$$\frac{p}{\eta} = \frac{\overline{v^2}}{\bar{v} \lambda} \approx \frac{\bar{v}}{\lambda}$$

which is the average number of collisions per unit time for a molecule.

(b) At STP, the pressure is $p = 1.013 \times 10^6$ dyn/cm^2. Hence the number of collisions per unit time is

$$\frac{p}{\eta} = \frac{1.013 \times 10^6}{1.8 \times 10^{-4}} = 5.63 \times 10^9 \text{ s}^{-1} .$$

2169

(a) Assuming moderately dilute helium gas so that binary collisions of helium atoms determine the transport coefficients, derive an expression for the thermal conductivity of the gas.

(b) Estimate the ratio of the thermal conductivity of gaseous ^{3}He to that of gaseous ^{4}He at room temperature.

(c) Will this ratio become different at a temperature near 2 K? Why?

(*Wisconsin*)

Solution:

(a) Consider an area element dS of an imaginary plane at $z = z_0$ which divides the gas into two parts A and B (see Fig. 2.35). We assume that the temperatures of A and B are T_A and T_B respectively. In the case of a small temperature difference, we can take approximately $n_A \bar{v}_A = n_B \bar{v}_B = n \bar{v}$, then the number of molecules exchanged between A and B through dS in time interval dt is $n\bar{v}dS\,dt/6$. According to the principle of equipartition of energy, the average kinetic energy of the molecules of A is $\frac{l}{2}kT_A$, and that of B is $\frac{l}{2}kT_B$ (l is the number of degrees of freedom of molecule). Therefore, the net energy transporting through dS in dt (or the heat transporting along the positive direction of the z-axis) is

$$dQ = lk(T_A - T_B)n\bar{v}dS\,dt/12 \ .$$

The temperature difference can be expanded in terms of the temperature gradient:

$$T_A - T_B = -2\bar{\lambda}\left(\frac{dT}{dz}\right)_{z_0} \ .$$

So

$$dQ = -\frac{1}{3}n\bar{v}\bar{\lambda}\frac{l}{2}k\left(\frac{dT}{dz}\right)_{z_0}dS\,dt \ ,$$

giving the thermal conductivity

$$\kappa = \frac{1}{3}n\bar{v}\bar{\lambda}\frac{l}{2}k = \frac{1}{3}\rho\bar{v}\bar{\lambda}c_v \ , \tag{*}$$

where c_v is the specific heat at constant volume.

(b) Since $\rho\bar{\lambda} \propto m/\sigma^2$, $\bar{v} \propto m^{-1/2}$, $c_v \propto 1/m$, with the formula (*), we have $\kappa \propto \frac{1}{\sqrt{m}}\sigma^{-2}$, where σ is the atomic diameter. For ^{3}He and ^{4}He, σ can be taken as the same, giving

$$\frac{\kappa_3}{\kappa_4} = \left(\frac{m_3}{m_4}\right)^{-1/2} = \left(\frac{3}{4}\right)^{-1/2} \approx 1.15 \ .$$

(c) When the temperature is near 2K, ^{3}He is in liquid phase and ^{4}He is in superfluid phase, so that the above model is no longer valid. The ratio of the thermal conductivities changes abruptly at this temperature.

2170

A certain closed cell foam used as an insulating material in houses is manufactured in such a way that the cells are initially filled with a poly-atomic gas of molecular weight ~ 60. After several years the gas diffuses out of the foam and is replaced by dry air (mean molecular weight ~ 30). Assuming that the insulating property arises largely from the thermal conductivity of the gas.

Discuss the factors which influence the thermal conductivity of the gas. For each factor make an argument for whether the insulating ability increases or decreases. What is the overall effect upon the insulating ability?

(*MIT*)

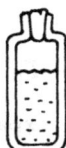

Fig. 2.36.

Solution:

The thermal conductivity is $\kappa \sim \lambda \bar{v} n C_v$, where λ is the mean free path, $\bar{v}$ the mean speed and n the number density of the gas molecules and c_v is the thermal capacity per molecule. We have $\bar{v} \propto \sqrt{T/A}$, where A is the molecular weight. $n\lambda \propto 1/\sigma$, where σ is the cross section of a molecule, being $\propto A^{2/3}$. So

$$\kappa \propto A^{-2/3}\sqrt{T/A} = \sqrt{T}/A^{7/6} \ .$$

Thus the molecular weight is the most important factor. The overall insulating ability decreases when the poly-atomic gas is replaced by dry air.

2171

Thermos Bottle.

(a) State and justify how the thermal conductivity of an ideal gas depends on its density at fixed temperature.

(b) A thermos (Dewar) bottle is constructed of two concentric glass vessels with the air in the intervening space reduced to a low density. Why can it act as an insulating container even though the vacuum is not perfect?

(*MIT*)

Solution:

(a) The mean speed of the air molecules is constant at fixed temperature. However the greater the gas density is, the more frequent will be the collisions and the transmission of energy will become faster. Hence the thermal conductivity is higher for greater gas density.

(b) As the air density is low, the thermal conductivity is also low. It can therefore enhance heat insulation.

2172

Sketch the temperature dependence of the heat conductivity of an insulating solid. State the simple temperature dependencies in limiting temperature ranges and dervie them quantitatively.

(*Chicago*)

Solution:

The thermal conductivity of a solid is $\kappa = c v_s \lambda / 3$, where c is thermal capacity per unit volume, v_s is velocity of sound, λ is mean free path of phonons. κ versus T is shown in Fig. 2.37.

(a) At low temperatures the heat capacity $c \propto T^3$, v_s and λ are constants, hence $\kappa \propto T^3$.

(b) At high temperatures v_s is constant and $\lambda \propto \dfrac{1}{T}$, the thermal capacity c is constant, hence $\kappa \propto \dfrac{1}{T}$.

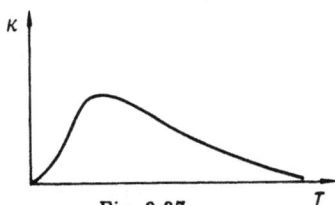

Fig. 2.37.

2173

Express the equilibrium heat flow equation in terms of the heat capacity, excitation or particle velocity, mean free path, and thermal gradient. Discuss the manifestation of quantum mechanics and quantum statistics in the thermal conductivity of a metal.

(*Wisconsin*)

Solution:

The equilibrium heat flow equation is

$$\mathbf{j} = -\lambda \nabla T ,$$

where $\lambda = \frac{1}{3} C_v v l$ is the thermal conductivity, C_v being the heat capacity per unit volume, v the average velocity and l the mean free path of the particles. Here we have used the assumption that the electronic conductivity is much larger than the lattice conductivity (correct at room temperature). The metallic lattice has attractive interaction with the electrons, of which the potential can be considered uniform inside the metal, and zero outside. Hence we can consider the valence electrons as occupying the energy levels in a potential well. Then the probability of occupying the energy level ε is given by the Fermi distribution:

$$f(\varepsilon) = \frac{1}{\exp[(\varepsilon - \varepsilon_F)/kT] + 1} .$$

The Fermi energy ε_F of metals are usually very large (of the order of magnitude $\sim$ eV). Since $1\, kT = 0.025$ eV at room temperature, ordinary increases of temperature have little effect on the electronic distribution. At ordinary temperatures, the electronic conductivity is contributed mainly by electrons of large energies i.e., those above the Fermi surface, which represent a fraction of the total number of electrons, $\dfrac{kT}{\varepsilon_F}$. Thus we must choose for the average velocity the Fermi velocity

$$v = v_F = (2\varepsilon_F/m_e)^{1/2} .$$

Then $l = v_F \tau$, where τ is the relaxation time of the electron. The difference of the quantum approach from classical statistics is that here the increase of temperature affects only the electrons near the Fermi surface. Using the approximation of strong degeneracy, the heat capacity is

$$C_v = \frac{\pi^2}{2} nk \left(\frac{kT}{\varepsilon_F} \right) ,$$

where n is the electron number density, giving

$$\lambda = \pi^2 nk^2 T\tau/3m .$$

This formula agrees well with experiments on alkali metals.

2174

A liquid helium container, shown in Fig. 2.38, contains 1000cm^3 of liquid helium and has a total wall area of 600cm^2. It is insulated from the surrounding liquid nitrogen reservoir by a vacuum jacket of 0.5 cm thickness. Liquid helium is at 4.2 K, while liquid nitrogen is at 77 K. If the vacuum jacket is now filled with helium at a pressure of 10 μm Hg, estimate how long it will take for all the 1000 cm^3 liquid helium in the container to disappear. (As a crude approximation, we can assume a constant temperature gradient across the vacuum jacket and evaluate the thermal conduction of the He-filled jacket at its mean temperature.

(*UC, Berkeley*)

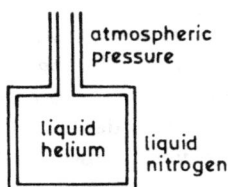

atmospheric pressure

liquid helium

liquid nitrogen

Fig. 2.38.

Solution:

When the liquid helium absorbs heat, it vaporizes and expands to escape. We can consider, approximately, the vaporization latent heat and the work done during expansion to be of the same order of magnitude, i.e., we take the phase transition curve as given by

$$\frac{dp}{dT} \sim \frac{p}{T} .$$

Therefore, the heat that is needed for the helium to escape is

$$Q \sim pV = nRT .$$

Assuming that n is comparable to the mole number of the same volume of water, we make the estimate

$$n \approx 1000 \times \frac{1}{18} = 56 \text{ mol.}$$

Hence $Q \approx 56 \times 8.3 \times 4.2 \approx 2 \times 10^3$ J.

Consider now the heat transfer. Since the molecular mean free path in the jacket is

$$l = \frac{1}{\overline{n\sigma}} = \frac{k\overline{T}}{p\rho} \approx 1 \text{ m} \gg 0.5 \text{ cm} ,$$

the heat transferred is

$$q \sim A(\overline{n}\ \overline{v}) \cdot (k\Delta T) \approx 30 \ \text{J/s} \ .$$

Thus the time for the helium to escape is

$$t = \frac{Q}{q} \approx 10^2 \ \text{s} \ .$$

2175

Transport properties of a simple gas.

Many properties of a gas of atoms can be estimated using a simple model of the gas as an assembly of colliding hard spheres. The purpose of this problem is to derive approximate expressions for a number of coefficients that are used to quantitatively describe various phenomena. For each of the coefficients below state your answer in terms of: k = Boltzmann's constant, T = temperature, R = radius of atom, m = mass of atom, c = heat capacity per gram, ρ = density. You may neglect factors of order unity. (Hint: First derive expressions for the mean free path between collisions, λ, and the root-mean-square speed, $\overline{v}$).

(a) Derive the coefficient of thermal conductivity, κ (units: g·cm/s³· K). This occurs in the relation between the heat flux and the temperature gradient.

(b) Derive the coefficient of viscosity, η (units: g/cm·s). This occurs in the relation between the tangential force per unit area and the velocity gradient.

(c) Derive the diffusion coefficient, D (units: cm²/s). This characterizes a system containing gases of two species. It relates the time rate of change of the density of one species to its inhomogeneity in density.

(*MIT*)

Solution:

The mean free path length is

$$\lambda \sim \frac{m}{R^2 \rho} \ .$$

The root-mean-square speed is

$$\overline{v} \sim \left(\frac{kT}{m}\right)^{1/2} \ .$$

(a) Suppose a temperature gradient $\frac{\partial T}{\partial x}$ exists along the x-direction and consider a unit area perpendicular to it. The net heat flow resulting from exchanging a pair of molecules across the unit area is

$$q \approx mc\left[T - \left(T + \lambda\frac{\partial T}{\partial x}\right)\right] = -\lambda mc\frac{\partial T}{\partial x} .$$

The number of such pairs exchanged per unit time is $\frac{\bar{v}\rho}{m}$, so the heat flux is

$$J_H \approx -\bar{v}\rho c\lambda\frac{\partial T}{\partial x}$$

and the thermal conductivity is

$$\kappa \sim \lambda\bar{v}\rho c \sim \frac{c}{R^2}(mkT)^{\frac{1}{2}} .$$

(b) The change of the component v_y of the average velocity in the exchange of a pair of molecules as mentioned in (a) is $-\lambda\frac{\partial}{\partial x}v_y(x)$, so that the tangential force on a unit area perpendicular to the x-direction is

$$F_y \sim -m\lambda\frac{\partial v_y}{\partial x}\cdot\frac{\bar{v}\rho}{m} .$$

Hence the coefficient of viscosity is

$$\eta \sim \lambda\bar{v}\rho \sim \frac{(mkT)^{\frac{1}{2}}}{R^2} .$$

(c) Suppose the mass density $\rho(z)$ is inhomogeneous in the z-direction. The mass flux in this direction is

$$J_\rho \approx \left[\rho - \left(\rho + \lambda\frac{\partial\rho}{\partial z}\right)\right]\bar{v}$$

$$= -\lambda\bar{v}\frac{\partial\rho(z)}{\partial z} ,$$

so the diffusion coefficient is

$$D = \lambda\bar{v} \sim (mkT)^{1/2}/R^2\rho .$$

2176

The speed of sound (c_s) in a dilute gas like air is given by the adiabatic compressibility:

$$c_s^2 = \left[\left(\frac{\partial \rho}{\partial p} \right)_s \right]^{-1} = \gamma \frac{kT}{M} ,$$

where M is the mean molecular weight, k is Boltzmann's constant and γ is the ratio of principal specific heats.

(a) Estimate numerically for air at room temperature,

(1) the speed of sound;
(2) the mean molecular collisions frequency;
(3) the molecular mean free path;
(4) the ratio of mean free path to typical wave length;
(5) the ratio of typical wave frequency to collision frequency. (Use $\nu = 300$ Hz as typical wave frequency.)

(b) Use the ratios found above to explain why adiabatic conditions are relevant for sound.

(*UC, Berkeley*)

Solution:

(a) (1) $c_s = \sqrt{\gamma \frac{kT}{M}} = 350$ m/s.

(2), (3) The mean free path is

$$l = \frac{1}{n\sigma} = \frac{kT}{p\sigma} = 4 \times 10^{-6} \text{ m} .$$

The mean collision frequency is

$$f = \frac{\overline{v}}{l} = 1.2 \times 10^8 \text{ s}^{-1} .$$

(4) $\dfrac{l}{\lambda} = \dfrac{l\nu}{c_s} = 3.4 \times 10^{-6} .$

(5) $\nu/f = 2.5 \times 10^{-6}.$

(b) As sound waves compress the air in a scale of the wavelength λ, we shall estimate the ratio of the time for heat, which is transferred by the motion of the molecules, to travel the distance λ to the period of the sound waves. A molecule of mean free path l makes N collisions during the displacement λ, where N is given by

$$\lambda^2 = Nl^2 .$$

The ratio we require is

$$\frac{t_H}{\tau} = \left(\frac{N}{f}\right) \Big/ \left(\frac{1}{300}\right)$$

$$= \frac{300\lambda^2}{fl^2} = \frac{300}{1.2 \times 10^8 \times (3.4 \times 10^{-6})^2} = 2.2 \times 10^5 .$$

Since $t_H \gg \tau$, the oscillation of the air is too fast for heat transfer to take place, adiabatic conditions prevail.

2177

The speed of sound in a gas is calculated as

$$v = \sqrt{\text{adiabatic bulk modulus/density}} .$$

(a) Show that this is a dimensionally-correct equation.

(b) This formula implies that the propagation of sound through air is a quasi-static process. On the other hand, the speed of sound for air is about 340 m/sec at a temperature for which the rms speed of an air molecule is about 500 m/sec. How then can the process be quasi-static?

(Wisconsin)

Solution:

The speed of sound is $v = \sqrt{B/\rho}$, where B is the adiabatic bulk modulus and ρ is the density.

(a) $[B] = \left[\frac{V \Delta p}{\Delta V}\right] = [\Delta p] = ML^{-1}T^{-2}$,

$[\rho] = ML^{-3}$,

thus,

$$\left[\sqrt{\frac{B}{\rho}}\right] = LT^{-1} = [v] .$$

(b) While under ordinary conditions the rms speed of a gas molecule is about 500 m/s, its mean free path is very short, about 10^{-5} cm, which is much smaller than the wavelength of sound waves. Therefore, the propagation of sound through air can be considered adiabatic, i.e., a quasi-static process.

2178

Consider a non-interacting relativistic Fermi gas at zero temperature.

(a) Write down expressions for the pressure and the energy density in the rest frame of the gas. What is the equation of state?

(b) Treating the system as a uniform static fluid, derive a wave equation for the propagation of small density fluctuations, and hence deduce an expression for the velocity of sound in the gas.

(SUNY, Buffalo)

Solution:

(a) The relation between the momentum and energy of a relativistic particle is given by $\varepsilon = pc$. The energy density is

$$u = \left(\frac{4\pi}{h^3}\right)(2J + 1)\int_0^{p_F} \varepsilon p^2 dp = (2J + 1)\pi\frac{cp_F^4}{h^3} ,$$

where J is the spin quantum number of Fermions and p_F is given by the equation

$$N = (2J + 1)\frac{V}{h^3}\frac{4}{3}\pi p_F^3 .$$

Hence

$$u = \hbar c\left[\frac{81\pi^2}{32(2J + 1)}\right]^{1/3}\left(\frac{N}{V}\right)^{4/3} .$$

The pressure is

$$p = -\frac{\partial(uV)}{\partial V} = -u + V\frac{\partial u}{\partial V} = \frac{u}{3} ,$$

i.e., the equation of state is

$$pV = \frac{E}{3} .$$

with $E = uV$.

(b) Let $\rho = \rho_0 + \delta\rho$ and $p = p_0 + \delta p$, where ρ_0 and p_0 are the density and pressure of the fluid respectively, and $\delta\rho$ and δp are the corresponding fluctuations. For a static fluid, $\mathbf{v} \simeq \delta\mathbf{v}$, and the continuity equation is

$$\frac{\partial\delta\rho}{\partial t} + \rho_0\nabla \cdot \delta\mathbf{v} = 0 .$$

In the same approximation Euler's equation can be reduced to

$$\frac{\partial \delta \mathbf{v}}{\partial t} = -\frac{\nabla \delta p}{\rho_0} .$$

The motion of an ideal fluid is adiabatical, thus

$$\delta p = \left(\frac{\partial p}{\partial \rho}\right)_S \delta \rho = \left(\frac{\partial p_0}{\partial \rho_0}\right)_S \delta \rho .$$

Combining these we obtain the wave equation

$$\frac{\partial^2}{\partial t^2} \delta \rho - v^2 \nabla^2 \delta \rho = 0$$

where $v = \sqrt{\left(\dfrac{\partial p_0}{\partial \rho_0}\right)_S}$ is the velocity of sound in the gas. As $\rho_0 = \dfrac{mN}{V}$

and $p_0 = \dfrac{\alpha}{3}\rho_0^{4/3}$, where

$$\alpha = \frac{\hbar c}{m} \left[\frac{81\pi^2}{32(2J+1)m}\right]^{1/3}$$

m being the mass of a particle, we have

$$v = \frac{2}{3}\sqrt{\alpha}\rho_0^{1/6} .$$

2179

A beam of energetic (> 100 eV) neutral hydrogen atoms is coming through a hole in the wall of plasma confinement device. Describe the apparatus that you would use to measure the energy distribution of these atoms.

(*Wisconsin*)

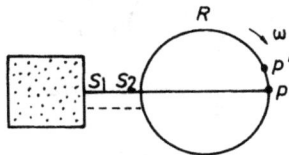

Fig. 2.39.

Solution:

An apparatus that could be used is shown in Fig. 2.39. Atomic beam enters into the cylinder R of diameter D after it passes through slits S_1 and S_2. The cylinder R rotates about its axis with angular velocity ω. Suppose an atom arrives at point p' on the cylinder, $\widehat{pp'} = s$. The time taken for the atom to travel from S_2 to p' is $t = \dfrac{D}{v}$, where v is its velocity.

During this time, the cylinder has rotated through an angle $\theta = \omega t$. Thus

$$s = \frac{D}{2} \cdot \theta = \frac{1}{2} \frac{D^2 \omega}{v} \ .$$

The energy of the atom is therefore $\varepsilon = \dfrac{m}{2} v^2 = mD^4\omega^2/8s^2$. Hence there is a one-to-one correspondence between s and ε. By measuring the thickness distribution of the atomic deposition on the cylinder we can determine the energy distribution of the atoms.

2180

Write the Maxwell distribution, $P(v_x, v_y, v_z)$, for the velocities of molecules of mass M in a gas at pressure p and temperature T. (If you have forgotten the normalization constant, derive it from the Gaussian integral,

$$\int_{-\infty}^{\infty} \exp(-x^2/2\sigma^2)dx = \sqrt{2\pi}\sigma \ .$$

When a clean solid surface is exposed to this gas it begins to absorb molecules at a rate W (molecules/s·cm^2).

A molecule has absorption probability 0 for a normal velocity component less than a threshold v_T, and absorption probability 1 for a normal velocity greater than v_T. Derive an expression for W.

(Wisconsin)

Solution:

The Maxwell distribution of velocity is given by

$$P(v_x, v_y, v_z) = \left(\frac{M}{2\pi kT}\right)^{3/2} e^{-\frac{M}{2kT}(v_x^2 + v_y^2 + v_z^2)} \ .$$

We take the x-axis normal to the solid surface. Then the distribution of the component v_x of velocity is

$$P(v_x) = \left(\frac{M}{2\pi kT}\right)^{\frac{1}{2}} e^{-\frac{M}{2kT}v_x^2} \ .$$

Hence

$$W = \int_{v_T}^{\infty} n v_x P(v_x) dv_x = n \left(\frac{kT}{2\pi M} \right)^{1/2} \exp\left(-\frac{M v_T^2}{2kT} \right) ,$$

where n is the molecular number density.

2181

A gas in a container consists of molecules of mass m. The gas has a well defined temperature T. What is

(a) the most probable speed of a molecule?
(b) the average speed of the molecules?
(c) the average velocity of the molecues?

(*MIT*)

Solution:

The Maxwell velocity distribution is given by

$$dw = \left(\frac{m}{2\pi kT} \right)^{3/2} \exp[-mv^2/2kT] dv_x dv_y dv_z .$$

(a) Let $f(v) = v^2 \exp(-mv^2/2kT)$. The most probable speed is given by $\frac{df(v)}{dv} = 0$, as

$$v = \left(\frac{2kT}{m} \right)^{1/2} .$$

(b) The average speed is

$$\bar{v} = 4\pi \left(\frac{m}{2\pi kT} \right)^{3/2} \int_0^{\infty} v f(v) dv = (8kT/\pi m)^{1/2} .$$

(c) The average velocity $\bar{\mathbf{v}} = (\bar{v}_x, \bar{v}_y, \bar{v}_z)$ is given by

$$\bar{v}_x = (m/2\pi kT)^{3/2} \int\!\!\!\int\!\!\!\int_{-\infty}^{\infty} v_x \exp[-mv^2/2kT] dv_x dv_y dv_z = 0 ,$$

and $\bar{v}_y = \bar{v}_z = 0$. Thus $\bar{\mathbf{v}} = 0$.

2182

Find the rate of wall collisions (number of atoms hitting a unit area on the wall per second) for a classical gas in thermal equilibrium in terms of the number density and the mean speed of the atoms.

(*MIT*)

Solution:

Take the z-axis perpendicular to the wall, pointing towards it. The rate of collision is

$$\Gamma = \int_{-\infty}^{\infty} dv_x \int_{-\infty}^{\infty} dv_y \int_{0}^{\infty} v_z f\, dv_z \ ,$$

where $f = n \left(\dfrac{m}{2\pi kT} \right)^{3/2} \exp\left[-\dfrac{m}{2kT}\left(v_x^2 + v_y^2 + v_z^2\right) \right]$, and n is the number density of the atoms. Integrating we obtain $\Gamma = \dfrac{1}{4} n\bar{v}$, where $\bar{v}$ is the mean speed,

$$\bar{v} = (8kT/\pi m)^{1/2} \ .$$

2183

At time $t = 0$, a thin walled vessel of volume V, kept at constant temperature, contains N_0 ideal gas molecules which begin to leak out through a small hole of area A. Assuming negligible pressure outside the vessel, calculate the number of molecules leaving through the hole per unit time and the number remaining at time t. Express your answer in terms of N_0, A, V, and the average molecular velocity, $\bar{v}$.

(*Wisconsin*)

Solution:

From the Maxwell velocity distribution, we find the number of molecules colliding with unit area of the wall of the container in unit time to be $\dfrac{nv}{4}$, where n is the number density of the molecules. Therefore, the number of molecules escaping through the small hole of area A in unit time is

$$-\frac{dN}{dt} = \frac{A}{4} n\bar{v} = \frac{A}{4V} N\bar{v} \ .$$

Using the initial condition $N(0) = N_0$, we obtain by integration

$$N(t) = N_0 e^{-\frac{A\bar{v}t}{4V}} \ ,$$

which gives the number of molecules remaining in the container at time t.

2184

A beam of molecules is often produced by letting gas escape into a vacuum through a very small hole in the side of the container confining the gas. The total intensity of the beam is defined as the number of molecules escaping from the hole per unit time. Find the change in total intensity of the beam if:

(a) the area of the hole is increased by a factor of 4;

(b) the absolute temperature is increased by a factor of 4, the pressure being maintained constant;

(c) the pressure in the container is increased by a factor of 4, the temperature remaining constant,

(d) at the original temperature and pressure, a gas of 4 times the molecular weight of the original gas is used.

(UC, Berkeley)

Solution:

The total intensity of the beam is

$$I = \frac{1}{4}n\bar{v}A \; ,$$

where

$$n = \frac{p}{kT}, \quad \bar{v} = \sqrt{\frac{8kT}{\pi m}} \; ,$$

or

$$I = \frac{1}{4}Ap\sqrt{\frac{8}{\pi mkT}} \; .$$

(a) If $A \rightarrow 4A$, then $I \rightarrow 4I$.

(b) If p is constant and $T \rightarrow 4T$, then $I \rightarrow I/2$.

(c) If T is constant and $p \rightarrow 4p$, then $I \rightarrow 4I$.

(d) If T and p are both constant and $m \rightarrow 4m$, then $I \rightarrow I/2$.

2185

Derive a rough estimate for the mean free path of an air molecule at STP. What is the path length between collisions for

(a) a slow molecule?

(b) a fast molecule?

(*Wisconsin*)

Solution:

Consider the motion of a molecule A. Only those molecules whose centers separate from the center of A by distances smaller than or equal to the effective diameter of the molecule can collide with A. We can imagine a cylinder, whose axis coincides with the orbit of the center of A, with a radius equal the effective diameter d of the molecule. Then all the molecules whose centers are in the cylinder will collide with A. The cross section of this cylinder is $\sigma = \pi d^2$.

In the time interval t, the path length of A is $\bar{u}t$ ($\bar{u}$ is the average relative speed), which corresponds to a volume $\sigma \bar{u}t$ of the cylinder. The number of collisions A suffers with other molecules is $n\sigma\bar{u}t$ (n is the number density). The frequency of collisions is therefore

$$\bar{z} = \frac{n\sigma\bar{u}t}{t} = n\sigma\bar{u} \ .$$

Hence the mean free path is $\bar{\lambda} = \dfrac{\bar{v}}{\bar{z}} = \dfrac{\bar{v}}{n\sigma\bar{u}}$. From the Maxwell distribution, we can show that $\bar{u} = \sqrt{2}\bar{v}$. Thus

$$\bar{\lambda} = \frac{1}{\sqrt{2}\sigma n} = \frac{1}{\sqrt{2}\pi n d^2} \ .$$

A more precise calculation from the Maxwell distribution gives the mean free path of a molecule whose speed is $\bar{v} = \sqrt{\dfrac{2kT}{m}}\, x$ as

$$\bar{\lambda} = \frac{1}{nd^2} \frac{x^2}{\sqrt{\pi}\psi(x)} \ ,$$

where $\psi(x) = x\exp(-x^2) + (2x^2 + 1)\displaystyle\int_0^x \exp(-y^2)dy$.

(a) For a slow molecule, $\bar{v} \to 0$, or $x \to 0$,

$$\bar{\lambda} \approx \frac{x}{\sqrt{\pi}nd^2} = \frac{mv}{nd^2\sqrt{2\pi mkT}} \ .$$

(b) For a fast molecule, $\bar{v} \to \infty$, or $x \to \infty$,

$$\bar{\lambda} \approx \frac{1}{\pi nd^2} \ .$$

2186

A simple molecular beam apparatus is shown in Fig. 2.40 The oven contains H_2 molecules at 300 K and at a pressure of 1 mm of mercury. The hole on the oven has a diameter of 100 μm which is much smaller than the molecular mean free path. After the collimating slits, the beam has a divergence angle of 1 mrad. Find:

(a) the speed distribution of molecules in the beam;

(b) the mean speed of molecules in the beam;

(c) the most probable speed of molecules in the beam;

(d) the beam power (number of molecules passing through the last collimating split per unit time);

(e) the average rotational energy of H_2 molecules.

(UC, Berkeley)

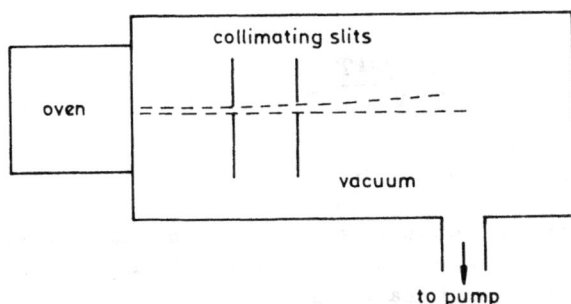

Fig. 2.40.

Solution:

(a) The Maxwell distribution is given by

$$n \left(\frac{m}{2\pi kT} \right)^{3/2} e^{-\frac{m}{2kT} v^2} dv .$$

The speed distribution of molecules in the beam is given by

$$nv \left(\frac{m}{2\pi kT} \right)^{3/2} e^{-\frac{m}{2kT} v^2} v^2 dv d\Omega .$$

(b) The mean speed is

$$\langle v \rangle = \frac{\displaystyle \int v \cdot v^3 e^{-\frac{m}{2kT} v^2} dv}{\displaystyle \int v^3 e^{-\frac{m}{2kT} v^2} dv} = \frac{3}{2} \sqrt{\frac{\pi kT}{2m}} .$$

(c) The most probable speed v_p satisfies

$$\frac{\partial}{\partial v}\left(v^3 e^{-\frac{m}{2kT}v^2}\right) = 0 \ .$$

Hence

$$v_p = \sqrt{\frac{3kT}{m}} \ .$$

(d) The beam power is

$$N = nA\left(\frac{m}{2\pi kT}\right)^{3/2}\left(\int_0^\infty e^{-\frac{mv^2}{2kT}}v^3\,dv\right)\Delta\Omega$$

$$= \frac{A}{2}n\sqrt{\frac{2kT}{\pi m}}(\Delta\theta)^2$$

$$= \frac{A}{2}\frac{p}{kT}\sqrt{\frac{2kT}{\pi m}}(\Delta\theta)^2$$

$$= 1.1 \times 10^{11}\text{s}^{-1} \ .$$

(e) The average rotational energy of the molecules in the beam is the same as that in the oven. From the theorem of equipartition of energy, we obtain the average rotational energy as kT.

2187

Consider a gas at temperature T and pressure p escaping into vacuum through a hole of area A which is in the wall of its container. Assume the radius of the hole is much less than the mean free path for the gas in the container.

(a) Roughly, what is the mass-rate of escape of the gas?

(b) If the gas is a mixture, is the relative mass-rate of escape of a component dependent only upon its relative concentration?

(Wisconsin)

Solution:

As the mean free path of the particles of the gas is much greater than the diameter of the hole, we can assume that the gas in the container is in thermal equilibrium and hence follows the Maxwell velocity distribution. If n is the number of particles per unit volume in the container at the moment t, the number of particles in a cylindrical volume of base area A and height v_x is

$$dN' = An \left(\frac{m}{2\pi kT}\right)^{\frac{1}{2}} \exp\left(-\frac{mv_x^2}{2kT}\right) v_x dv_x ,$$

Hence the mass-rate of escape of the gas as a fraction of the original mass is

$$\frac{M'}{M} = \frac{N'}{Vn} = \frac{A}{V} \int_0^\infty \left(\frac{m}{2\pi kT}\right)^{\frac{1}{2}} \exp\left(\frac{-mv_x^2}{2kT}\right) v_x dv_x = \frac{A}{4V}\bar{v} ,$$

where $\bar{v} = \sqrt{\dfrac{8\pi kT}{\pi m}}$ is the average speed.

(b) If the gas is a mixture, then each component by itself satisfies the Maxwell distribution. From the above result, we see that the relative mass-rate of escape is dependent on the molecular mass of the component through the average speed $\bar{v}$.

2188

Consider a two-dimensional classical system with Hamiltonian

$$H = \frac{1}{2m}(P_1^2 + P_2^2) + \frac{1}{2}\mu^2(x_1^2 + x_2^2) - \frac{1}{4}\lambda(x_1^2 + x_2^2)^2 .$$

A system of N particles of mass m each is in thermal equilibrium at temperature T within the potential well that appears in the Hamiltonian. T is small enough so that an overwhelming majority of the particles reside within the quadratic part of the well. However, some particles will always possess enough thermal energy to escape from the well by passing over the "top" of the well; in the one-dimensional slice of $V(x)$ shown in Fig. 2.14, this occurs at $x_1 = b$, where b can be determined from the above equation.

Calculate the escape rate for particles to leave the well by passing over the top.

(*Princeton*)

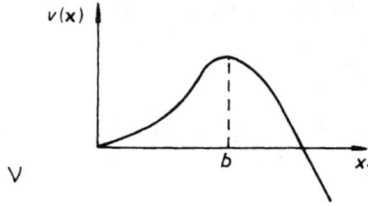

Fig. 2.41.

Solution:

Putting $x_2 = 0$ and $\dfrac{\partial H}{\partial x_1} = 0$, we obtain $b = \mu/\sqrt{\lambda}$ corresponding to the peak of potential barrier. Assume $b \gg l$, where l is the mean free path of the particles, so that even near the peak the particles are in thermal equilibrium. We need consider only the escape rate near the peak:

$$\dot{N} = \int 2\pi b v_x n(b) f d\mathbf{v} \Big/ \int f d\mathbf{v}$$

$$= 2\pi b n(b) \frac{\displaystyle\int_0^\infty v_x e^{-\frac{m}{2kT}v_x^{\,2}} dv_x}{\displaystyle\int_0^\infty e^{-\frac{m}{2kT}v_x^{\,2}} dv_x}$$

$$= \pi b n(b) \sqrt{\frac{2kT}{\pi m}} \; ,$$

where $n(b)$ is the number density at the peak. To find $n(b)$ we note that

$$n(r) = ce^{-\frac{V(r)}{kT}}$$

$$= ce^{-\frac{1}{kT}\left(\frac{1}{2}\mu^2 r^2 - \frac{1}{4}\lambda r^4\right)} \; ,$$

where $r^2 = x_1^2 + x_2^2$, and c is a normalizing factor defined by the following equation:

$$N = \int 2\pi r n(r) dr = 2\pi c \int r e^{-\frac{1}{kT}\left(\frac{1}{2}\mu^2 r^2 - \frac{1}{4}\lambda r^4\right)} dr \; .$$

As the majority of the particles reside within the quadratic part of the potential well, the above integral can be approximated as follows:

$$N = 2\pi c \int_0^\infty r \left(1 + \frac{\lambda r^4}{4kT}\right) e^{-\frac{\mu^2 r^2}{2kT}} \, dr$$

$$= 2\pi c \frac{2kT}{\mu^2} \left[\int_0^\infty t e^{-t^2} \, dt + \frac{\lambda}{4kT} \left(\frac{2kT}{\mu^2}\right)^2 \int_0^\infty t^5 e^{-t^2} \, dt\right]$$

$$= \frac{4\pi kT}{\mu^2} \left(\frac{1}{2} + \frac{\lambda kT}{\mu^4}\right) c \, ,$$

thus

$$c = \frac{N\mu^2}{4\pi kT} \cdot \frac{1}{\dfrac{1}{2} + \dfrac{\lambda kT}{\mu^4}} \, ,$$

and

$$n(b) = \frac{N\mu^2}{2\pi kT} \cdot \frac{1}{1 + \dfrac{2\lambda kT}{\mu^4}} e^{-\frac{\mu^4}{4kT\lambda}} \, .$$

Hence the escape rate is

$$\dot{N} = \frac{N\mu^3}{\sqrt{2\pi m \lambda kT}} \cdot \frac{1}{1 + \dfrac{2\lambda kT}{\mu^4}} e^{-\frac{\mu^4}{4kT\lambda}}$$

$$\approx \frac{N\mu^3}{\sqrt{2\pi m \lambda kT}} \cdot e^{-\frac{\mu^4}{4kT\lambda}} \, .$$

2189

A sealed $\frac{1}{4}$ litre bottle filled with oxygen at a pressure of 10^{-4} atmospheres is left on the surface of the moon by an astronaut. At a time when the temperature of the bottle is 400 K the jar develops a leak at a thin part of a wall through a small hole of diameter 2 microns. How will the amount of gas in the bottle depend on time and about how long will it take for the gas to decrease to $\frac{1}{10}$ of its original amount?

Show your work, estimate any constants needed besides Boltzmann's constant. You may assume that the temperature is maintained constant by the sunlight of the lunar day.

$$k = 1.38 \times 10^{-16} \text{ erg/K} \, .$$

(Wisconsin)

Solution:

The mean free path of the gas is

$$\bar{\lambda} = \frac{1}{\sqrt{2}\pi d^2 n} = \frac{kT}{\sqrt{2}\pi d^2 p} \ .$$

With $T = 400$ K, $d = 3.6 \times 10^{-10}$ m and $p = 10^{-4}$ atm ≈ 10 N/m^2, we get $\bar{\lambda} \approx 10^{-3}$ m. Thus we can consider the gas in the bottle as being in thermal equilibrium at any instant. From the Maxwell distribution, we find the rate of decrease of molecules in the bottle to be

$$\frac{dN}{dt} = -\frac{AN}{4V}\bar{v} \ , \quad \bar{v} = \sqrt{\frac{8kT}{\pi m}} \ ,$$

subject to the initial condition

$$N(t = 0) = N_0 \ .$$

Hence

$$N(t) = N_0 \exp\left(-\frac{A}{4V}\bar{v}t\right) \ .$$

That is, the number of particles in the bottle attenuates exponentially with time.

The number of particles in the bottle is N at time τ given by

$$\tau = \frac{4V}{A\bar{v}} \ln \frac{N_0}{N} \ .$$

For $N = N_0/10$, with

$$\bar{v} = \sqrt{\frac{8kT}{\pi m}} \approx 514 \text{ m/s} \ ,$$

$V = 0.25 \times 10^{-5}$ m^3, $A = \pi(10^{-6})^2$m^2, we find

$$\tau \approx 1.43 \times 10^6 \text{s} \ .$$

2190

(a) What fraction of H_2 gas at sea level and $T = 300$ K has sufficient speed to escape from the earth's gravitational field? (You may assume an ideal gas. Leave your answer in integral form.)

(b) Now imagine an H_2 molecule in the upper atmosphere with a speed equal to the earth's escape velocity. Assume that the remaining atmosphere above the molecule has thickness $d = 100$ km, and that the earth's entire atmosphere is isothermal and homogeneous with mean number density $n = 2.5 \times 10^{25}/m^3$ (not a very realistic atmosphere).

Using simple arguments, estimate the average time needed for the molecule to escape. Assume all collisions are elastic, and that the total atmospheric height is small compared with the earth's radius.

Some useful numbers: $M_{earth} = 6 \times 10^{24}$ kg,
$$R_{earth} = 6.4 \times 10^3 \text{ km.}$$

(*Princeton*)

Solution:

(a) The Maxwell velocity distribution is given by

$$4\pi \left(\frac{m}{2\pi kT} \right)^{3/2} e^{-\frac{mv^2}{2kT}} v^2 dv .$$

The earth's escape velocity is

$$v_e = \sqrt{\frac{2GM}{R}} = 7.9 \times 10^3 \text{m/s} .$$

H_2 molecules with velocities greater than v_e may escape from the earth's gravitational field. These constitute a fraction

$$f = \left(\frac{4}{\sqrt{\pi}} \right) \int_a^\infty x^2 \exp(-x^2) dx ,$$

where $a = v_e/v_0$ with $v_0 = \sqrt{\frac{2kT}{m}} = 2.2$ km/s. Hence

$$f = \frac{2}{\sqrt{\pi}} \left[ae^{-a^2} + \frac{2}{\sqrt{\pi}} \int_a^\infty e^{-x^2} dx \right]$$

$$= 1.4 \times 10^{-5} + 1.13 \int_{3.55}^\infty e^{-x^2} dx = 6 \times 10^{-5} .$$

(b) The average time required is the time needed for the H_2 molecules to diffuse through the distance d with a significant probability. The mean free path is

$$l = \frac{1}{n\sigma} = 4 \times 10^{-6} \text{ m}.$$

The time interval between two collisions is

$$\tau = \frac{1}{v_0} = 5 \times 10^{-10} \text{s}.$$

After N collisions, the mean-square of the diffusion displacement is

$$\langle z^2 \rangle = N l^2.$$

Putting $\langle z^2 \rangle = d^2$, so that $N = d^2/l^2$, we have

$$t = N\tau = \frac{\tau d^2}{l^2} = 3 \times 10^{11} \text{s} \approx 10^4 \text{ years},$$

i.e., it takes about 10^4 years for a H_2 molecule to escape from the atmosphere.

2191

(a) Consider the emission or absorption of visible light by the molecules of hot gas. Derive an expression for the frequency distribution $F(\nu)$ expected for a spectral line of central frequency ν_0 due to the Doppler broadening. Assume an ideal gas at temperature T with molecular mass M. Consider a vessel filled with argon gas at a pressure of 10 Torr (1 Torr = 1 mm of mercury) and a temperature of 200°C. Inside the vessel is a small piece of sodium which is heated so that the vessel will contain some sodium vapor. We observe the sodium absorption line at 5896 Å in light from a tungsten filament passing through the vessel.
Estimate:

(b) The magnitude of the Doppler broadening of the line.
(c) The magnitude of the collision broadening of the line.

Assume here that the number of sodium atoms is very small compared to the number of argon atoms. Make reasonable estimates of quantities that you may need which are not given and express your answers for the broadening in angstroms.

Atomic weight of sodium $= 23$.

(*CUSPEA*)

Solution:

(a) We take observations along the z-direction. The Maxwell-Boltzmann distribution for v_z is given by

$$dP = \left(\frac{M}{2\pi kT}\right)^{1/2} e^{-\frac{Mv_z^2}{2kT}} dv_z .$$

The Doppler shift of frequency is given by

$$\nu = \nu_0 \left(1 + \frac{v_z}{c}\right) .$$

Thus

$$v_z = c\left(\frac{\nu - \nu_0}{\nu_0}\right)$$

and

$$dP = \left(\frac{M}{2\pi kT}\right)^{1/2} e^{-\frac{Mv_z^2}{2kT}} dv_z$$

$$= \frac{1}{\nu_0}\left(\frac{Mc^2}{2\pi kT}\right)^{1/2} e^{-\frac{Mc^2}{2kT}\left(\frac{\nu-\nu_0}{\nu_0}\right)^2} d\nu .$$

(b) The magnitude of the Doppler broadening is

$$\Delta\nu \approx \sqrt{\frac{kT}{Mc^2}}\,\nu_0 ,$$

$$\Delta\lambda \approx \sqrt{\frac{kT}{Mc^2}}\,\lambda_0 = \sqrt{\frac{kT}{M}}\frac{1}{c}\lambda_0$$

$$= \sqrt{\frac{8.3 \times 473}{40 \times 10^{-3}}}\frac{\lambda_0}{3 \times 10^8} \simeq 1.04 \times 10^{-6}\lambda_0$$

$$= 1.04 \times 10^{-6} \times 5896\text{Å} = 6.13 \times 10^{-3}\text{Å} .$$

(c) The broadening due to collisions is

$$\Delta\nu \approx \frac{1}{\tau} ,$$

where τ is the mean free time between two successive collisions (of a Na atom). We have $\tau = \frac{\Lambda}{v}$, where v is the average velocity of Na atom and Λ

is its mean free path. As $\Lambda = \frac{1}{n\sigma}$, where n is the number density of argon molecules, σ is the cross section for scattering:

$$\sigma = \pi r^2 \approx 3 \times 10^{-20} \mathrm{m}^2 \ .$$

We have

$$n = \frac{p}{RT} \times N_A = 1.01 \times 10^5 \times 10 \times 6.02 \times 10^{23} / (760 \times 8.3 \times 473)$$
$$= 2.04 \times 10^{23} \mathrm{m}^{-3} \ ,$$
$$\Lambda \approx 1.7 \times 10^{-14} \mathrm{m} \ ,$$
$$v = \sqrt{\frac{kT}{M}} = \left(\frac{8.3 \times 473}{23 \times 10^{-3}} \right)^{1/2} \approx 413 \ \mathrm{m/s} \ .$$
$$\tau \approx 4 \times 10^{-7} \mathrm{s} \ ,$$

and hence

$$\Delta\lambda = \frac{\lambda}{v}\Delta\nu = \frac{\lambda^2}{c\tau} \approx 3 \times 10^{-5} \mathrm{\AA} \ .$$

2192

A gas consists of a mixture of two types of molecules, having molecular masses M_1 and M_2 grams, and number densities N_1 and N_2 molecules per cubic centimeter, respectively.

The cross-section for collisions between the two different kinds of molecules is given by $A|V_{12}|$, where A is a constant, and V_{12} is the relative velocity of the pair.

(a) Derive the average, over all pairs of dissimilar molecules, of the center-of-mass kinetic energy per pair.

(b) How many collisions take place per cubic centimeter per second between dissimilar molecules?

(UC, Berkeley)

Solution:
According to the Maxwell distribution,

$$f dv = N \left(\frac{M}{2\pi kT} \right)^{3/2} e^{-\frac{M}{2kT} v^2} dv \ .$$

(a) $\varepsilon = \dfrac{1}{N_1 N_2} \displaystyle\iint \dfrac{1}{2(M_1 + M_2)}(M_1\mathbf{v}_1 + M_2\mathbf{v}_2)^2 f_1 f_2 dv_1 dv_2$.

Note that the integral for the cross term $\mathbf{v}_1 \cdot \mathbf{v}_2 f_1 f_2$ is zero. Hence

$$\varepsilon = 3kT/2 .$$

(b) The number of collisions that take place per cubic centimeter per second between dissimilar molecule is

$$J = \iint A|V_{12}| \cdot |V_{21}| f_1 f_2 dv_1 dv_2$$

$$= A \iint (\mathbf{v}_1 - \mathbf{v}_2)^2 f_1 f_2 dv_1 dv_2$$

$$= 3AN_1 N_2 kT \left(\dfrac{1}{M_1} + \dfrac{1}{M_2} \right) .$$

2193

Consider air at room temperature moving through a pipe at a pressure low enough so that the mean free path is much longer than the diameter of the pipe. Estimate the net flux of molecules in the steady state resulting from a given pressure gradient in the pipe. Use this result to calculate how long it will take to reduce the pressure in a tank of 100 litres volume from 10^{-5} mm of Hg to 10^{-8} mm of Hg, if it is connected to a perfect vacuum through a pipe one meter long and 10 cm in diameter. Assume that the outgassing from the walls of the tank and pipe can be neglected.

(UC, Berkeley)

Solution:

(a) Assume that the length of the pipe is much longer than the mean free path, then we can regard the gas along the pipe as in localized equilibrium at different pressures but at the same temperature. From the Maxwell distribution we obtain the mean velocity:

$$v_0 = \dfrac{\displaystyle\int_0^\infty v_z e^{-\frac{m}{2kT}v_z^2} dv_z}{\displaystyle\int_0^\infty e^{-\frac{m}{2kT}v_z^2} dv_z} = \sqrt{\dfrac{2kT}{\pi m}} .$$

Thus the molecular flux along the pipe is

$$\phi = -Av_0 \cdot \Delta n$$

$$= -Av_0 l \frac{\Delta n}{l} = -Av_0 l \frac{1}{kT}\left(\frac{\Delta p}{l}\right) \ .$$

Since

$$l = \frac{1}{n\sigma} = \frac{kT}{p\sigma} \ ,$$

we have

$$\phi = -\frac{Av_0}{\sigma}\frac{1}{p}\frac{dp}{dz} \ .$$

(b) As given, $p \le 10^{-5}$ mmHg, we have

$$l = \frac{kT}{p\sigma} \ge 3 \times 10^2 \text{m} \gg 1\text{m} \ .$$

That is, the mean free path is much longer than the pipe and the above expression for ϕ is not valid. However, as the diameter of the pipe is much smaller than its length, we have $\phi = Av_0 n$.

Assume that both the initial and final states are in thermal equilibrium at temperature T, then

$$V\frac{dn}{dt} = -Av_0 n \ .$$

Hence

$$t = \frac{V}{Av_0}\ln\frac{n_i}{n_f} = \frac{V}{Av_0}\ln\frac{p_i}{p_f}$$

$$= \frac{V}{A}\sqrt{\frac{\pi m}{2kT}}\ln\frac{p_i}{p_f} = 0.4\text{s} \ .$$

2194

Consider the hydrodynamical flow conditions. The cooling of the gas during expansion can be expressed as follows, $\dfrac{T_0}{T} = 1 + \dfrac{M^2}{3}$, where T_0 is the temperature before expansion, T is the temperature after expansion, and M is the ratio of the flow velocity v to the velocity of sound c at temperature T.

(a) Derive the above expression.

(b) Derive a corresponding expression for $\frac{p_0}{p}$, and calculate the value of M for a condition where $\frac{p_0}{p} = 10^4$.

(c) Calculate the value of T for $\frac{p_0}{p} = 10^4$ and $T_0 = 300$ K.

(d) Find the maximum value of v in the limit $T \to 0$.

(UC, Berkeley)

Solution:

(a) Consider the process of a small volume of gas consisting of N molecules passing through a small hole. When it enters the hole it carries internal energy $N c_v T_0$ and the bulk of the gas does work on it to the amount of $p_0 V_0 = N k T_0$. When the volume of gas leaves the hole, its internal energy is $N c_v T$ and it does work $pV = N k T$ on the external gas. Its kinetic energy is now $N m v^2/2$. Thus we have for each molecule of the volume $c_p T_0 = c_p T + m v^2/2$, where

$$c_p = c_v + k = \frac{\gamma}{\gamma - 1} k, \quad \gamma = c_p/c_v .$$

Noting that the velocity of sound is $c = \sqrt{\dfrac{\gamma k T}{m}}$, we have

$$\frac{T_0}{T} = 1 + \frac{\gamma - 1}{2} M^2 .$$

For air, $\gamma = \dfrac{5}{3}$, and $\dfrac{T_0}{T} = 1 + \dfrac{M^2}{3}$.

(b) From the adiabatic relation,

$$\frac{p_0}{p} = \left(\frac{T_0}{T} \right)^{\frac{\gamma}{\gamma - 1}} = \left(1 + \frac{M^2}{3} \right)^{\frac{5}{2}} ,$$

we have

$$M = \sqrt{3 \left[\left(\frac{p_0}{p} \right)^{\frac{2}{5}} - 1 \right]} .$$

When $\dfrac{p_0}{p} = 10^4, M = 11.$

(c) $T = T_0 \left(\dfrac{p}{p_0}\right)^{\frac{2}{5}} = 7.5$ K.

(d) When $T \to 0$, we have

$$\frac{1}{2}mv_M^2 = c_pT_0 .$$

Hence $v_M = \sqrt{\dfrac{2c_pT_0}{m}} = \sqrt{\dfrac{5kT_0}{m}} = 557$ m/s.

2195

The schematic drawing below (Fig. 2.42) shows the experimental set up for the production of a well-collimated beam of sodium atoms for an atomic beam experiment. Sodium is present in the oven S, which is kept at the temperature $T = 550$ K. At this temperature the vapor pressure of sodium is $p = 6 \times 10^{-3}$ torr. The sodium atoms emerge through a slit in the wall of the oven. The hole is rectangular, with dimensions 10 mm $\times$ 0.1 mm. The collimator C has a hole of identical size and shape, and the sodium atoms which pass through C thus constitute the atomic beam under consideration. The atomic mass of sodium is 23. The distance d in the figure is 10 cm.

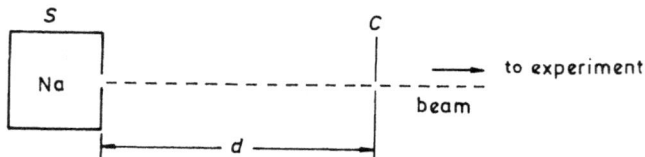

Fig. 2.42.

(a) Compute the number ϕ of sodium atoms which pass through the slit in C per second.

(b) Derive an expression for the function $D(v)$ which describes the distribution of velocities of the particles in the beam in the sense that $D(v)dv$ is the probability that an atom passing through C has a velocity in the range $(v, v + dv)$.

(c) The region in which the beam propagates must, of course, be a reasonably good vacuum. Estimate (and give answer in torr) just how good the vacuum ought to be if the beam is to remain collimated for at least 1 meter. (1 torr = 1mmHg).

Solution:

(a) The Maxwell distribution is given by

$$f \, dv_x \, dv_y \, dv_z = \left(\frac{m}{2\pi kT}\right)^{3/2} e^{-\frac{m}{2kT}(v_x^2 + v_y^2 + v_z^2)} \, dv_x \, dv_y \, dv_z \ .$$

There are $nAv_x f \, dv_x \, dv_y \, dv_z$ atoms in the velocity interval $\mathbf{v} \sim \mathbf{v} + d\mathbf{v}$ that escape through the area A of the hole. The number of atoms that pass through the second hole (the collimator C) is

$$\phi = nA \int \left(\frac{m}{2\pi kT}\right)^{3/2} v_x e^{-\frac{m}{2kT}(v_x^2 + v_y^2 + v_z^2)} \, dv_x \, dv_y \, dv_z$$

$$= nA \left(\frac{m}{2\pi kT}\right)^{3/2} \int_0^\infty v^3 e^{-\frac{mv^2}{2kT}} \, dv \cdot \int \cos\theta \, d\Omega$$

$$= nA \left(\frac{m}{2\pi kT}\right)^{3/2} \frac{1}{2} \left(\frac{2kT}{m}\right)^2 \frac{A}{d^2}$$

$$= \frac{A^2}{2\pi d^2} \cdot \frac{p}{kT} \sqrt{\frac{2kT}{\pi m}} \ .$$

With $A = 10 \times 0.1 = 1.0$ mm^2 $= 10^{-6}$ m^2,

$\quad d = 10$ cm $= 1.0 \times 10^{-1}$m ,

$\quad p = 6 \times 10^{-3}$ torr $= 0.80 N/m^2$,

$\quad T = 550$ K

we have $\phi = 6 \times 10^{11}$ s^{-1} .

(b) $D(v)dv = Cv^3 e^{-\frac{m}{2kT}v^2} \, dv$, where C is the normalizing factor given by

$$1 = \int D(v)dv = C \cdot \left(\frac{2kT}{m}\right)^2 \cdot \frac{1}{2} \ .$$

Hence

$$D(v) = 2v^3 \left(\frac{m}{2kT}\right)^2 e^{-\frac{m}{2kT}v^2} \ .$$

(c) Assume that the vacuum region is at room temperature $T = 300$ K. Since the mean free path is $l = 1$ m, we have

$$p \sim \frac{kT}{l\sigma} = \frac{1.38 \times 10^{-23} \times 300}{1 \times 10^{-20}} = 0.414 \text{ Pa}$$

$$= 3 \times 10^{-3} \text{ torr} \ .$$

2196

An insulated box of volume $2V$ is divided into two equal parts by a thin, heat-conducting partition. One side contains a gas of hard-sphere molecules at atmospheric pressure and $T = 293$ K.

(a) Show that the number of molecules striking the partition per unit area and unit time is $n\bar{v}/4$.

(b) A small round hole of radius r is opened in the partition, small enough so that thermal equilibrium between the two sides is maintained via heat conduction through the partition. Calculate the pressure and temperature as functions of time in both halves of the box.

(c) Suppose the partition is a non-conductor of heat. Discuss briefly and qualitatively any deviations from the time-dependence of temperature and pressure found in part (b).

(*UC, Berkeley*)

Solution:

(a) The Maxwell distribution is given by

$$f dv = \left(\frac{m}{2\pi kT}\right)^{3/2} e^{-m(v_x^2 + v_y^2 + v_z^2)/2kT} dv_x dv_y dv_z .$$

Among the molecules which strike on unit area of the partition in unit time, the number in the velocity interval $\mathbf{v} \sim \mathbf{v} + d\mathbf{v}$ is $n v_x f dv_x dv_y dv_z$. Integrating, we get the number of molecules striking the partition per unit area per unit time:

$$\frac{n}{4}\sqrt{\frac{8kT}{\pi m}} = \frac{n}{4}\bar{v} .$$

(b) Take the gas as an ideal gas whose internal energy is only dependent on temperature. As the box is insulated, the temperature of the gas is constant. Then we need only obtain the molecular number densities as functions of time in the two parts. Let n_1, n_2 be the molecular number densities of the left and right parts respectively at time t, V the volume of each part and A the area of the small hole. Then from the equations

$$V\frac{dn_1}{dt} + \frac{A\bar{v}}{4}(n_1 - n_2) = 0,$$

$$V\frac{dn_2}{dt} + \frac{A\bar{v}}{4}(n_2 - n_1) = 0,$$

$$n_1(0) = \frac{N}{V} = n_1 + n_2$$

$$n_2(0) = 0,$$

we get

$$n_1 = \frac{N}{2V}(1 + e^{-at}),$$

$$n_2 = \frac{N}{2V}(1 - e^{-at})$$

where

$$a = \frac{A\bar{v}}{2V} = \frac{\pi r^2 \bar{v}}{2V}.$$

From $p = nkT$ we have

$$p_1 = p_0[1 + \exp(-at)]/2$$
$$p_2 = p_0[1 - \exp(-at)]/2,$$

where $p_0 = NkT/V$.

(c) When the partition is a thermal insulator, we can still assume that each part is in thermal equilibrium by itself. At the beginning, molecules of higher energies in the left side will more readily enter the right side than molecules of lower energies. Therefore, the temperature of the right side will be slightly higher than the initial value, and correspondingly that of the left side will be slightly lower. The process being adiabatic, the change in pressure will be faster than that given by (b). The behaviors of the temperatures and pressures are shown in Fig. 2.43 and Fig. 2.44.

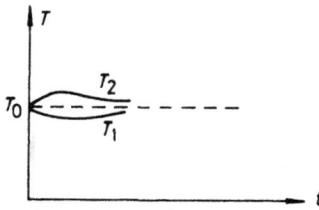

Case (a) : solid curves
Case (b) : dashed curve

Fig. 2.43.

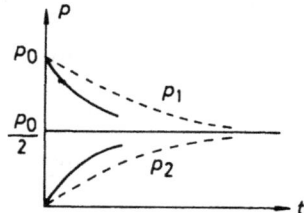

Case (b): dashed curves
Case (c): solid curves

Fig. 2.44.

2197

Consider a two-dimensional ideal monatomic gas of N molecules of mass M at temperature T constrained to move only in the xy plane. The usual volume becomes in this case an area A, and the pressure p is the force per unit length (rather than the force per unit area).

(a) Give an expression for $f(v)dv$, the total number of molecules with speeds between v and $v + dv$. (Assume that the classical limit is applicable in considering the behavior of these molecules).

(b) Give the equation of state (relating pressure, temperature etc.).

(c) Give the specific heats at constant area (two dimensional analogue of specific heat at constant volume) and at constant pressure.

(d) Derive a formula for the number of molecules striking unit length of the wall per unit time. Express your result in terms of N, A, T, M and any other necessary constants.

(UC, Berkeley)

Solution:

(a) From the Maxwell velocity distribution

$$f dv \propto e^{-\frac{M}{2kT}(v_x^2 + v_y^2)} dv_x dv_y ,$$

we have $f dv = ce^{-\frac{Mv^2}{2kT}} v dv$, where c is the normalizing factor given by

$$N = \int_0^\infty f dv .$$

Thus $f dv = \dfrac{MN}{kT} v e^{-\frac{Mv^2}{2kT}} dv.$

(b), (c), (d). The above can be written as

$$f \, dv_x dv_y = \frac{MN}{2\pi kT} e^{-\frac{M}{2kT}(v_x^2+v_y^2)} dv_x dv_y .$$

We first calculate the number of molecules which collide with unit length of the "wall" per unit time:

$$n = \int \int_{v_x>0} \frac{v_x}{A} \cdot \frac{MN}{2\pi kT} e^{-m(v_x^2+v_y^2)/2kT} dv_x dv_y$$

$$= \frac{N}{A} \cdot \sqrt{\frac{kT}{2\pi M}} ,$$

where A is the area of the system. Then we calculate the pressure:

$$p = \int \int \frac{2M v_x^2}{A} f \, dv_x dv_y = \frac{N}{A} kT ,$$

which gives the equation of state $pA = NkT$. From the theorem of equipartition of energy, we know that

$$c_v = Nk ,$$

and $c_p = c_v + Nk = 2Nk$.

2198

A parallel beam of Be $(A = 9)$ atoms is formed by evaporation from an oven heated to 1000 K through a small hole.

(a) If the beam atoms are to traverse a 1 meter path length with less than $1/e$ loss resulting from collisions with background gas atoms at room temperature (300 K), what should be the pressure in the vacuum chamber? Assume a collision cross-section of 10^{-16} cm^2, and ignore collisions between 2 beam atoms.

(b) What is the mean time $(\bar{\tau})$ for the beam atoms to travel one meter? Show how the exact value for $\bar{\tau}$ is calculated from the appropriate velocity distribution. Do not evaluate integrals. Make a simple argument to get a numerical estimate for $\bar{\tau}$.

(c) If the Be atoms stick to the far wall, estimate the pressure on the wall due to the beam where the beam strikes the wall. Assume the density of particles in the beam is $10^{10}/\text{cm}^3$. Compare this result with the pressure from the background gas.

collimating slits

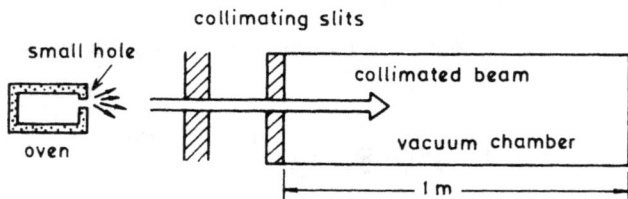

Fig. 2.45.

Solution:

(a) The fractional loss of atoms in the beam is $1 - \exp(-x/l)$, where l is the mean free path. If the loss is to be less than $1/e$ after travelling a distance L, L must be less than $l \ln \left(\dfrac{e}{e-1} \right)$. As

$$n = \frac{1}{l\sigma} < \ln \left(\frac{e}{e-1} \right) \Big/ L\sigma \ ,$$

we require

$$p = nkT < kT \ln \left(\frac{e}{e-1} \right) \Big/ L\sigma = 0.18 N/m^2 \quad \text{for } L = 1 \text{ m} \ .$$

(b) If the velocity in x-direction is v_x, the flying time is L/v_x. Since the distribution of the particle number in the beam is

$$f dv_x \propto v_x e^{-\frac{mv_x^2}{2kT}} dv_x \ ,$$

we have

$$\bar{\tau} = \frac{\displaystyle\int_0^\infty \frac{L}{v_x} \cdot v_x e^{-\frac{mv_x^2}{2kT}} dv_x}{\displaystyle\int_0^\infty v_x e^{-\frac{mv_x^2}{2kT}} dv} = L\sqrt{\frac{\pi m}{2kT}} = 1.3 \times 10^{-3} \ s \ .$$

(Note that scattering from background gas atoms has been neglected.)

(c) The pressure exerted by particles in the velocity interval $v_x \sim v_x + dv_x$ on the wall is proportional to

$$N \frac{v_x A m v_x f dv_x}{A} \ ,$$

where N is the particle number density in the beam and A is the cross section of the beam. Hence

$$p_0 = N \frac{\int m v_x^2 \cdot v_x e^{-\frac{m v_x^2}{2kT}} dv_x}{\int v_x e^{-\frac{m v_x^2}{2kT}} dv_x} = 2NkT_0 = 3 \times 10^{-4} \text{ N/m}^2 \, ,$$

which is much less than the pressure from the background gas.

2199

A quantity of argon gas (molecular weight 40) is contained in a chamber at $T_0 = 300$ K.

(a) Calculate the most probable molecular velocity.

A small hole is drilled in the wall of the chamber and the gas is allowed to effuse into a region of lower pressure.

(b) Calculate the most probable velocity of the molecules which escape through the hole.

The pressures of the chamber and the region outside the hole are adjusted so as to sustain a hydrodynamic flow of gas through the hole, such that viscous effects, turbulence, and heat exchange with the wall of the hole may be neglected. During this expansion the gas is cooled to a temperature of 30 K.

(c) Calculate the velocity of sound c at the lower temperature.

(d) Calculate the average flow velocity $\bar{v}$ at the lower temperature, and compare the distribution of velocities with the original distribution in the chamber.

(*UC, Berkeley*)

Solution:

(a) The Maxwell distribution is given by

$$f \, dv = \left(\frac{m}{2\pi kT} \right)^{3/2} e^{-\frac{m}{2kT} v^2} dv \, ,$$

or

$$f \, dv = 4\pi \left(\frac{m}{2\pi kT} \right)^{3/2} v^2 e^{-\frac{m v^2}{2kT}} dv \, .$$

The most probable velocity is the value of v corresponding to maximum f. From $\dfrac{\partial f}{\partial v} = 0$, we get

$$v_p = \sqrt{\frac{2kT_0}{m}} = 352 \text{ m/s} .$$

(b) The velocity distribution of the escaping molecules is given by

$$F dv = N v^3 e^{-\frac{m}{2kT} v^2} dv ,$$

where N is the normalizing constant. From $\dfrac{\partial F}{\partial v} = 0$, we get

$$v_m = \sqrt{\frac{3kT_0}{m}} = 431 \text{ m/s} .$$

(c) The velocity of sound is

$$c = \sqrt{\left(\frac{dp}{d\rho}\right)_S} .$$

Using the adiabatic relation $p = \rho^\gamma \cdot \text{const.}$, we get

$$c = \sqrt{\gamma \frac{kT}{m}} ,$$

where $\gamma = c_p/c_v$. For argon gas, $\gamma = 5/3$, and $c = 101$ m/s.

(d) The average flow velocity is

$$\bar{v} = \int v \cdot v^3 e^{-\frac{m}{2kT_0} v^2} dv \Big/ \int v^3 e^{-\frac{m}{2kT_0} v^2} dv$$

$$= \frac{3}{2} \sqrt{\frac{\pi kT_0}{2m}} = 468 \text{ m/s} .$$

2200

Estimate, to within an order of magnitude, on the basis of kinetic theory the heat conductivity of a gas in terms of its temperature, density, molecular weight, and heat capacity at constant volume. Make your own estimates of collision cross-sections and molecular mean free paths. You may restrict your attention to pressures near atmospheric, temperatures near room temperature and dimensions of the order of centimeters or meters. Do not concern yourself with heat transfer by convection. $(k = 1.38 \times 10^{-16} \text{ erg/K})$.

(*UC, Berkeley*)

Solution:

Assume that a temperature gradient $\dfrac{dT}{dx}$ exists in the gas and molecules drift from region of higher temperature to that of lower temperature. The number crossing unit area perpendicular to the drift in unit time is $n\bar{v}/4$. Each molecule, on the average, makes a collision in travelling a distance l, the mean free path, and transfers an energy $c_v \Delta T \sim c_v l \dfrac{dT}{dx}$. The heat flow per unit area per unit time is therefore $q = \dfrac{1}{4} n\bar{v} c_v l \dfrac{dT}{dx} = \kappa \dfrac{dT}{dx}$, where

$$\kappa = \frac{1}{4} n l \bar{v} c_v = \frac{c_v}{4\sigma} \sqrt{\frac{3kT}{M}}$$

is the thermal conductivity of the gas. Taking air as an example, with $M = 29 \times 1.67 \times 10^{-27}$ kg, $\sigma = 10^{-20} \text{m}^2$, $c_v = 5k/2$, $T = 300$ K, we have

$$\kappa = 0.44 \text{ J/mK} .$$

2201

A propagating sound wave causes periodic temperature variations in a gas. Thermal conductivity acts to remove these variations but it is generally claimed that the waves are adiabatic, that is, thermal conductivity is too slow.

The coefficient of thermal conductivity for an ideal gas from kinetic theory is $k \approx 1.23 C_v \bar{v} l$ where C_v is the heat capacity per unit volume, $\bar{v}$ is the mean thermal speed, and l is the mean free path.

What fraction of the temperature variation ΔT will be conducted away vs λ and what is the condition on λ for thermal conductivity to be ineffective?

(*Wisconsin*)

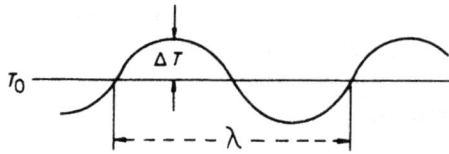

Fig. 2.46.

Solution:

The temperature at x can be written as

$$T = T_0 + \Delta T \cos\left(\frac{2\pi x}{\lambda} + \varphi_0\right) \; .$$

Then the change of temperature due to thermal conduction is

$$\delta T = -2l\left(\frac{dT}{dx}\right)$$
$$= \frac{4\pi l}{\lambda}\Delta T \sin\left(\frac{2\pi x}{\lambda} + \varphi_0\right) \; .$$

Thus

$$\frac{\delta T}{\Delta T} = \frac{4\pi l}{\lambda}\sin\left(\frac{2\pi x}{\lambda} + \varphi_0\right) \; .$$

This is the fraction of ΔT which results from thermal conduction. The condition for thermal conduction to be ineffective is

$$\frac{\delta T}{\Delta T} \ll 1 \; , \quad \text{that is } \lambda \gg l \; .$$

2202

Give a qualitative argument based on the kinetic theory of gases to show that the coefficient of viscosity of a classical gas is independent of the pressure at constant temperature.

(UC, Berkeley)

Solution:

Consider the flow of gas molecules along the x-direction whose average velocity $\bar{v} = v_x$ has a gradient in the y-direction. The number passing through unit area perpendicular to the x-direction in unit time is $n\bar{v}/4$. In a collision the momentum transfer in the y-direction is $m\Delta v_x$. Since a collision occurs over a distance $\sim l$, the molecular mean free path, for an order of magnitude calculation we can take

$$\Delta v_x = \Delta y \frac{\partial v_x}{\partial y} \sim l \frac{\partial v_x}{\partial y} .$$

Thus the shearing force across the unit area or the viscous force is

$$f = \frac{nm\bar{v}\Delta v_x}{4} = \frac{1}{4}nm\bar{v}l\frac{\partial v_x}{\partial y} .$$

Hence the coefficient of viscosity is

$$\eta = \frac{nm\bar{v}l}{4} = \frac{m}{4\sigma}\sqrt{8kT/\pi m} ,$$

where σ is the molecular collision cross section. η is seen to be independent of pressure at constant temperature.

2203

Consider a dilute gas whose molecules of mass m have mean velocity of magnitude $\bar{v}$. Suppose that the average velocity in the x-direction u_x increases monotonically with z, so that $u_x = u_x(z)$ with $|u_x| \ll \bar{v}$ and all gradients small. There are n molecules per unit volume and their mean free path is l where $l \gg d$ (molecular diameter) and $l \ll L$ (linear dimension of enclosing vessel).

(a) The viscosity η is defined as the proportionality constant between the velocity gradient and the stress in the x-direction on an imaginary plane whose normal points in the z-direction. Find an approximate expression for η in terms of the parameters given.

(b) If the scattering of molecules is treated like that of hard spheres, what is the temperature dependence of η? The pressure dependence? Assume a Maxwellian distribution in both cases.

(c) If the molecular scattering cross section $\sigma \propto E_{cm}^2$, where E_{cm} is the center-of-mass energy of two colliding particles, what is the temperature dependence of η? Again assume a Maxwellian distribution.

(d) Estimate η for air at atmospheric pressure ($10 \, \mathrm{dyn/cm^2}$) and room temperature. State clearly your assumptions.

(Princeton)

Solution:

(a) The number of molecules passing through a unit area perpendicular to the z-direction per unit time is $n\bar{v}/4$. Each particle makes a collision after travelling a distance of the order of magnitude of l in which a momentum in the x-direction is transferred of the amount $m\left(u_x + \Delta z \dfrac{\partial u_x}{\partial z} - u_x\right) = m\Delta z \dfrac{\partial u_x}{\partial z}$. For an approximate estimate, we take $\Delta z \sim l$, so that the viscous force and the viscosity are respectively

$$\tau = \frac{1}{4} n\bar{v} m l \frac{\partial u_x}{\partial z} \ ,$$

$$\eta = \frac{1}{4} m\bar{v} n l \ .$$

(b) As $l = \dfrac{1}{n\sigma}, \eta = \dfrac{m\bar{v}}{4\sigma}$. The hard sphere model gives $\sigma = $ const., then as $\bar{v} \propto \sqrt{T}$, $\eta \propto \sqrt{T}$ and is independent of pressure.

(c) If $\sigma \propto E_{cm}^2 \propto T^2$, then $\eta \propto T^{-3/2}$ and is independent of pressure.

(d) For room temperature, we can approximately take the molecular weight of air to be 30, $\bar{v}$ to be the speed of sound and $\sigma \sim 10^{-20}$ m^2. Then $\eta \approx 1.3 \times 10^{-6}$ kg/ms.

2204

Electrical conductivity. Derive an approximate expression for the electrical conductivity, σ, of a degenerate electron gas of density n electrons/cm^3 in terms of an effective collision time, τ, between the electrons.

(MIT)

Solution:

For a degenerate electron gas, the velocity is uniform on the Fermi surface. Then the net current in any direction is zero. Under the effect of an electrical field E_z, the electrons move as a whole in the z-direction, forming an electrical current. We have

$$E_z e = m\frac{dv_z}{dt} \ , \quad \Delta v_z \approx \tau \frac{dv_z}{dt} \ ,$$

giving the current density

$$j_z = en\Delta v_z \approx enE_z e\tau/m \ ,$$

where m is the mass of the electron and e is its charge. Comparing it with the relation between electrical current density and electrical conductivity, we get

$$\sigma = e^2 n\tau/m \ .$$

2205

Consider a system of charged particles confined to a volume V. The particles are in thermal equilibrium at temperature T in the presence of an electric field E in the z-direction.

(a) Let $n(z)$ be the density of particles at the height z. Use equilibrium statistical mechanics to find the constant of proportionality between $\dfrac{dn}{dz}$ and n.

(b) Suppose that the particles can be characterized by a diffusion coefficient D. Using the definition of D find the flux J_D arising from the concentration gradient obtained in (a).

(c) Suppose the particles are also characterized by a mobility μ relating their drift velocity to the applied field. Find the particle flux J_μ associated with this mobility.

(d) By making use of the fact that at equilibrium the particle flux must vanish, establish the Einstein relation between μ and D:

$$\mu = \frac{e}{k}\frac{D}{T} \ .$$

(*Wisconsin*)

Solution:

(a) The particle is assumed to have charge e, its potential in the electric field E being

$$u = -eEz \ .$$

Then the concentration distribution at equilibrium is

$$n(z) = n_0 \exp\left(\frac{eEz}{kT}\right) \ ,$$

where n_0 is the concentration of particles at $z = 0$, whence we get

$$\frac{dn(z)}{dz} = \frac{eE}{kT}n(z) .$$

(b) By definition,

$$J_D = -D\frac{dn(z)}{dz} = -D\frac{eE}{kT}n(z)$$
$$= -D\frac{eE}{kT}n_0 \exp(eEz/kT) .$$

(c) The particle flux along the applied electric field is

$$J_\mu = n(z)\bar{v} = n(z)\mu E = \mu E n_0 \exp\left(\frac{eEz}{kT}\right) .$$

(d) The total flux is zero at equilibrium. Hence $J_D + J_\mu = 0$, giving

$$\mu = \frac{eD}{kT} .$$

2206

Consider a system of degenerate electrons at a low temperature in thermal equilibrium under the simultaneous influence of a density gradient and an electric field.

(a) How is the chemical potential μ related to the electrostatic potential $\phi(x)$ and the Fermi energy E_F for such a system?

(b) How does E_F depend on the electron density n?

(c) From the condition for μ under thermal equilibrium and the considerations in (a) and (b), derive a relation between the electrical conductivity σ, the diffusion coefficient D and the density of states at the Fermi surface for such a system.

(SUNY, Buffalo)

Solution:

(a) From the distribution $n_\epsilon = \{\exp[(\epsilon - \mu - e\phi(x))/kT] + 1\}^{-1}$, we obtain $E_F = \mu_0 + e\phi(x)$, where $\mu_0 = \mu(T = 0)$.

(b) $N = 2 \cdot \dfrac{V}{h^3}\dfrac{4}{3}\pi p_F^3 = 2 \cdot \dfrac{V}{h^3} \cdot \dfrac{4}{3}\pi(2mE_F)^{3/2}$

(c) The electric current density **j** under the electric field **E** is

$$\mathbf{j} = \sigma \mathbf{E} = -\sigma \nabla \phi(x) .$$

The diffusion current density is $\mathbf{J} = -D\nabla\rho$. In equilibrium we have

$$\mathbf{j}/e = -\mathbf{J}/m \quad \text{i.e.,} \quad \sigma\nabla\phi(x)/e = -D\nabla\rho/m = -D\nabla n .$$

The density of states at the Fermi surface is

$$N_F = \frac{dN}{dE_F} = 4\pi V (2m)^{3/2} \frac{E_F^{1/2}}{h^3} .$$

The electric chemical potential $\tilde{\mu} \equiv \mu + e\phi(x)$ does not depend on x in equilibrium. Thus

$$\nabla\tilde{\mu} = \nabla\mu + e\nabla\phi(x) = 0 ,$$

i.e.,

$$e\nabla\phi(x) + \left(\frac{\partial\mu}{\partial n}\right)_T \nabla n = 0 .$$

Hence,

$$D = \frac{\sigma}{e^2}\left(\frac{\partial\mu}{\partial n}\right)_T \approx \left(\frac{\sigma}{e^2}\right)\frac{\partial E_F}{\partial n} = \frac{2\sigma E_F}{3e^2 N}$$

$$= \frac{2\sigma}{3e^2 n}\left[\frac{h^3}{4\pi(2m)^{3/2}}\frac{N_F}{V}\right]^2 .$$

2207

Consider a non-interacting Fermi gas of electrons. Assume the electrons are nonrelativistic.

(a) Find the density of states $N(E)$ as a function of energy ($N(E)$ is the number of states per unit energy interval) for the following cases:

    1) The particles are constrained to move only along a line of length L.

    2) The particles move only on a two dimensional area A.

    3) The particles move in a three dimensional volume V.

(b) In a Fermi electron gas in a solid when $T \ll T_F$ (the gas temperature is much less than the Fermi temperature), scattering by phonons and impurities limits electrical conduction. In this case, the conductivity σ can be written as

$$\sigma = e^2 N(E_F) D \ ,$$

where e is the electron charge, $N(E_F)$ is the density of states, defined above, evaluated at the Fermi energy and D is the electron diffusivity. D is proportional to the product of the square of the Fermi velocity and the mean time, τ_e, between scattering events $(D \sim v_F^2 \tau_e)$.

1) Give a physical argument for the dependence of the diffusivity on $N(E_F)$.

2) Calculate the dependence of σ on the total electron density in each of the three cases listed in part (a). The electron density is the total number of electrons per unit volume, or per unit area, or per unit length, as appropriate.

(*CUSPEA*)

Solution:

(a) 1) Motion along length L. The wave eigenfunction of a particle is

$$\sin\left(\frac{n\pi x}{L}\right) \ , \quad n = 1, 2, \ldots \ .$$

The Schrödinger equation gives the quantum energy levels as

$$E_n = \frac{\hbar^2}{2m}\left(\frac{\pi n}{L}\right)^2 \ ,$$

i.e.,

$$n = \frac{2L}{h}\sqrt{2mE} \ .$$

The number of states $\overline{N}$ for each n is 2 to account for spin degeneracy. Thus

$$N(E) = \frac{d\overline{N}}{dE} = \frac{d\overline{N}}{dn}\cdot\frac{dn}{dE} = 2\frac{dn}{dE} = L\left(\frac{8m}{h^2 E}\right)^{1/2} \ .$$

2) Motion in a square of side L $(L^2 = A)$. The eigenfunction of a particle is

$$\sin\frac{n_x \pi x}{L}\sin\frac{n_y \pi y}{L} \ , \quad \begin{array}{l} n_x = 1, 2, \ldots \\ n_y = 1, 2, \ldots \end{array}$$

with energy

$$E(n_x, n_y) = \frac{h^2(n_x^2 + n_y^2)}{8mL^2} = \frac{h^2 n^2}{8mL^2} = E(n) \ .$$

Thus the number of states between n and $n + dn$ is

$$2 \cdot \frac{1}{4} \cdot 2\pi n dn = \pi n dn \ .$$

Hence,

$$N(E) = \pi n \frac{dn}{dE} = \frac{4\pi m L^2}{h^2} \ ,$$

$$= \frac{4\pi m A}{h^2} \ ,$$

where we have used

$$n = \frac{L\sqrt{8mE}}{h} \ .$$

3) Motion in volume $V (L^3 = V)$.

$$E(n) = \frac{h^2}{8m} \left(\frac{n}{L}\right)^2 \ ,$$

with

$$n^2 = n_x^2 + n_y^2 + n_z^2 \ ,$$

where $n_x = 1, 2, \ldots , n_y = 1, 2, \ldots$, and $n_z = 1, 2 \ldots$.
The number of states between n and $n + dn$ is

$$2 \cdot \frac{1}{8} \cdot 4\pi n^2 dn = \pi n^2 dn \ .$$

Hence $N(E) = \pi n^2 \dfrac{dn}{dE} = 4\pi V \left(\dfrac{2m}{h^2}\right)^{3/2} E^{1/2}.$

(b) 1) The mean free time is inversely proportional to the probability of collision, and the latter is proportional to the density of states on the Fermi surface (as $T \ll T_F$, scatterings and collisions only occur near the Fermi surface and we assume that elastic scatterings are the principal process). Thus

$$\tau_e \sim \frac{1}{N(E_F)} \ .$$

Hence

$$D \sim \frac{v_F^2}{N(E_F)} \ .$$

2) We have $\sigma = e^2 D N(E_F) \sim e^2 v_F^2 \sim E_F$. Let the total number of electrons be z, and the number density of electrons be μ, then

$$\mu = \begin{cases} z/L & \text{(one-dimensional)} \\ z/A & \text{(two-dimensional)} \\ z/V & \text{(three-dimensional)} . \end{cases}$$

As $\quad E_F = \dfrac{h^2}{8m} \left(\dfrac{n_F}{L}\right)^2$ for all the three cases, and

$$z = \begin{cases} 2n_F & \text{(one-dimensional)} \\ \pi n_F^2/2 & \text{(two-dimensional)} \\ \pi n_F^3/3 & \text{(three-dimensional)} , \end{cases}$$

we have

$$\sigma \sim E_F \sim \begin{cases} \mu^2 & \text{(one-dimensional)} \\ \mu & \text{(two-dimensional)} \\ \mu^{2/3} & \text{(three-dimensional)} . \end{cases}$$

This results differ greatly from those of the classical theory. The reason is that only the electrons near the Fermi surfaces contribute to the conductivity.

2208

(a) List and explain briefly the assumptions made in deriving the Boltzmann kinetic equation.

(b) The Boltzmann collision integral is usually written in the form

$$(\partial f(\mathbf{r}, \mathbf{v}_1, t)/\partial t)_{\text{coll}} = \int d^3\mathbf{v}_2 \int d\Omega \sigma(\Omega) |\mathbf{v}_1 - \mathbf{v}_2| (f_1' f_2' - f_1 f_2) ,$$

where $f_1 = f(\mathbf{r}, \mathbf{v}_1, t)$, $f_2' = f(\mathbf{r}, \mathbf{v}_2', t)$ and $\sigma(\Omega)$ is the differential cross section for the collision $(\mathbf{v}_1, \mathbf{v}_2) \rightarrow (\mathbf{v}_1', \mathbf{v}_2')$. Derive this expression for the collision integral and explain how the assumptions come in at various stages.

(SUNY, Buffalo)

Solution:

(a) The assumptions made are the following.

1) A collision can be considered to take place at a point as the collision time is generally much shorter than the average time interval between two collisions.

2) The particle number density $f(\mathbf{r}, \mathbf{v}, t)$ is the same throughout the interval $d^3r\, d^3\mathbf{v}$.

3) Particles of different velocities are completely independent, i.e., the distribution for particles of different velocities $\mathbf{v}_1, \mathbf{v}_2$ can be expressed as

$$f(\mathbf{r}, \mathbf{v}_1, t) \cdot f(\mathbf{r}, \mathbf{v}_2, t) \ .$$

4) Only central forces and two-body elastic collisions need to be considered.

(b) By assumption 1), we need consider only the collision rate $(\partial f_1 / \partial t)_{\text{coll}}$ of particles with $\mathbf{v}_1$ in the interval $d^3r\, d^3\mathbf{v}_1$.

By assumption 2), the probability for a particle with $\mathbf{v}_1$ is to be scattered by a particle with $\mathbf{v}_2$ into solid angle Ω in time interval dt can be written as

$$|\mathbf{v}_1 - \mathbf{v}_2| \sigma(\Omega) f(\mathbf{r}, \mathbf{v}_2, t) d\Omega d^3\mathbf{v}_2 dt \ .$$

By assumption 3) we can write the number density of particles with $\mathbf{v}_1$ scattered into solid angle Ω as

$$|\mathbf{v}_1 - \mathbf{v}_2| \sigma(\Omega) f(\mathbf{r}, \mathbf{v}_2, t) f(\mathbf{r}, \mathbf{v}_1, t) d\Omega d^3\mathbf{v}_2 d^3\mathbf{v}_1 dt \ .$$

Similarly, the increase in the number density of particles with $\mathbf{v}_1$ in the space $d^3r\, d^3\mathbf{v}_1$ after the collision $(\mathbf{v}_1', \mathbf{v}_2') \to (\mathbf{v}_1, \mathbf{v}_2)$ is given by

$$|\mathbf{v}_1' - \mathbf{v}_2'| \sigma(\Omega) f(\mathbf{r}, \mathbf{v}_2', t) f(\mathbf{r}, \mathbf{v}_1', t) d\Omega d^3\mathbf{v}_2' d^3\mathbf{v}_1' dt \ .$$

Assumption 4) gives $d^3\mathbf{v}_1 d^3\mathbf{v}_2 = d^3\mathbf{v}_1' d^3\mathbf{v}_2'$ and

$$|\mathbf{v}_1 - \mathbf{v}_2| = |\mathbf{v}_1' - \mathbf{v}_2'| \ .$$

Hence

$$\left(\frac{\partial f_1}{\partial t} \right)_{\text{coll}} = \int d^3\mathbf{v}_2 \int d\Omega \sigma(\Omega) |\mathbf{v}_1 - \mathbf{v}_2| (f_1' f_2' - f_1 f_2) \ .$$

www.ingramcontent.com/pod-product-compliance
Lightning Source LLC
Chambersburg PA
CBHW061615220326
41598CB00026BA/3765